Advances in Geological Science

Studies in the twentieth century uncovered groundbreaking facts in geophysics and produced a radically new picture of the Earth's history. However, in some respects it also created more puzzles for the research community of the twenty-first century to tackle. This book series aims to present the state of the art of contemporary geological studies and offers the opportunity to discuss major open problems in geosciences and their phenomena. The main focus is on physical geological features such as geomorphology, petrology, sedimentology, geotectonics, volcanology, seismology, glaciology, and their environmental impacts. The monographs in the series, including multi-authored volumes, will examine prominent features of past events up to their current status, and possibly forecast some aspects of the foreseeable future. The guiding principle is that understanding the fundamentals and applied methodology of overlapping fields will be key to paving the way for the next generation.

Mitsuhiro Toriumi

Physics of Geochemical Mechanics and Deep Neural Network Modeling with Diffusion Augmentation

Applications to Earthquake Prediction

Springer

Mitsuhiro Toriumi
Japan Agency for Marine-Earth Science
and Technology
Yokohama, Kanagawa, Japan

ISSN 2524-3829 ISSN 2524-3837 (electronic)
Advances in Geological Science
ISBN 978-981-97-9375-4 ISBN 978-981-97-9376-1 (eBook)
https://doi.org/10.1007/978-981-97-9376-1

This Springer imprint is published by the registered company Springer Nature Singapore Pte Ltd.
The registered company address is: 152 Beach Road, #21-01/04 Gateway East, Singapore 189721, Singapore

Preface

During the last decades, natural sciences have changed drastically in their depth of recognition, width of field, and method of observation. The astro-sciences have been enlarged their field from cosmology and stellar sciences to the astrobiology and exoplanetary sciences with huge amounts of monitoring data using satellite observation systems. On the other hand, earth sciences developed extensively in the experimental framework in the ultrahigh-pressure reaching to the condition of the planetary center and in the high-energy spectrographic investigations, data-processing methods of the neural network modeling and generative AI, and in the satellite monitoring of geophysical and geological data involving exoplanets.

On the other hand, in the solid earth sciences, the comprehensive investigations among nanoscale materials science of minerals in the earth interiors, microscale mechanics and thermodynamics of rocks, macroscale continuum mechanics and fluid dynamics with reaction, and planetology and celestial mechanics are much concerned for us to understand the nature of dangerous giant earthquake, giant volcanic eruption, and global warming. Therefore, it has been proposed in the previous book that geochemical mechanics involving dynamics of various sized cracks and pores–aqueous fluid interaction, time-scale problem of crystal growth in plate boundary rocks, permeable convection of geofluid in the plate boundary zone, and the short-term cyclic change, with several to several ten years periodicity of the global and regional seismicity and the plate velocity are modeled by the mechanical network dynamics of the solid earth process. Furthermore, it was pointed out that the global correlated seismicity is strongly controlled by the earth rotation rate and by the change of ellipticity of the earth appeared in the variation of the satellite gravity.

Therefore, it will be worth thinking that the background correlated seismicity showing the pitchfork bifurcation of the nonlinear dynamics is strongly related with the somewhat change in fluid earth affecting earth's angular moment change. Moreover, it should be needed for human community that the practical prediction of forthcoming large earthquakes becomes installation in the safety system of our society. To obtain the available system of the practical prediction, in this book, the author intends to propose an augmentation method of the multivariate timeseries by applying the diffusion type augmentation of rather rare large earthquake events involving characteristic timeseries of the correlated seismicity and local seismicity, and augmentation-based full-prediction model using the deep neural network of the large earthquake in the regional and global seismic system.

On the other hand, micro- and nanoprocesses of geochemical mechanics relevant to various crack dynamics with large amount of dissipation mechanism at the crack tip may be discussed, based on the physical modeling, and they probably suggest the observable precursors before the large earthquake as proposed in this book. For the foundation of physical modeling of the process sequence from earthquake triggering to generation in the earth interior, it should be required that the nano- and microscale defects structure, interacting with the environmental stress, temperature change, fluid flow, and mineral reaction, should be understood by construction of their physical model. It is obvious that the large earthquake is responsible for the explosive high-speed slippage by propagation mode of the very large shear cracks embedded in the solid earth interior, which are surely under the critical state of the propagation mode, although the size of the present shear crack is enough larger than the critical Griffith crack size. The conditions that large shear cracks are still in the state under the critical condition before propagation are seemingly due to the large amount of dissipation at the shear crack tip, and therefore, it needs that the subsidiary microcracks cluster formation, plastic deformation, and mineral reaction controlling the fluid flow take place in the stress-concentrated tip of large growing shear crack.

On the other hand, the large shear crack grown from small shear crack through the above dissipation process seems to have the rough-shaped shear plane but not sharply flat surface depending on the dissipation mechanisms such as subsidiary crack formation and plastic deformation with mineral reactions. These dissipation mechanisms may give arise the viscous behavior to the surrounding rocks in the deep crustal conditions near the plate boundary. The Kelvin–Helmholtz instability of the neutral interface of the shear slippage interface is possibly expected in the above dissipated large shear crack surface for the physical modeling of the rough-shaped shear plane surface, of which interface shows the characteristic waveform governed by the shear slip velocity. Then, it seems that the elastic rock mass of the hanging wall should generate the elastic wave at the rough-shaped bumps on the wavy shear plane, and its frequency is controlled by the shear crack speed and the wavelength and spacing of the bump.

Moreover, the various dissipation mechanisms controlling the size of large growing shear cracks occur heterogeneously in the rocks because of heterogeneous distribution of the shear cracks and fluid flow paths, and then the distribution of the constituent mineral grains may change depending on the deformation rate of the rocks. In the previous book, it showed that albite grains in the plate boundary metamorphic rocks occur as large cluster and its annealed large grain (porphyroblastic grain), thereby suggesting the deformation-induced grain clustering in the geochemical mechanics process in the plate boundary rocks as proposed in this book.

On the other hand, recent global seismicity appears somewhat different from that of this year in the point that the large earthquakes over magnitude 7.5 in the region apart from the subduction boundary occurred at Noto Peninsula earthquake, Turkey-Syria earthquake, and East Taiwan earthquake. Last year, the subduction boundary earthquake took place at the Mindanao, Philippines. In the previous book, that the global correlated seismicity tends to keep active after Sumatra and Tohoku-Oki giant earthquakes in 2004 and 2011 has been noticed, and thus, it needs to be investigated whether this trend continues still in the period from 2021 to 2023 or not. If this is the case, it, therefore, shows that the recent active seismicity is surely responsible for the frequent occurrence of the large earthquakes noted above.

By the way, it has been pointed out that the global and regional correlated seismicity dynamics contains in part the nonlinear dynamics with bifurcation, thereby showing that the practical prediction method of the large earthquake is necessary for the availability of neural network modeling of the artificial earthquake event above the multivariate timeseries inferred from the nonlinear dynamical equations. In this book, the author tries to study the deep neural network modeling in the cases of artificial earthquake associated with the two-state and Lorenz-type nonlinear differential equations, and it suggests that the deep CNN modeling with large earthquake labeling is available for actual prediction.

In summary, the content in this book may be available for studying the recent trends of the research in global and regional seismicity dynamics relevant to the practical prediction of the large earthquake, and the comprehensive approaches to the earth mechanics integrating the nano- to microscale materials sciences, geochemistry, seismology, and geodynamics.

Kamakura, Japan — Mitsuhiro Toriumi

Acknowledgements

Metadata and data products for this study were accessed through NOAA, IERS, GRACE, and USGS data centers for SLR, LOD, gravity data, and global seismic data. The author sincerely thanks to these organizations of the global monitoring of geophysical data for open science usage. The author intends to express sincerely thanks for Google Collaboratory to use the cloud processing by DNN modeling on KERAS with TENSORFLOW.

The author very thanks to Data Center of Earthquake Research Institute of University of Tokyo for helpful use of database of JMA-1. He also thanks much to National Institute of Earthquake Disaster for helpful use of earthquake database. He thanks greatly Japan Agency of Marine Science and Technology for providing the MCS data of the Nankai Trough and the facility of the institute, and Japanese Center of Nuclear Power Control for providing the data of boring cores.

The author is also very indebted to Prof. Yoshiyuki Kaneda of University of Kagawa for many critical discussions and continuous supports for this research. He also gives me many comments of next generation of geophysical research and their social missions.

He is also indebted to Dr. Tatsu Kuwatani of Japan Agency of Marine Science and Technology for many supports on writing this book. He thanks much Dr. No of Japan Agency of Marine Science and Technology for providing MCS image. He thanks greatly Japan Agency of Marine Science and Technology.

Finally, I am deeply indebted to Etsuko Toriumi for many supports and discussions, and Drs. Yosuke Nishida, Madanagopal D Deenadayalan, and Niraja Deshmukh of Springer Nature Publisher for many encouragements and comments.

About This Book

In this book, physical processes of rapid growth of minerals in the jog and cluster of the micro-open cracks during the shear crack growth and explosive propagation are discussed, and their meanings for the fluid flow and stable shear crack enlargement in Chap. 2. In the same chapter, jog wave on shear crack and interface development of the shear plane associated shear crack growth will be noted in detail to apply for the geological structure of the large fault zone with slightly viscous materials. In Chap. 3, structure evolution concerning grain distribution and clustering of minerals in shear deformation and chemical reactions, decompression growth of mineral grains in the shear crack jog, and shape change annealing of fluid inclusion array formed by healing of micro-open crack in single crystals are modeled to discuss the characteristic distribution of mineral grains in plate boundary metamorphic rocks. Decompression fragmentation and rapid explosive extrusion mechanism will be noted in Chap. 3.

Furthermore, in Chap. 4, it shows that the recent space-time variation of the global and regional correlated seismicity appears as the extension of the timeseries from 1990 to 2021 in the global system, and from 1998 to 2021 in Japanese islands, though in the last decade, the global correlated seismicity displays a sharp plateau during the period of 2018–2022, being derived from the giant volcanic eruption of Hunga-Tonga-Hunga volcano in Tonga. Besides, it is noted that the secular trend of the global correlated seismicity, which is strongly correlated with the secular change of the rate of earth rotation, will be explained by change of angular moment due to the elliptical change of the earth on the timing of spin-up to spin-down. In addition, the physical model of non-volcanic tremor along on the subduction plate boundary will be presented and discussed possibility of precursor of the large earthquake in this chapter.

On the other hand, in Chap. 5, the nonlinear behavior of the global correlated seismicity from 1990 to 2023, which is not possibly interpreted by the simple celestial mechanics of the earth rotation rate and angular moment change, may be discussed by the mechanical network transition of the subduction zones. The seismicity network of the subduction boundary zones is characterized by the three-arm chain of slider-spring block connection system in the plate tectonic system of the earth, and the mechanical behavior of the proposed three-arm chain system activated by periodic external force will be simulated to be likely to the behavior of the global correlated seismicity.

In Chap. 6, the author presents the new model of diffusion augmentation of the multivariate timeseries with rare events such as large earthquakes in the plate boundary zones. To evaluate the availability of the deep machine learning by neural network modeling in the case of nonlinear dynamics system, it suggests that the artificial earthquakes on the multivariate timeseries simulated by two types of nonlinear differential equations and their combinations with random noisy drift will be tested and evaluated in this chapter. Furthermore, it will be presented that the full-prediction method based on 2dCNN modeling is founded by using input data of augmented multivariate timeseries of correlated seismicity and local seismicity of the target locality, and the timeseries of labels of large earthquakes.

In conclusive remarks of Chap. 7, the author discusses the future development of the practical prediction of the large earthquakes in the plate boundary regions, being applied by the deep machine learning with diffusion augmentation of rare events arising the giant disasters to the human society.

Contents

Chapter 1
Introduction

Abstract Natural sciences are now enlarging the width and depth of understanding the world, together with rapid developing observation facilities and mathematical basis of data-driven sciences with machine learning. In earth sciences, the satellite observation, global network observation of seismicity, positioning, geodesical displacement, geomagnetic variation, oceanic and atmospheric climate changes, and celestial mechanics variation are monitored in wide and narrow bands' scales. The physics of geochemical mechanics in this book introduces mechanisms combining geochemical, geological, seismological, and global tectonic processes based on the data-driven and numerical sciences.

Keywords Geochemical mechanics · Shear crack jog · Data-driven sciences

1.1 Physics of Geochemical Mechanics and Global Seismicity

Recent progress in natural sciences is greatly owed to the comprehensive investigations using various deep machine learning and neural network modeling of multivariate timeseries and high dimension image data equipped from high-density and high-accurate multi-sensors network system in the nanoscale to global scale. As a result, it leads that for the large-scale and inhomogeneous complicated materials, and huge amounts of boundary and initial conditions, characteristic features, physical parameters of response function, inversion of the initial and boundary conditions, finding of the physical rules hidden in the complicated system, are possibly formulated and understood, and applied to the development of various technologies. Especially, it is greatly developed that the generative autoencoder with neural network model and the transformer with multi-attention model are available for characterization of the high dimension image data and of the language text data, which involve very complicated, recursive, and redundant information. Furthermore, it has become the next stage to understand the complexity hidden in the natural system that the connected nonlinear systems including the various tipping points can be investigated

M. Toriumi, *Physics of Geochemical Mechanics and Deep Neural Network Modeling with Diffusion Augmentation*, Advances in Geological Science,
https://doi.org/10.1007/978-981-97-9376-1_1

by using dense machine learning method in the noise-induced nonlinear dynamical tipping model.

Neural network modeling (NNM) with many layers and nodes, which are often connected bidirectionally and/or unidirectionally, and in the modes of dense connection and/or of the partial connection among nodes is characterized by its huge number of parameters, and by the minimum approximation of the objective function (evaluation function) between the model output and natural input data as the machine deep learning. In this NNM system, it is very important that the objective function can involve the many types of constraint conditions and physical rules; for examples, these are normalization (L1 norm, L2 norm, e.g., Chollet and Allaire 2018), regularization, total energy constraints, total number of constituent elements, equation of state, continuity equation, positive entropy production, etc. It is also possible that the constraints are discontinuity between the present elements under consideration and other types of chemical equations.

In addition, it is possible that the generative variational autoencoder (GVAE) modeling makes a synthesized universal image from huge amounts of images with high dimensions data involving space, time, and color indexes, and even the interconnection relations of the partial objects in the various images based on the characterization in the latent space mapping. In the previous book of the author (Toriumi 2022), it shows that the multivariate timeseries of the global seismicity rates of the various localities along the plate boundaries (subduction, spreading) are possibly transformed to high dimension vectors of the global correlated seismicity, and the variation of those global correlated seismicity by segmentation of timeseries seems to be possibly represented by the transition of the seismicity network of the global plate boundary and by the nonlinear dynamical equations of the correlated seismicity.

By the way, it shows that the network dynamics of the proteins is formulated by the minimalization of total elastic strain energy of atomic bond defined by deviation of standard atomic structure with a special network structure represented by Kirchhoff matrix. On the other hand, the author proposed a model of seismicity network dynamics that the network potential is defined by the total sum of the byproduct of logarithmic seismicity rates of which the gridded locality displays the high intensity of their correlated seismicity of the present characteristic seismicity network nodes, that is represented by Kirchhoff matrix of actual network nodes as shown in previous book (Toriumi 2022). The present type of network dynamics is primarily given by the transition from the one network geometry to another network one, and then it may be suggested that the kinetics of the network transition is formulated by the gradient of network potential difference concerning the seismicity rates of the key localities and noisy drift term. In this case, the first term of network potential is to be zero because of near equilibrium, but the potential function difference can be expanded to the Taylor series by small fluctuation of the intensity of the correlated variable of nodes, which is probably driven by the global parameters such as flattening rate of the earth and tidal force in the present case.

In the above-mentioned network dynamics, it seems that the networks should be decomposed by basis Kirchhoff matrices with related node intensity and therefore that the kinetics of the network dynamics is represented by the change of weight parameters controlled by the external variables such as time derivative of the earth rotation (length of day) and of tidal force. It follows that the total network potential of the system is to be summation of every contribution of basis network potential with contribution factors, which are governed by the geometry of network and intensive variable-dependent factor of network components. Therefore, it seems that the transition kinetics of network system depends on the weight parameters of the basis matrix components, governed by the operated external variables such as plate velocity and elliptical strain of the whole earth. Furthermore, it is possible to consider that the multiple stable modes of the present kinetics proposed in the previous book are probably derived from time variation of the weight parameters having the different relation against the external variable. For the change in the above-mentioned variables with advancing the time, the different styles of the functional relation mentioned above are seemingly responsible for the different modes of the network.

As noted above, the natural phenomena represent apparently complicated large amounts of variables including diverse initial and boundary conditions, together with heterogeneous internal structure of the system, and thereby it seems that observables of natural phenomena should be decomposed into the local state variables, such as the stress, temperature, fracture and defect density, fluid content, mineral composition, and their spatial extension. Otherwise, it is possible to consider that the functional relation of the field and related variables representing the characteristic parameters of the lithospheric plates and underlying mantle and core governs the stable states and their evolution in the long- and short-term phenomena. In this case, the controlling equations are probably decomposed into the basic processes of the physically simple mechanisms in the solid earth, for examples, these are various crystalline defects (vacancy, dislocations, boundaries) and their dynamics, cracks, minerals, and fluid (aqueous solution, magma, oil, gases) and their dynamics in massive rocks, and elastic waves, boundary wave, and faults motion in earth interior, thereby controlling the elastic and plastic deformation, phase transition and reactions, and fluid flow.

By the way, it has been discussed that the rocks of the subduction plate boundary are identified to be the metamorphic schists and gneisses that have been brought into the earth interior by plate subduction, and the characteristic processes in the metamorphic rocks are hydration and dehydration mineral reactions, ductile deformation and brittle deformation, fluid flow and diffusion, oxidation and reduction, and their combination. Until now, it has been discussed and established that the metamorphic belt extensively exposed in the plate boundary region is composed of various metamorphic grades from zeolite facies, prehnite-pumpellyite facies, greenschist facies, glaucophane-schist (blueschist) facies, epidote–amphibolite facies, eclogite facies, and granulite facies, in the ascending order of metamorphic temperature. Within these metamorphic facies, the zeolite facies, prehnite-pumpellyite facies, greenschist, and glaucophane-schist facies rocks are very important concerns because they are formed in the conditions from 300 to 600 °C of temperature and pressure of 0.1–3 GPa, being associated with dehydration and hydration mineral reactions, and abundant sealed shear and open cracks. Furthermore, it seems to be potential objects that the metamorphic rocks mentioned above remain the physical footprints of the characteristic processes of various size and speed shear slip events in the microstructures of metamorphic minerals and their grain boundaries (e.g., Karato and Toriumi 1989).

On the other hand, for the disciplinary sciences in metamorphic petrology, it is probable that paradigm change should be needed, from the static and mineral facies-based phase petrological characterization and estimation of the pressure–temperature path of rocks, to the various mechano-chemical kinetics from very short (milliseconds) to much long terms (ten million years), showing rate of fluid flow, rate of crystal growth, rate of shear slip, rate of temperature and pressure change, and their recovery (annealing) reaction rates. In other words, it is required to be discussed that the metamorphic rocks mentioned above accumulate the physical footprints of the divergent processes with various time scales. Therefore, it needs that we are much concerned with investigation of the fundamental physical mechanisms relating to these mechano-chemical processes in the low-grade metamorphic rocks mentioned above, those are called as the 'Physical Process of Geochemical Mechanics' in this book. From the viewpoint of physical process of geochemical mechanics, it is quite important that scale-invariant periodic structure of the shear slip surface appears as a waveform from 1 mm of microscopic size to 10 km of multichannel seismic survey (MCS) as shown in Fig. 1.1.

In addition, global mechanics of the solid earth is becoming the important theme of the earth sciences field because it seems to be responsible for the dynamics of the fluid earth composed of atmosphere, ocean, and outer core, and the former two

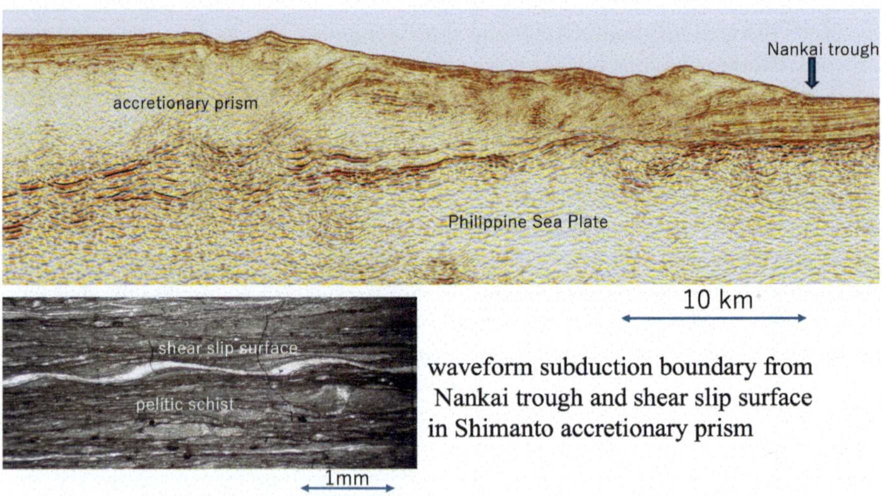

Fig. 1.1 Cross section across the Nanai trough to Off-Muroto in Shikoku by multichannel seismic image (MCS image; provided from JAMSTEC seismic survey database site https://doi.org/10.17596/0002069), showing wavy subducting plate boundary with strong reflection, and wavy shear slip surface filled with quartz (white thin layer) of sheared low-grade politic schists of Shimanto belt in western Kyushu under optical microscope

are strongly controlled by the celestial mechanics and by the climate change as proposed in the previous book. Furthermore, it has been suggested that the recent global seismic activity increased obviously rather than the former century, thereby suggesting the synchronous change of global seismicity and the global warming of the earth surface. On the other hand, the author proposed the systematic but nonlinear progressive change of the global correlated seismicity associated with the changing rate of earth rotation, that is the rate of length of day (LOD), and it is necessary that the rate of rotation change should give rise to change in earth ellipticity as well as the solid tide of the earth, thereby suggesting the reason of the increase in the global seismic activity. Therefore, it should be needed that what is the reason of the nonlinear relation between the global correlated seismicity and the rate of earth rotation change. Recent detailed investigation of the geomagnetic change in the core clarified the earth rotation change is synchronous with the zonal geostrophic flow of the outer core electromagnetic fluid (Gillet et al. 2010), thereby suggesting the influence on the LOD change from outer core zonal flow. Thus, the fluid earth surely gives rise to change in the rate of LOD, but it should remain still that the simple relation of the LOD changing rate and zonal flow of the outer core, the stratosphere, and the ocean does not possibly interpret yet the nonlinearity of relationship between the LOD rate and global correlated seismicity clarified in the previous book (Toriumi 2022). In this book, the author, therefore, intends to propose a model that explains the above-observed relationship between them.

1.2 Large Earthquake Prediction by DNN Modeling with Augmentation

As stated previously, it is obvious that the probability prediction of large earthquake events in the locality and the o'clock concerned is very difficult because the large earthquake and giant earthquake events are very rare, but huge amounts of small earthquakes of the present locality. For examples, there are only two events of giant earthquakes (Sumatra Earthquake 2004 and Tohoku-Oki Earthquake 2011) after the foundation of global network of the seismic observatory at 1980, thereby showing these extremely rare occurrences of events impossible to learn by DNN with huge amounts of coeval and global seismic data of small earthquakes. It, therefore, needs that both timeseries of large earthquakes as target of prediction and small earthquakes as background seismicity is to be augmented in the reasonable form as natural cases.

There are several augmentation models for the use of regression to timeseries: one is the total variance keeping method, and the other is the mean-keeping reversal method. On the other hand, the generative variational autoencoder-decoder modeling (GVAE) with CNN is one of augmentation model using neural network machine learning, for this GVAE modeling of the timeseries generates the new timeseries from the characteristic latent space of the given timeseries, and random fluctuation around the low dimension latent space plots. However, these augmentation models are not favorable for application to the timeseries of large earthquake associated with small earthquakes data sets, because these are based on the characteristic features deducing from the total timeseries, despite the necessity of time-dependent features sets corresponding to the large/giant earthquakes.

In this work, it is required that the simple physical model of the large earthquake associated with unknown activities of small earthquakes and other geophysical signals should be formulated by Langevin-type differential equation with some degree of Gaussian noise drift, and that the augmentation model, therefore, is based on the available fluctuation-induced and time-dependent synthetic timeseries. The favorable augmentation model, thus, seems to be diffusion augmentation method in this case, referring to the generative autoencoder model and state-space modeling of the timeseries by Kitagawa (2009). In this book, the author tries to propose a diffusion augmentation model using the terms containing the drift and fluctuation of Gaussian noises, associated with successive timeseries of characteristic features. The present augmentation can be applied for labeling of the large earthquake at the concerned locality corresponding to large number of sets of the characteristic features data matrixes augmented enough for deep machine learning by deep neural network model (DNN).

It remains that the deep convolution neural network (DCNN) modeling in the testing was not enough accurate from the training data sets in the previous investigations (Toriumi 2021), suggesting the effect of random change between several states of the seismic activity both in the global and regional cases. Thus, it should be clarified that whether the artificial events of the earthquake along on the simple nonlinear dynamics with chaotic and bifurcation behavior were possibly predicted by the CNN modeling or not. Therefore, the author intends to study the behavior of the logistic and Lorentz dynamics with chaotic timeseries added by drift terms of Gaussian noises and related episodic events. The artificial earthquake events analogous to the natural earthquake on the artificial seismicity background are generated on the two-state nonlinear dynamics timeseries (one dimension) and Lorenz timeseries (three dimensions) by the signal level over the given value. It will be concluded that even in the case of bifurcated and chaotic change of the background timeseries, the training sets of artificial earthquakes give rise possibly to predict the accurate testing sets by using DCNN modeling.

Until now, the full prediction of the near-future earthquake by means of DNN modeling has not been carried out, but many works of the prediction study of earthquake are concerned with the testing research after training of DNN models using various methods, those are recurrent neural network (RNN), long- and short-term memory modeling (LSTM), convolution neural network (CNN), and transformer with multiple attention. In the previous book, the prediction testing that the author tried limited in such framework of prediction study, because number of large earthquake events is too small, and not available for the machine learning necessary for training the DNN modeling as mentioned previously. However, in the present book, it may be ready to investigate the full prediction of large earthquake that the diffusion augmentation of the global and regional correlated seismicity timeseries labeled by large earthquake event makes possible to infer the near-future extensions of them.

In this book, we will call the probability prediction of near-future large earthquake, using the future extensions of the correlated seismicity and local logarithmic seismicity timeseries by DNN modeling, as the full-prediction model of large earthquake event. In this full-prediction model, there are three modules: the first is the correlated seismicity module, the second is the augmentation module, and the third is the near-future modeling module. The first module is the decomposition layer of the seismicity data matrix by SVD (singular value decomposition), the second is the augmentation facility of the multivariate correlated seismicity timeseries, and the third means the 2dCNN with large earthquake labeling prediction facility. In the full-prediction model, it will be tried that the one-year predictions of the large earthquake of several localities in the global and regional cases are investigated for the period from 2023/3/30 to 2024/3 from the referred data matrix from 1990/1/1 to 2023/3/29.

In the present investigation, it seems that the full-prediction modeling of the large earthquakes which will take place near-future is probably still under the construction to be available for the practical application in the social system, and thus, it will be noticed that thus, in this study, the full prediction as the timeseries of large earthquake probability for the anonymous localities is carried out. In near future, the author hopes that the more available full-prediction system will be founded and installed in the human society.

References

Chollet F, Allaire JJ (2018) Deep learning with R. Manning Publishing, USA

Gillet N, Jault D, Canet E, Fournier A (2010) Fast torsional waves and strong magnetic field within the Earth's core. Nature 465/6. https://doi.org/10.1038/nature0901

Karato S, Toriumi M (1989) Rhelogy of solids and of the earth. Oxford Science Press

Kitagawa G (2009) Introduction to time series modelling. CRC Press

Toriumi M (2021) Global seismicity dynamics and data-driven science. Springer

Toriumi M (2022) Geochemical mechanics and neural network modelling: application to earthquake prediction. Springer

Chapter 2
Physics of Geochemical Mechanics

Abstract The shear crack jog should be open at the time of shear crack growth and propagation, of which the shear slip velocity varies in the very wide range from μm/s to km/s, suggesting that fluid pressure in the jog should decrease rapidly by small amount of fluid influx from the surrounding matrix. Thus, it seems that albite and quartz are highly oversaturated in the jog, thereby being consistent with the occurrence of dendritic quartz and ameboid-shaped albite in the jog. On the other hand, the boundary surface of the low-fluid pressure jog experiences the decompression breakout forming systematic dent structure. On the other hand, the size of large shear crack is possibly enlarged by dissipation of strain energy derived from formation of microcrack cluster and fluid flow into shear crack. Based on the dissipation, the jog density wave appears in the shear crack growth process by numerical simulation. In addition, it is shown that the shear crack interface in the ductile materials appears to be the characteristic waveform obtained by neutral stability condition of Kelvin–Helmholtz instability.

Keywords Shear crack jog · Rapid growth of crystals in jog · KH instability of shear surface

2.1 Seismic-Crystal Diagram in Shear Crack Jog of the Plate Boundary Metamorphic Rocks

Recent investigations of the sealed cracks and related shear crack jog in the plate boundary metamorphic rocks manifest various morphology of filling minerals related to the seismic shear crack activity in the ancient plate boundary rocks of the Sanbagawa and Shimanto metamorphic belts (Toriumi 2022). On the line of these investigations, it is suggested that the morphological map of sealing minerals in the shear jog and subsidiary open crack is possible for various degrees of oversaturation of minerals by the regime of dendritic, acicular, flower-like and euhedral shapes, corresponding to the seismic to non-seismic shear slip velocities.

M. Toriumi, *Physics of Geochemical Mechanics and Deep Neural Network Modeling with Diffusion Augmentation*, Advances in Geological Science,
https://doi.org/10.1007/978-981-97-9376-1_2

In addition, it has been clarified that the shear crack jog is forcibly open by the shear crack motion even in the condition of imbalance between newly open jog volume and inflow aqueous solution. In that extreme situation, the open jog volume may be occupied by vapor of H_2O with small amount of CO_2 (or their two-phase fluid), and the gas pressure decreases very rapidly. Then, in the jog, the saturated minerals should very rapidly nucleate and crystallize to form the dendritic aggregates and/or acicular crystal. Therefore, it may be straightforward that appearance of the dendritic and acicular rapid growth of minerals manifests the rapid decrease of fluid pressure by entrainment in the shear crack jog.

By the way, the author pointed out the occurrence of flower-like grains of albite, quartz, garnet, and chlorite, showing the waveform outline of their grains in the low-grade metamorphic rocks of the Sanbagawa metamorphic belt (Toriumi 2022). The present grain shape of sugar candy is called as 'konpeito' which is originated in classic Holland, and the waveform of the konpeito has been discussed by means of one of universality in the nonlinear dynamics (Sakai and Hayakawa 2006). On the other hand, the author proposed the model that the konpeito-like growth of minerals in natural metamorphic rocks is responsible for the capillary effect controlled by the Mullins-Sekerka instability and the wavelength is governed by the growth and diffusion rates at the growth front (Toriumi 2022). Furthermore, in that paper, it is suggested that the growth rate is a function of the inverse of the wavelength of the flower-like grain (konpeito-like grain).

In addition, the long acicular grains of actinolite appear commonly in the albite grains sealing the open crack and shear crack jog (Toriumi 2022). The growth form of the present acicular shape with large aspect ratio is probably derived from the rapid growth in the largely oversaturated conditions from aqueous solution/vapor. It seems that the acicular grains of amphiboles should be formed in the rapid crack jog opening process as well as the flower-like grains.

By the way, it has been discussed that the texture of the metamorphic rocks is divided into several types: these are equigranular, granoblastic, porphyroblastic, poikiloblastic, and oikoblastic textures, and of the deformed rocks, these are mylonitic, porphyroclastic, cataclastic textures (Spry 1969). These textures are seemingly obvious under the microscope in the high-grade metamorphic rocks and mantle peridotites, for these are recrystallized enough in the high temperature and pressure conditions ($T > 500$ °C, $P > 1$ GPa). It is the reason why the texture is strongly controlled by the interfacial free energy and plastic deformation of the constituent mineral grains in the high temperature and non-hydrostatic conditions. On the other hand, the low-grade metamorphic rocks show various textures controlled by dissolution and growth, deformation, and fracturing of minerals.

However, in the plate boundary metamorphic rocks, it is obvious that the mineral reactions proceed at the grain boundary, but the mineral fracturing and deformation occur in the trans-granular volume. The pressure and temperature conditions of the reaction front between constituent minerals are not homogeneous in the whole rock volume, but they vary in the rate-limiting processes coupled with deformation/fracturing and mineral reactions associated with fluid. Furthermore, it is sure that these textures formed by inhomogeneous rate processes above should change gradually after their formation in the annealing and reconstructive processes in the metamorphism process. Therefore, considering the plural events of reaction and deformation/fracturing in metamorphism, the textures formed by these events should change successively in different duration and time scales through post-event annealing process.

On the other hand, the plate boundary zone is characterized by very active seismicity showing the huge observable range of the moment release rates, and the seismic shear plane reaches the area exceeding 1000 km × 500 km and 100 m displacement. Furthermore, as suggested by Aki and Richard (1980), there are abundant barriers on the seismic shear plane, which control the displacement velocity of the seismic shear, meaning the mechanical obstacles (Toriumi 2022). In the context of the geology, the mechanical obstacle corresponds to the subducted seamounts (Kodaira et al. 2004) and the shear crack jog/jog cluster (Toriumi 2022). Even if the case is the former, the subducted seamount just on the shear plane should be associated with the large-scale jog pairs at the time of rapid slip motion, and thus the jog formation may be important in the inhomogeneous short-term fluid migration and mineral-precipitation process in the plate boundary zone. In this paper, the author would like to propose a model of the short-term and intermittent events of jog formation and rapid precipitation of minerals in the plate boundary metamorphic rocks.

There are probably two types of jog formation: one is the obstacles on the shear slip plane of the shear cracks, and the other is the jog by the Riedel and penetrative shear crossing the main shear plane of the shear cracks as shown in Fig. 2.1. Both jog volumes are formed by the case of rapid shear slippage, and if the slow slip is the case, the surrounding rocks may fill the jog volume by plastic and brittle deformation immediately. In this study, the rapid case of shear crack jog formation should be discussed as later sections.

As is well-known (e.g., Wei 2010), there are three types of cracks: one is the open crack (type I), second is the in-plane shear crack (type II), and the third is the out-of-plane shear crack (type III). In the three-dimensional space, a shear crack is

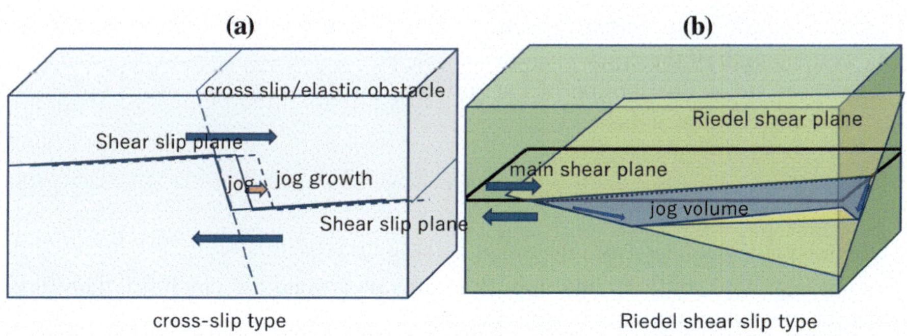

Fig. 2.1 Two types of shear crack jog. A; shear crack slip across cross slip or obstacle makes jog and jog growth, B; shear crack slip and associated Riedel shear makes jog

composed of type II and type III segments as like as crystal dislocation composed of screw and edge dislocations. On the other hand, the crack surface is defined as the discontinuous interface between two continuum blocks surrounded by the closed curvilinear circle, and displacement vector is almost constantly parallel to the shear plane surrounded by the crack tip outline. The change of displacement vector is concentrated near the crack tip, and thus the strain magnitude increases to the crack tip.

The shear slip velocity of the crack is represented by the velocity of the present crack tip, and thus the stress components around the crack tip change according to migration of the crack tip, as their time derivatives increase with increasing crack tip velocity. This process manifests the moment release by shear crack slippage and makes the emission of elastic waves of double couple type, which is the seismic wave. The frequency of the emitting elastic waves should be governed by the velocity of the shear crack.

By the way, the shear crack motion is divided into the crack growth and the crack propagation mode. The latter is the fast velocity mode and the former the slow mode, and the boundary of the modes is determined by the critical value of the differential stress and mechanical conditions of crack tip. The crack growth is so-called as the subcritical crack growth, and it is controlled by chemical and mechanical environments around crack tip and by repeated applied differential stress. On the other hand, the rapid crack propagation is governed strongly by the stress intensity factor, and thus, the crack tip velocity is proportional to the large power of the stress intensity factor K, being likely to the explosive fracturing (Atkinson 1984).

In the case of crack growth in the hydrous and repeated differential stress conditions, the crack tip velocity is proportional to the square of the stress intensity factor (Wei 2010) which is proportional to square root of the crack length and differential stress as,

$$\begin{aligned} V &= CK^2 \\ K &= \tau(\pi c)^{1/2} \end{aligned} \tag{2.1}$$

in which K is the stress intensity factor, τ is the shear stress, C is a constant term, and c is the length of the crack.

The shear stress distribution of the surrounding the crack tip toward the stress maximum direction becomes

$$\Delta\sigma = K/(2\pi r)^{1/2} \tag{2.2}$$

in which r is the distance from the crack tip (Fig. 2.2).

In the high-temperature condition, the minerals around the crack tip, therefore, should be over the yield stress, and deforms by moving dislocation and vacancy. Even in the rather low temperature, the mineral reaction with aqueous solution may be enhanced by the high stress condition (high equivalent pressure) and relaxes stress amplification around the tip. In this case, the stress accumulation at the crack

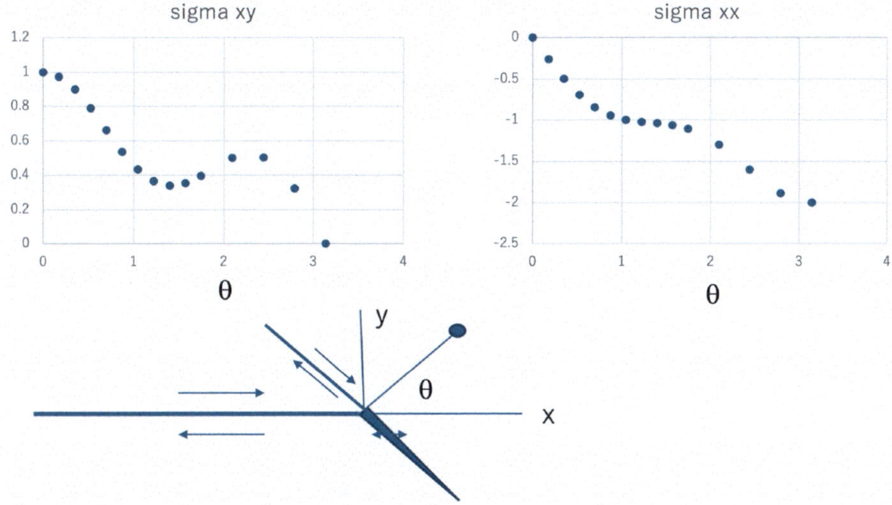

Fig. 2.2 Stress field around the shear crack tip showing σ_{xy} (upper left) and σ_{xx} (upper right) against angle from the shear plane at the distance = 1. Lower figure shows the coordination of the shear crack (type II) and secondary cross shear crack slip and open crack (broad line)

tip may be relaxed, but the velocity of crack growth is controlled by the strain rate of plastic deformation and by the reaction rate of water–rock interaction. Hydrogen embrittlement and hydrolytic weakening are such examples.

In the case of type II crack, the stress field manifests the dipole structure that the top region is the compressive field, and the bottom is the extensive field. Therefore, it is obvious that the equivalent pressure gradient from the upper to bottom across the shear crack tip makes the flux of H_2O and dissolved species of hydrous reaction. Applying that the logarithmic solubility products of albite and chlorite have positive pressure gradient (ΔVqz = 0.437 J/mol.bar, ΔVab = 7.4 J/mol.bar and $\Delta Vchl$ = 33.7 J/mol.bar calculated from Holland and Powell (1998), quartz, albite, and chlorite grains in the higher-pressure upper region should dissolve, and the ionic species may migrate, and they should precipitate in the bottom region. Considering that the ionic diffusion rate of Na^+ should be faster than that of Mg^{2+}, then the albite should precipitate earlier than chlorite in the lower pressure domain under the shear crack tip. On the other hand, the upper region may become deficient in albite component but enrich in epidote and amphiboles in the low-grade metamorphic rocks. Rate of this process depending on the shear crack tip velocity should control the width of the albite-rich and chlorite-rich regions, and the thin layers of the albite and chlorite layers are developed with advancing the shear crack. In this case, the shape of the albite and chlorite layers is strongly controlled by the geometry of the shear crack in the rocks.

As noted previously, the shear crack velocity should be proportional to the present shear crack length and the square of shear stress, and therefore, the crack size should increase with advancing time if the recovery is not the case. In this process, the velocity must increase proportional to the length of the crack (that is K^2) and equivalent fluid pressure gradient does increase too, but the rate of reaction may be proportional to the difference of chemical potential of relevant minerals (albite/quartz and chlorite) between the compression and depression areas around the crack tip. The velocity of crack tip (u_s) in the crack growth mode holds (Wei 2010),

$$u_s = AK^2 \tag{2.3}$$

in which A is activation energy term with Arrhenius relation and K is the stress intensity factor.

On the other hand, the rate of the precipitation (R) in the extensional region becomes (Toriumi 2022),

$$R = GK \big/ (2\pi r_o)^{1/2} P_s \tag{2.4}$$

in which P_s is the solid pressure in the rocks apart from the crack tip (= saturation pressure of albite/quartz/chlorite at the present temperature), r_o is the effective distance from the crack tip, and G is the coefficient term related to temperature.

Therefore, it holds,

$$R = \left(G \big/ (2\pi r_o)^{1/2} A^{1/2} P_s\right) u_s^{1/2}. \tag{2.5}$$

In this equation, G and A are parameters dependent on temperature and P_s is rock pressure. It shows that the rate of mineral precipitation gradually increases with increasing the shear crack velocity (Fig. 2.3). As a result, the mineral layers should be formed under the shear crack plane, and adjacently the residual mineral layer above the shear plane may be formed as a paired band as illustrated in Fig. 2.4.

The competitive process of precipitation of minerals against shear crack growth such as albite and quartz in low-grade metamorphic rocks should vary its mode of growth shape of minerals depending on the growth rate. In the rapid growth from the solution, the growth shape appears dendritic outline, but in the slow growth, it shows euhedral shape with singular surface. In the intermediate growth, the shape has wavy growth surface as shown in the previous book (Toriumi 2022). Therefore, the diagram of R and u_s is available for the classification of mode of crystal growth associated with shear crack growth (shear slip velocity).

There are well-known modes of grain growth in the solution and gaseous environment: one is euhedral growth having singular surfaces, second is the dendritic growth controlled by the ratio of diffusion of the components in the environment and on the crystal surface, acicular growth by large contrast of growth velocities of different singular surfaces. These grain shapes are governed by the growth condition determined by the temperature which controls strongly the diffusion, and by the degree of oversaturation which depends on the temperature, pressure, solubility product, and pH. In addition, the author proposes recently the waveform grain growth of minerals in the conditions between euhedral and dendritic growth by Mullins–Sekerka instability (Kurz and Fisher 1986). In this case, the wavelength of the waveform grain

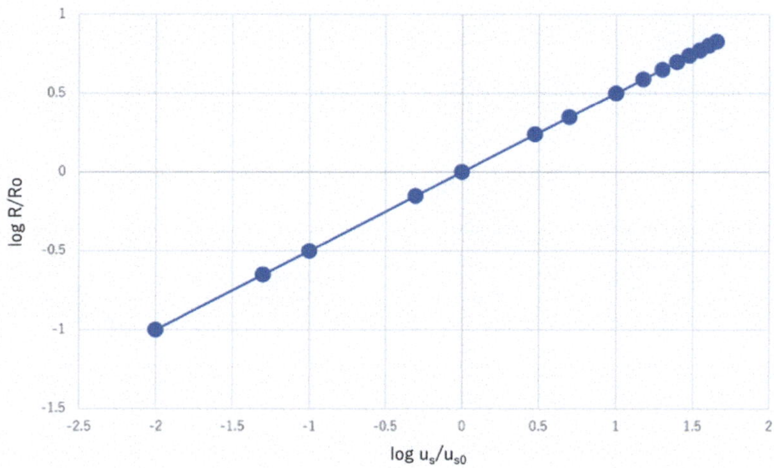

Fig. 2.3 Estimated relation between the ratio of mineral growth over shear crack growth rate and shear crack length, showing the continuous decrease with shear crack growth

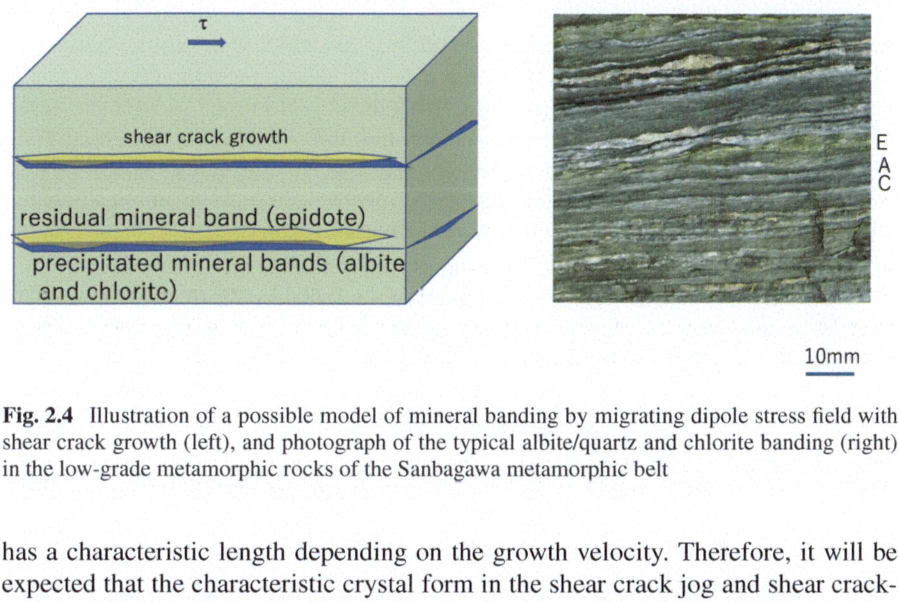

Fig. 2.4 Illustration of a possible model of mineral banding by migrating dipole stress field with shear crack growth (left), and photograph of the typical albite/quartz and chlorite banding (right) in the low-grade metamorphic rocks of the Sanbagawa metamorphic belt

has a characteristic length depending on the growth velocity. Therefore, it will be expected that the characteristic crystal form in the shear crack jog and shear crack-related banding could be mapped in the diagram of shear crack velocity and the temperature.

The high-speed shear slip may be associated with large rate of void generation in the rocks by the shear crack jog as shown in Fig. 2.5. In the high-temperature conditions, the jog motion may not create the void, but make the surrounding rock volume plastically deforms and fills the void. However, in the case that the rheology of the rocks is still hard in the high-speed jog motion, the newly generated void should be filled with fluid entrained from grain boundary fluid in the surrounding

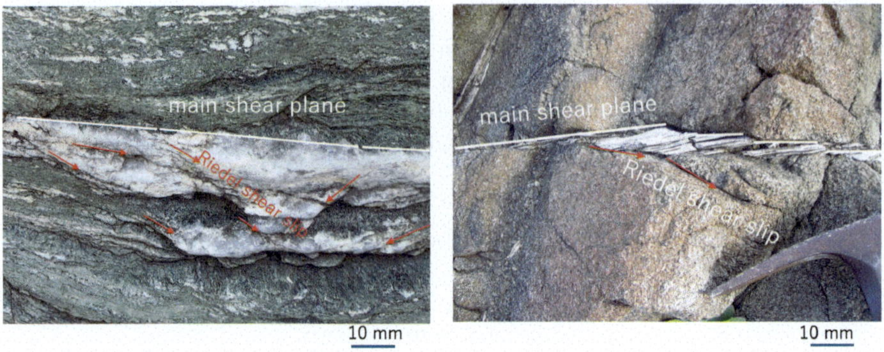

Fig. 2.5 Photographs representing typical structure of sealed shear crack jog filled with albite/quartz and chlorite by multiple Riedel shears (left), and sealed simple jog filled with elongated calcite grains in the Muroto gabbro of Shimanto belt

rocks. In the present case, the influx of the fluid component H_2O should be controlled by the increasing rate of the jog volume.

Assuming the molecular influx of the H_2O into the jog to be $m(t)$ and the volume increase of jog to be $V(t)$, then we have from the state equation of fluid,

$$P_f(t)V(t) = m(t)RT \tag{2.6}$$

in which P is the pressure, T is the temperature, and R is the gas constant. In this study, the temperature may decrease with sudden decrease of pressure under adiabatic condition, but the ratio of this temperature decrease is rather small in the effect on the mineral oversaturation than that of large pressure down.

The volume of the jog space is assumed to be governed by the macroscopic shear slip velocity $u(t)$, and influx of the H_2O molecules from grain boundary fluid through jog surface (Fig. 2.6), j, is followed by Darcy's law,

$$j = k \operatorname{grad} P_f \tag{2.7}$$

in which k is the permeability of the matrix, S is the jog surface, and P_f is the fluid pressure.

Here, $V(t)$ is the integral of the $S\,u(t)$, and grad P_f is approximated to be $(P_f - P_s)/r_o$, and $m(t)$ is expressed to be,

$$m(t) = Sj/v_o \tag{2.8}$$

in which v_o is the molar volume of fluid. Therefore, we obtain the following state equation,

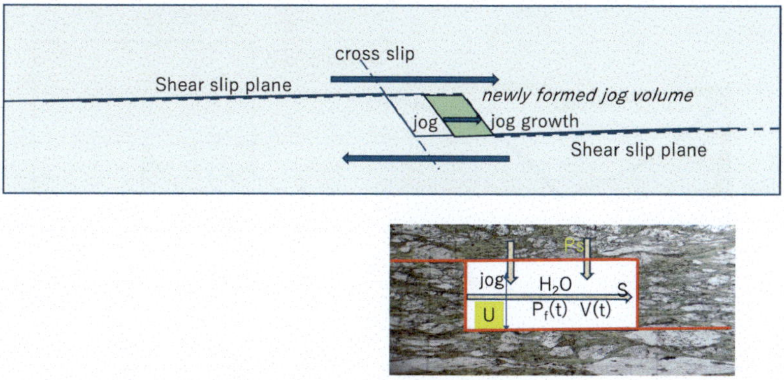

Fig. 2.6 A model of rapid growth of jog volume by shear crack growth, being associated with fluid influx to jog from surrounding rock. V; jog volume, S; jog surface, P_f; fluid pressure of jog, P_s; solid pressure of rock, u; shear crack velocity

$$P_f(t)\int_0^t Su(t)\mathrm{d}t = RT\int_0^t Sj(t)\mathrm{d}t/v_o \tag{2.9}$$

In which S is assumed to be constant, and $u(t)$ (m/s) ($< S$) is approximated to be the linear function of time like the seismic velocity as,

$$\begin{aligned} u(t) &= Ut(0 < t \le t_o,\ U \text{ is constant}) \\ u(t) &= 0(t_0 < t). \end{aligned} \tag{2.10}$$

$$P_f(t)\int_0^t Su\mathrm{d}t = k\mathrm{RT}\int_0^t S(P_s - P_f)/v_o r_o \mathrm{d}t \tag{2.11}$$

Differentiation of above equation, we obtain,

$$(\mathrm{d}P_f/\mathrm{d}t)\left(\int_0^t Su\mathrm{d}t\right) + SP_f u = (kRT/v_o r_o)S(P_s - P_f) \tag{2.12}$$

$$\begin{aligned} \mathrm{d}P_f(t)/\mathrm{d}t &= (BP_s - (B + Ut)P_f(t))/\left(V_o + Ut^2/2\right) \\ B &= k\mathrm{RT}/v_o r_o \sim 1 \end{aligned} \tag{2.13}$$

$$\mathrm{d}P_f(t)/\mathrm{d}t = (P_s - (1 + Ut)P_f)/\left(V_o + Ut^2/2\right) \tag{2.14}$$

In this equation, assuming $v_o = 0.01$ cc and $P_s = 1000$ MPa, the relations of P_f versus time and P_f versus volume of jog can be simulated in the simple simulation method. The results are shown in Figs. 2.7, 2.8, 2.9, 2.10.

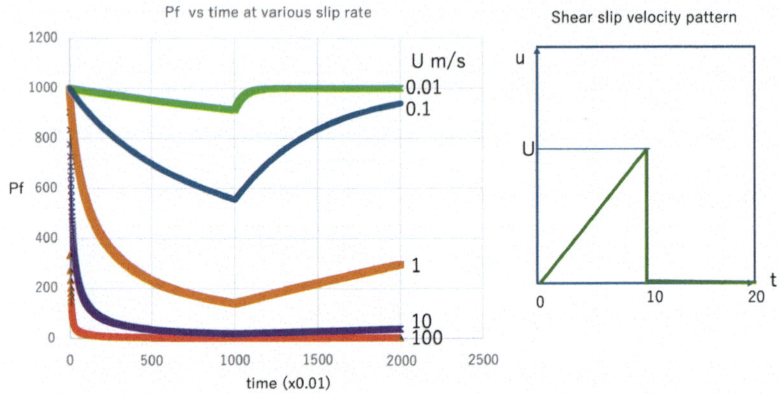

Fig. 2.7 Simulated fluid pressure changes with time in the various maximum velocity of shear crack growth (0.01–100 m/s). Right figure shows the pattern of shear crack velocity with maximum *U*

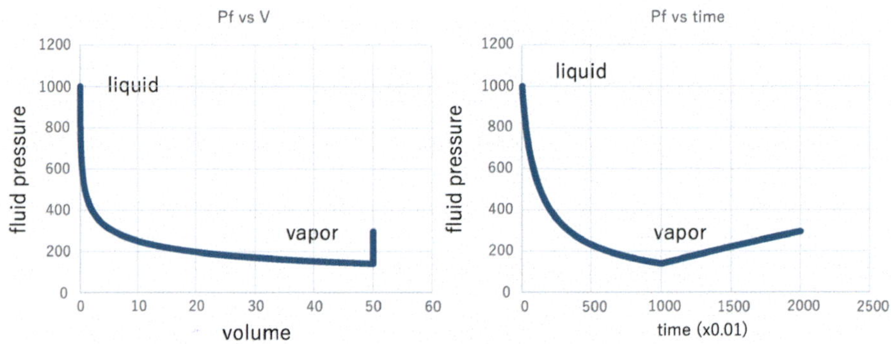

Fig. 2.8 Simulation results of P_f-volume of jog during the jog growth and P_f with time at the case of $U = 1$ m/s, showing the aqueous fluid is initially fluid but later it changes to gas

The temporal changes of the fluid pressure in the jog space, therefore, display the rapid decrease from the rock pressure to the very low pressure, indicating evaporation of H_2O molecules in the jog space in the case of high shear velocity. On the other hand, in the low shear velocity, it is obvious that decrease of the fluid pressure in the jog is small from surrounding solid pressure P_s.

The above scenario may be satisfied in the conditions that the surrounding rock is very hard like the artificial piston, but it should be modified in the cases of ductile rocks and low fracture strength of surrounding rocks. The ductile rocks in the high temperature may dissipate the high differential stress around the jog and crack tip, and jog space would be slowly deformed and filled with ductile rock materials, suggesting that the shear slip plane may deform into the curvilinear cross section. On the other hand, in the low fracture strength case, the surrounding rocks would be fragmented by rapid decompression and microflashing. Next, it should be discussed

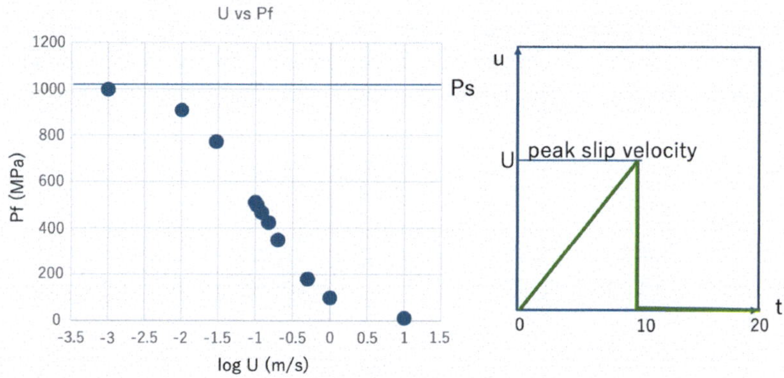

Fig. 2.9 Simulation results of relationship between lowest fluid pressure P_f (initial pressure $P_f =$ 1 GPa) and maximum shear crack velocity U shown in right, showing the sigmoidal relation of the P_f versus shear slip velocity

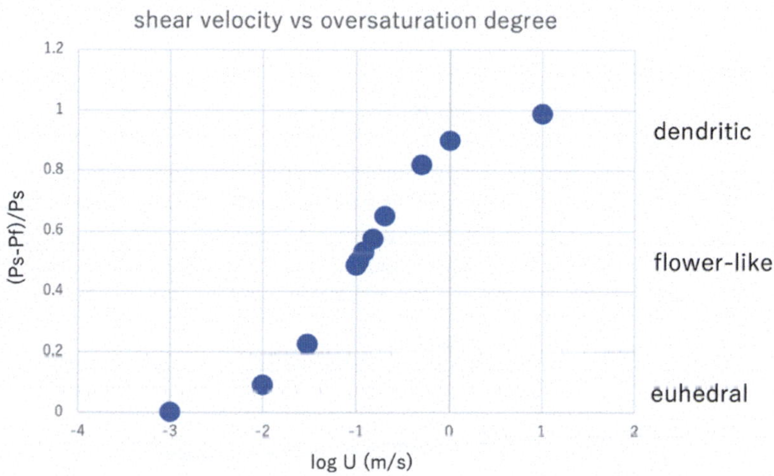

Fig. 2.10 Relation between the degree of oversaturation in pressure and logarithmic crack velocity showing the sigmoidal relation approximately to $1 + \tanh(\log U)$

whether jog volume formed by rapid shear slip event may be collapsed or not by decompression fracturing of surrounding rocks.

Microflashing as the decompression fracturing of jog wall should be due to the extensional fragmentation by multiple opening of microcracks network. The elementary process of this multiple opening should be governed by small increments of microcrack growth. Then, the growth velocity v_c of each microcrack under critical Griffith crack length) is possibly approximated (Wei 2010) by,

$$v_c = A(\pi c)\Delta\sigma^2 - B \tag{2.15}$$

in which A and B are parameters, c is the crack length near the grain size, and $\Delta\sigma$ is the differential stress. The differential stress should be equal to the fluid pressure difference between jog interior and surrounding rock, and thus, the above equation becomes,

$$v_c = A(\pi c)\Delta P_{\mathrm{f}}^2 - B. \tag{2.16}$$

On the other hand, the multiple opening of the microcrack network should take place at the time of reaching the jog size, L. Therefore, we obtain the time of jog collapse as,

$$t_c = L/v_c = 1/\left(k_1 \Delta P_{\mathrm{f}}^2 - k_2\right) \tag{2.17}$$

with

$$k_1 = A(\pi c)/L \sim< 1$$

and

$$k_2 = B/L \sim< 1.$$

Therefore, the time scales of the jog collapse can be estimated from the crossing points of Fig. 2.11, as shown in Fig. 2.12, indicating the fluid pressure at the jog collapse formed by shear slip event. It is obvious that the slow slip favors the stable jog growth, and the fast slip does short-term survival of the jog volume (Fig. 2.13).

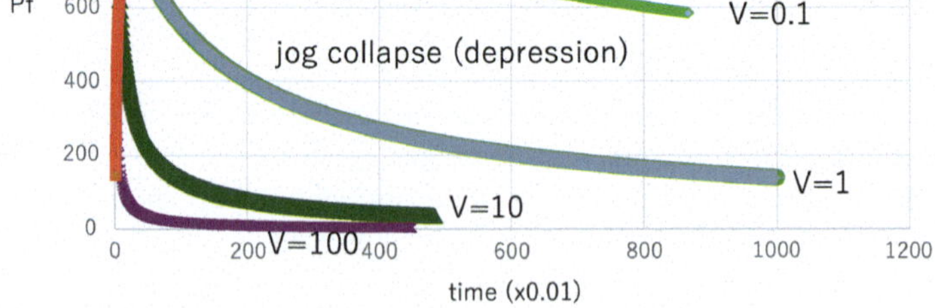

Fig. 2.11 Life time of jog space until collapse of jog by large decompression stress in the rock surrounding the jog in the diagram of the fluid pressure change in jog. The crossing points between the fluid pressure change and collapse time show the life time of jog

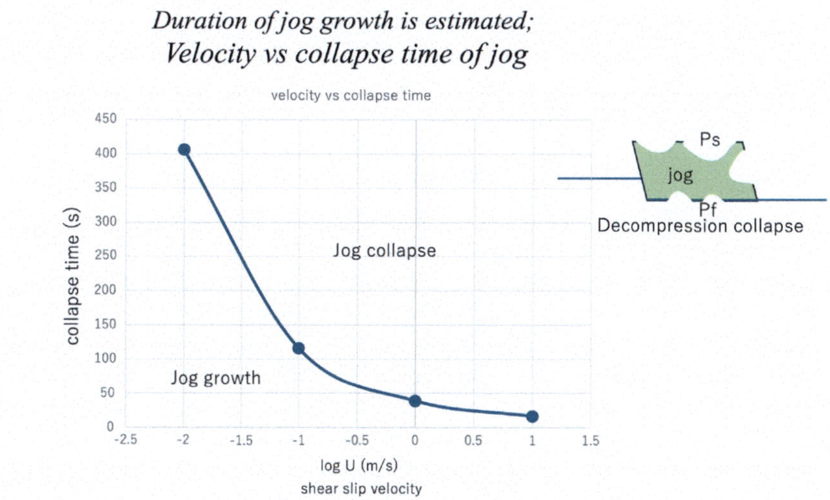

Fig. 2.12 Simulation of collapse time of jog and maximum slip velocity of shear crack, showing the jog growth regime in the left region and collapse regime in the right region

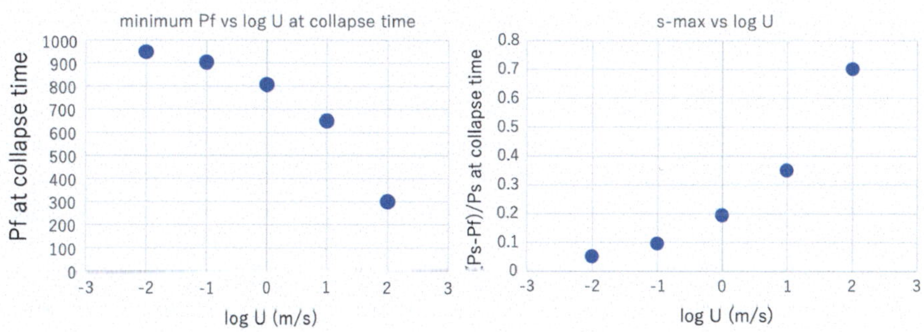

Fig. 2.13 Fluid pressure at the collapse time of jog (left), and the relation of degree of oversaturation in pressure and logarithmic crack velocity (right)

As discussed in the previous section, the growth velocity of the minerals from the aqueous fluid (liquid or gas) in the jog space varies in the wide range, depending on the degree of the oversaturation, and which is governed by the shear slip velocity and the surrounding rock toughness. The shear slip velocity should change with time as like as the seismic slip from start to stop through maximum velocity. Therefore, it seems that the starting phase of shear slip appears rapid decrease of fluid pressure in the jog volume, showing the rapid increase of the degree of oversaturation of metamorphic minerals, but after the peak shear velocity the degree of oversaturation should become low in the case of slow slip velocity. On the other hand, in the high slip velocity, the degree of the pressure oversaturation increases very rapidly, and it keeps very high level as shown in Fig. 2.10.

Growth of metamorphic minerals is controlled by diffusion onto and along the mineral surface and adsorption on the surface. The diffusion of constituent ions of mineral onto mineral surface is to be ballistic but not random process in the gaseous fluid of the jog space, and surface diffusion along on the mineral surface may be ignored because of rapid adsorption on the mineral surface. Therefore, the growth process is likely to the ballistic–dendritic growth from the jog surface against the rocks. On the other hand, if the fluid in the jog space is the aqueous solution as being in the late stage of shear crack slippage, the influx of ions on to the mineral surface is to be diffusion, and then the mineral growth process is controlled by capillary effect and diffusion rate as formulated by Mullins–Sekerka instability. In this case, the grain shape of minerals appears the waveform outline as like as flower-shape (or konpeito), showing the characteristic wavelength depending on the growth velocity and diffusion rate (Toriumi 2022).

On the other hand, considering the case of rather rapid surface diffusion, the linear growth velocities of the singular surfaces and their ratios of growing mineral are governed by the degree of oversaturation and temperature/pressure. For examples, actinolite has singular surfaces of (010), (110), and (011), and it displays acicular form elongated to *c*-axis in the high oversaturation degree in the low temperature, but in the low degree of oversaturation and high temperature it does tabular shape. On the other hand, epidote has (001), (100), and (011) and (111), and elongates to *b*-axis in the high oversaturation and low-temperature condition. In the case of prismatic quartz, growth velocity of *m*–surface is much lesser than *r*–surface, and thus it appears acicular quartz grain in the large degree of oversaturation.

Therefore, it is possible that the albite/quartz crystal form precipitated in the jog space and band zones noted previously is divided into dendrite, flower-like, acicular, and euhedral tabular shapes in the diagram of the logarithmic shear slip velocity related to the oversaturation degree and temperature as shown in Fig. 2.14, being likely to the Nakaya diagram of snow in the air. In this paper, the author would like to propose the shear velocity and mineral shape diagram in terms of jog space mineral growth as discussed previously. The degree of oversaturation, s, is the ratio of difference between equilibrium pressure and actual one against equilibrium one, and in the jog space, it becomes,

$$\begin{aligned} s &= \Delta v(-P_{\mathrm{f}}(t) + P_{\mathrm{s}})\big/\mathrm{RT} \\ &= \Delta v P_{\mathrm{s}}(\tanh(\log U) + 1)\big/2\mathrm{RT} \end{aligned} \tag{2.18}$$

in which c is the shear crack length, Δv is the partial molar volume difference between mineral and ionic species, and C is the constant term.

By the way, the linear growth velocity may be proportional to the degree of oversaturation in the low oversaturation conditions (Langer 1980), and thus it holds,

$$\begin{aligned} V &= \alpha s^2 \\ &= c\lambda^{-2} \end{aligned} \tag{2.19}$$

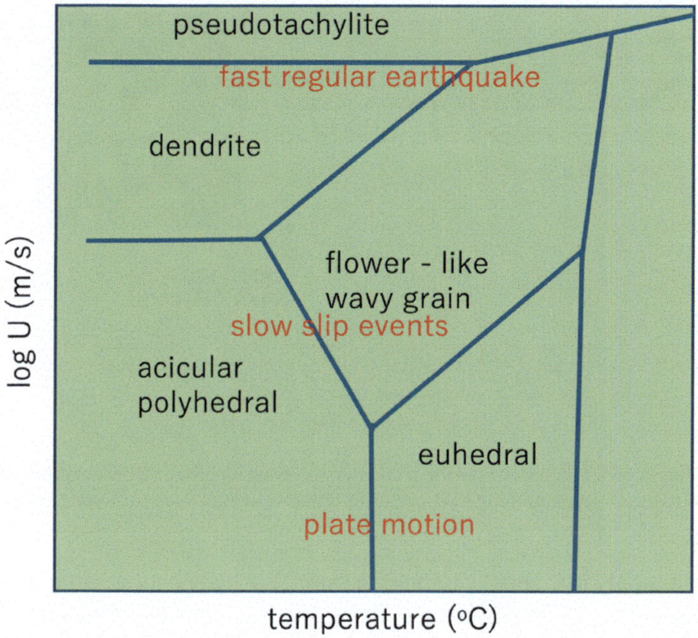

Fig. 2.14 Possible phase diagram of shear crack velocity and crystal habit of albite/quartz in jog

and, from above equation, we obtain

$$\tanh(\log U) + 1 = 2\mathrm{RT}(c/\alpha)^{1/2}\lambda^{-1}/\Delta\nu P_{\mathrm{s}}. \tag{2.20}$$

Thus, the relation of the slip velocity and wavelength of flower-grain becomes

$$\begin{aligned} &\log U - \tanh^{-1}\left(C\lambda^{-1}-1\right) \\ &C = 2\mathrm{RT}c^{1/2}/\alpha^{1/2}\Delta\nu P_{\mathrm{s}} \end{aligned} \tag{2.21}$$

Therefore, the shear slip velocity is a function of wavelength of the flower grains of mineral in the jog, and it is possible that the observable mineral shape in the jog is an indicator of shear slip velocity (Fig. 2.15).

On the other hand, increase of temperature should increase growth rate because of Arrhenius type activation process, and then the diagram of log U and temperature for various wavelength of flower-like grain growth can be inferred. In this diagram, the reason why the contour line of the iso-wavelength displays the rapid increase with increasing temperature is responsible for the asymptotic relation between log U and degree of oversaturation of minerals discussed previously.

By the way, the strongly anisotropic growth of minerals is possible in the low temperature and high oversaturation degree because of differential linear growth velocity on different low index surfaces of euhedral grain. For example, quartz

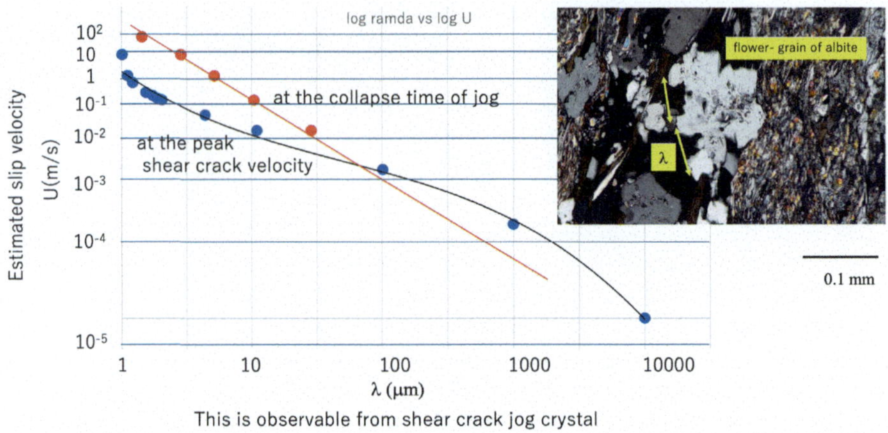

Fig. 2.15 Simulated relation between the characteristic wavelength of the wavy grain boundary of growing albite/quartz in jog and logarithmic shear crack velocity

displays commonly a prismatic grain in the vein, but it appears acicular in the sealed jog and open crack. The relation between the length and width of the tabular to acicular grain of quartz for various jog and open crack in the Shimanto rocks is shown in Fig. 2.16. In the same figure, acicular quartz grains in on-forming submarine chimney of Iheya, Okinawa Trough (Yamamura et al. 1997), and Dalnegorsk skarn of Russia are plotted, suggesting the large degree of oversaturation and low-temperature precipitation of them in various jog and open crack volume.

Therefore, we obtain the relation of crystal form and seismic shear velocity that called as the seismic-crystal diagram in this paper is shown in Fig. 2.17, indicating that the dendritic crystals would be formed in the conditions of high-speed seismic slip in the jog space, but the flower-like ones should be crystallized in the rather slow earthquake event. In addition, it is expected that the change of crystal morphology from

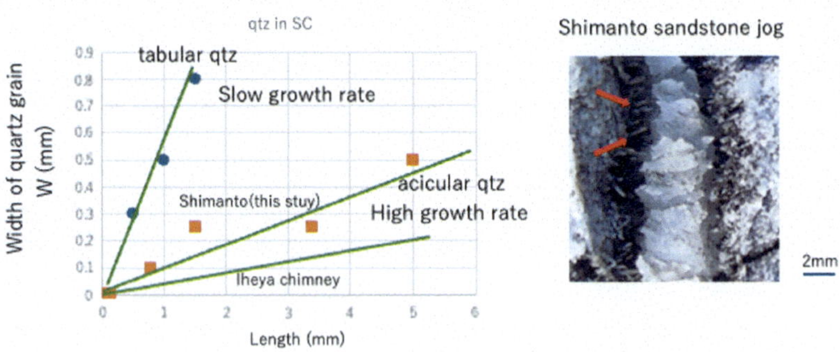

Fig. 2.16 Acicular euhedral grains of quartz showing the large aspect ratio in the sealed open crack of Shimanto sandstone and submarine chimney of the Iheya, Okinawa trough and common quartz veins in Shimanto sandstone

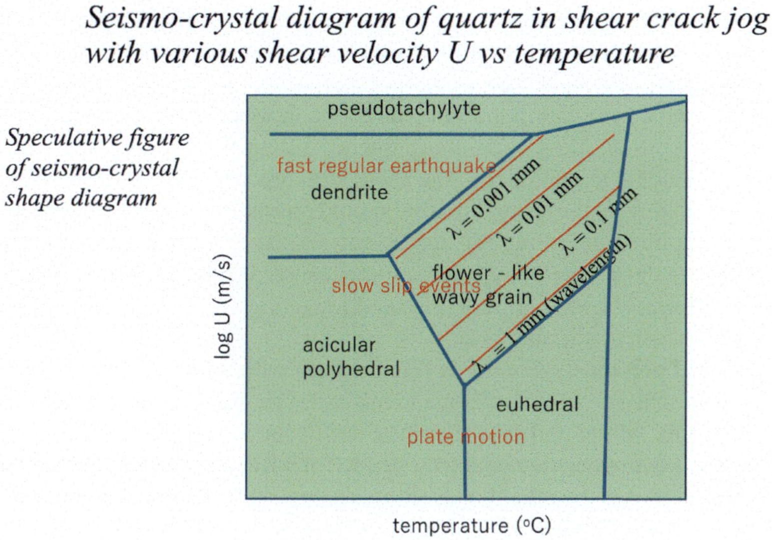

Fig. 2.17 Phase relation of the crystal morphology of albite/quartz in sealed shear crack jog in the diagram of logarithmic shear crack velocity and temperature called as Seismic-crystal diagram in this paper

dendritic to flower-like grains in the shear crack jog may take place with advancing seismic slip motion in the plate boundary metamorphic rocks.

In this section, the author intends to show the mineral morphology in the shear crack jog and mineral bands. To describe the characteristic features of minerals in jog, the shear crack jog should be described by its characteristic features in terms of petrography. Shear crack jogs appear commonly as lens-like shape under the thin section parallel to lineation and normal to schistosity. On the other hand, on the plane parallel to the schistosity, the jog displays the ellipsoidal outline with various long axis. The long axis orientation does not coincide with the mineral stretching lineation, and it varies angles from several to several ten-degree crossing with stretching lineation, suggesting the jog migration in the direction different with shear slip direction. It shows that the jog velocity is the sum of shear slip velocity and tangential velocity of the jog.

There are two patterns of the sealed jog in the metamorphic rocks: one is the symmetric type and the other is the asymmetric one as shown in Fig. 2.5. The symmetric type appears lens-like shape symmetric against the central plane of the jog as seen in the calcite filling jog in gabbro of Muroto peninsula, but in the asymmetric type, albite and quartz domain shows apparently concave and diffuse top feature different with bottom sharp feature against chlorite domain. The former displays the aligned mineral elongation oblique to the shear plane and jog boundary. On the other hand, the latter type shows the structure like the flame structure as seen in the layered soft sediments. It is noteworthy that the asymmetric sealed jogs show the similar configuration of the top and bottom described previously in the 10–100 m scale outcrops. This asymmetric type should be interpreted by the combination of the cross slip by Riedel and penetrative shear on main shear plane as illustrated in Figs. 2.1 and 2.5.

The jog boundary normal to the shear direction moves toward the shear direction under the thin section above, and then minerals filling the jog displays tabular to acicular grains elongated parallel to the shear slip direction. On the other hand, the jog boundary oblique to the shear direction should migrate parallel to the jog velocity. In this case, the minerals filling jog space grows to be elongated normal to the jog boundary that is the jog velocity orientation because of directed growth in the jog space. It, therefore, shows that both of sealed jog elongation and sealing grain may be oblique to the shear slip direction.

As described previously, the jog is filled mainly with albite, quartz, chlorite and calcite in basic schists, but minor constituents are very fine-grained epidote, actinolite, and goethite. Albite and quartz sometimes occur as dendritic and/or compact clusters of small grains near the boundary surface of jog against surrounding matrix as shown in Fig. 2.18. On the other hand, albite, calcite, and chlorite-filling jog volume appears as flower-like grains with characteristic spacing of the spikes of the wavy outline (Fig. 2.19). In addition, the elongated grains showing intergrowth texture of albite and quartz are often found in the jog.

The large lens-like jogs are not so common but found somewhere in the basic and pelitic schists. The size of them reaches about several 0.1 to 1 m length. In basic rocks, the large jogs are filled with intergrown albite and quartz, but in pelitic rocks, they are occupied by intergrowth aggregates of quartz and calcite. The intergrown albite and quartz appear to be like a dendritic feature as a single grain. In the large lens-like jogs of pelitic schist, the intergrown albite and calcite look like flower-like grains with rounded corner and spikes as shown in Fig. 2.20, suggesting the annealing or rather high growth velocity of albite and calcite intergrowth.

The seismic and non-seismic shear slip produce the jog space in which inflow of fluid phase should be imbalanced with the opening rate of the jog volume. Then, the

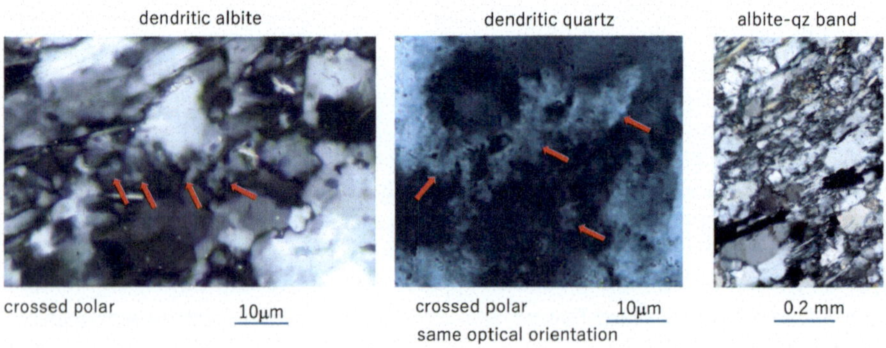

Fig. 2.18 Dendritic growth of fine-grained albite (left, red arrows) and quartz (right, red arrows), appearing in the jog filled with waveform grains of albite and quartz in the chlorite zone of the Asemigawa route, Sanbagawa metamorphic belt. Crossed polars

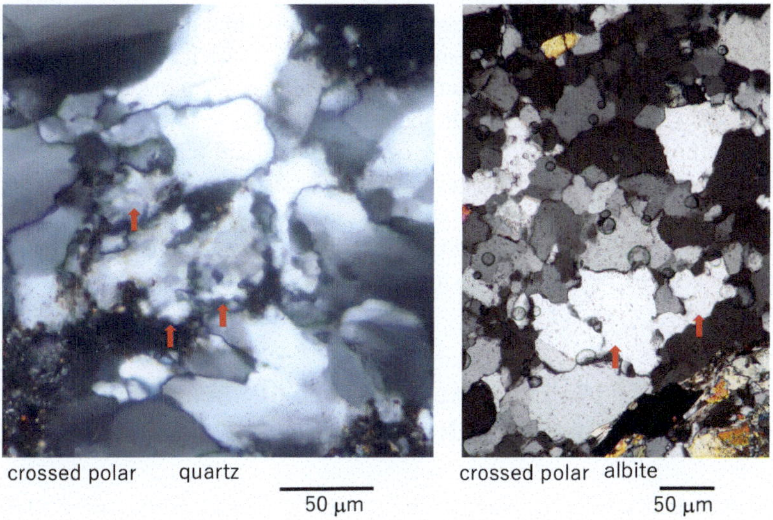

Fig. 2.19 Flower-like quartz (left) and albite (right) with wavy grain boundary in sealed jog in the chlorite zone rocks of the Asemigawa route. Crossed polars

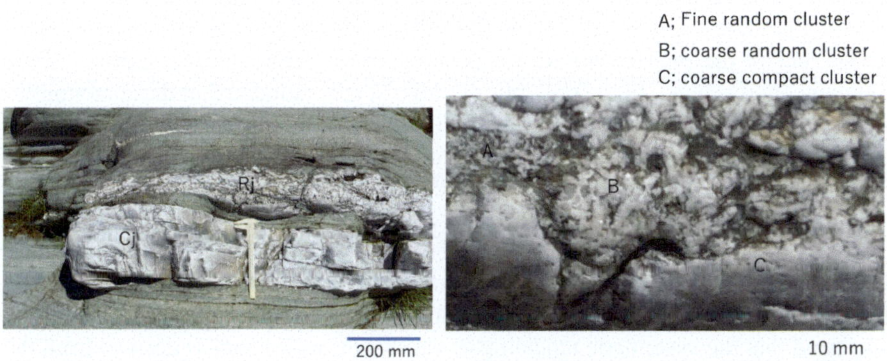

Fig. 2.20 Two types of large sealed jogs: one is compact jog filled with coarse-grained albite and quartz (CJ) and random clustered jog filled with flower-like grains of albite, quartz, and matrix calcite (RJ) which display the layered structure of fine-grained, medium-grained cluster and compact cluster

fluid pressure in the jog space decreases very rapidly at first and then increases with opening the jog, and the degree of oversaturation of minerals in the jog increases rapidly and then decreases later. The progressive change of fluid pressure and degree of oversaturation could be simulated by equation of state and mass conservation of fluid with Darcy law in this study. Furthermore, the large discrepancy of the fluid pressure between matrix and jog space makes the jog volume to collapse by fracturing of the matrix. As a result, the degree of oversaturation of minerals in aqueous fluid changes with sigmoidal curve against the logarithmic maximum velocity of shear slip. Because the growth velocity of minerals increases with increasing the degree of oversaturation, it seems that the growth velocity of minerals in the jog increases with increasing the shear slip velocity. On the other hand, the growth shape of minerals in aqueous fluid is controlled by the growth velocity and temperature to form the dendrite, flower-like, and acicular grains with decreasing growth velocity. Euhedral shape should be satisfied in the low growth velocity conditions. Then, it seems that various crystal shapes can be mapped in the diagram between the shear slip velocity and the temperature, being called as the seismic-crystal shape diagram in this paper.

2.2 Decompression Collapse Structure of Shear Crack Jog in the Plate Boundary Rocks

There is abundant schistosity-parallel monomineralic band in the low-grade metamorphic schists in the plate boundary zone: in basic schists, layers are albite and chlorite bands, and in the pelitic schists are quartz and muscovite band. Each band contains a few amounts of accessary minerals such as quartz, chlorite, and calcite in albite band. In these schists, the matrix is composed of amphiboles (actinolite, crossite), epidote, chlorite, albite, quartz, sphene, and magnetite in the basic greenschists, but in the pelitic schists, composed of muscovite, chlorite, quartz, albite, magnetite, and sphene. In the previous sections, it is concluded that these monomineralic bands have been formed from precipitation of minerals from aqueous solution in the moving and expanding shear crack jogs along the shear crack.

In the previous section, it shows that the simulation study of the rapid expansion of jog manifests the rapid decompression and successive pressure recovery of jog space, and thus does large degree of supersaturation of albite, chlorite, and other matrix minerals rather than the aqueous solution in the matrix, thereby suggesting the appearance of rapid grown grains of albite and quartz. Furthermore, it is also suggested that the systematic repeating of various mineral bands is derived from the difference of the strength of decompression in the jog.

By the way, the decompressed volume formed by the growth of jog along the shear crack should be mechanically unstable because of imbalance of stress between matrix and jog, and thus the jog should be collapsed by negative pressure against the matrix during the precipitation of minerals in jog. Therefore, it seems that the collapse structure should be made by swelling of the jog wall into jog interior, or the brecciation of the matrix occurs by decompression. In the next, the author intends to show the characteristic structure of decompression collapse of the shear crack jog.

The collapsed jog is commonly found in the low-grade basic and pelitic schists of the Sanbagawa metamorphic belt, and it is sealed by mainly albite in basic schists and by quartz in peltic schists, but minor components of them are quartz, chlorite, epidote, and actinolite. The characteristic petrography of the collapsed jog shows a wide range of grain size having the various appearance of spatial distribution of

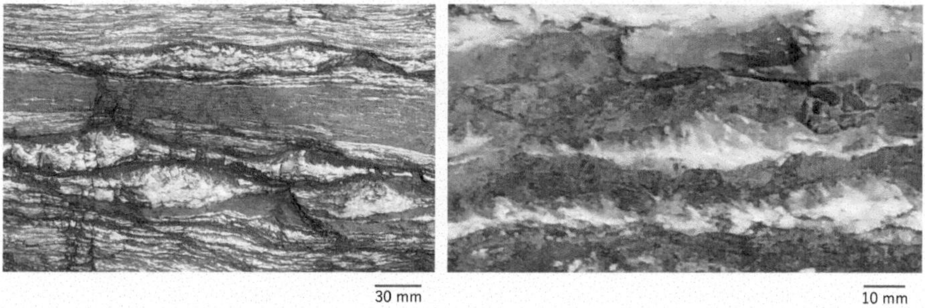

Fig. 2.21 Photographs showing decompression dent structure like sedimentary flame structure of the shear crack jog sealed by albite and quartz. In these photographs, the downward depression structure is seen

grain size, and various shapes of the grain ranging from euhedral to flower-like and ameboid shape, and rarely dendritic shape. It seems that this type of shape variation of the grains should be derived from crystallization from aqueous fluid in various degrees of oversaturation that means the various crystal growth velocity in the jog.

First, it should be noted that there is apparently strange occurrence of the collapsed structure in the outcrop scale. As shown in Fig. 2.21, upward-open and frame-like structure of white band (albite band) appears frequently in the basic schists. The length and thickness of the frame-like structure vary ranging from several mm to meters, and the wavelength of the characteristic frame-like structure ranges from mm to meters also. On the other hand, frame-like structure displays various patterns on the top surface of the frame structure as shown in Fig. 2.21: there are several fundamental patterns of the white and green parts in the top surface of the sealed jog, which are rounded–elliptical white domain, geometrical angular white domain, and parallel–tabular white domain, and they are surrounded by green domain. The white domain is mainly composed of albite and small amount of quartz and/or calcite, but the green one mainly of chlorite and small amount of albite.

It sounds rather strange that although apparent top of the sealed jog shows obvious and distinctive frame structure, apparent bottom feature displays different pattern with enough small wavelength of apparent frame structure. These apparent frame structures of both top and bottom of jogs appear the similar concave pattern against the sealed jog interior. However, it seems that the wavy pattern of apparent frame displays periodic one, and the size of wavelength can be identified: one of apparent top is much larger than that of bottom as shown in Fig. 2.22.

At this time, the author would like to identify the above-described apparent frame structure to be the collapsed swell structure of the jog, because it is not the sedimentary frame structure but one of the decompression patterns of the jog during the crystallization of sealing minerals such as albite, quartz, and chlorite. Later, this identification will be suggested from the results that the geometry of the shape of the swell itself is identified from the sedimentary frame shape and possible catenary curve.

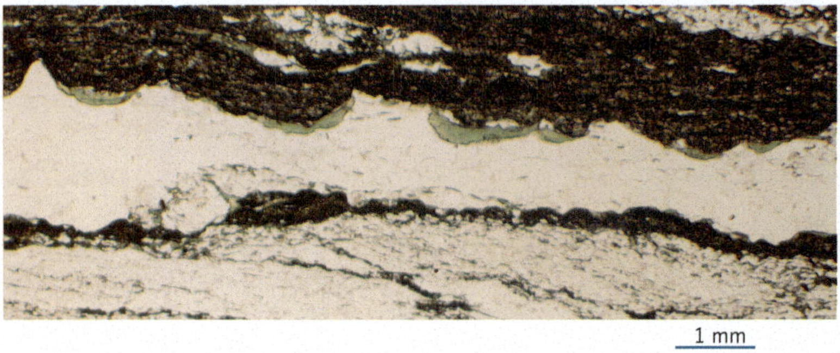

Fig. 2.22 Photograph showing periodic decompression dent structure sealed by albite and quartz under microscope

The top and bottom structure of the sealed jog representing the collapsed dent appears almost always the same orientation in the single outcrop of 10 m width, although rare case that two top and bottom structures with mutually reverse relation are rarely found in the several outcrops in the studied Nagatoro area of the Sanbagawa metamorphic belt, as seen in Fig. 2.23. From these occurrences, it should be obvious that the top and bottom structure of the sealed jogs in the very schistose and banded schists of the Sanbagawa metamorphic belt formed in the plate subduction zone has been derived from the decompression collapse of the shear crack jog with low pressure aqueous fluid in the crystallization of sealing minerals in the gravitational field. In this model, the top of the sealed jog having periodic collapsed swell was a large volume composed with low pressure fluid, and even at the bottom layer, the low-pressure fluid remains to be collapsed upward in the mode of small wavelength.

In this section, the relationship between the wavelength and amplitude of the swell structure of the top and bottom of the jog will be measured in the outcrops of the Nagatoro area of the Sanbagawa metamorphic belt, Kanto Mountains. In the studied

Fig. 2.23 Photographs showing upward geometry of decompression dent structure which appear upward concave in the outcrop scales of Sanbagawa low grade greenschist

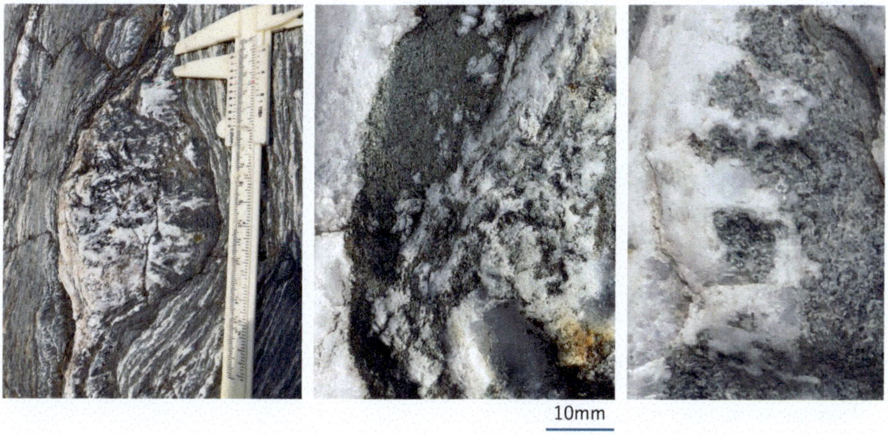

Fig. 2.24 Top view of the decompression dents on the plane parallel to schistosity from Sanbagawa low grade greenschist

area, there are abundant various sized dented jogs of the basic and pelitic schistose low-grade metamorphic rocks even in the single outcrop. So, it is possible to obtain the systematic change of the dent geometry ranging in size from 1 to 100 mm scales.

The results for the top features of the jog are shown in Fig. 2.24, and it seems that the wavelength of the jog dent is proportional to the thickness of jog and amplitude of wavy dent structure. On the other hand, these values of the bottom feature of the jog cannot be measured because of lack of the periodic dent structure as shown in Fig. 2.24, thereby showing the direct contact between massive layer of sealed jog and the chlorite thin band of the matrix part. In the case of the clear periodic dent in the bottom of the jog, the results are shown in Fig. 2.24, thereby suggesting the large discrepancy of wavelength and amplitude between the top and bottom

By the way, it is well-known that the soft materials make the repeating pattern of swelling structure during the decompression process by nonlinear kinetics of transformation (Guvendiren et al. 2009). In their study, it has been obtained experimentally that the thickness d of the plastic film is proportional to the wavelength λ of the dent structure (collapse structure) as,

$$\lambda(\mu\mathrm{m}) = 0.5d(\mu\mathrm{m}). \tag{2.22}$$

On the other hand, in this study of natural decompression dent of the jog, the relation between wavelength and thickness becomes,

$$\lambda(\mathrm{mm}) \sim 0.6d(\mathrm{mm}). \tag{2.23}$$

It seems strange that the decompression behavior of the soft matters quite resembles that of the shear crack jog in the highly sheared metamorphic rocks. However, thinking for the elastic–plastic behavior during decompression process, the soft

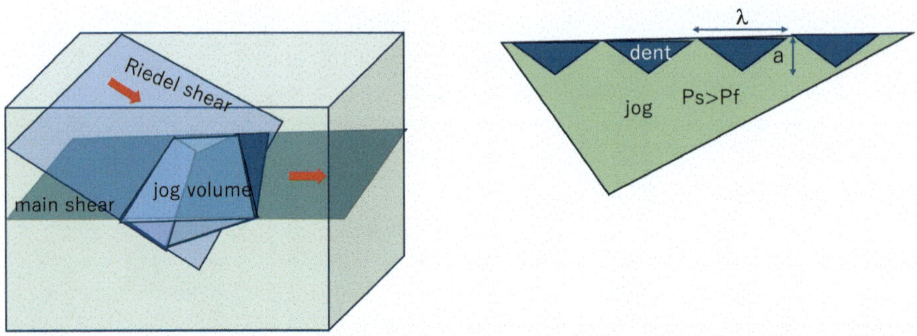

Fig. 2.25 A model of shear crack jog volume formed by cross-slip with Riedel shear (left) and dent of jog surface formed by decompression buckling (right). λ and a are wavelength and amplitude of the decompression dent on jog surface

matters may be governed by the similar universal relation in nonlinear dynamics, that is the macroscopic pattern formation consisting with wavy decompression collapse structure like present jog collapse structure.

Therefore, the author intends to propose a new model for the depression structure formed in the decompression process of the shear crack jog in the metamorphism.

First, let us assume a model of the decompression jog with tabular shape as shown in Fig. 2.25.

In this model, the shape of depression is approximated to be triangular instead of sinusoidal shape, because of simple estimation of the energy increase by periodic depression of the jog. In this model, wavelength and amplitude of the depression are λ and a, respectively.

The total strain energy E_s for triangular depression is obtained by average strain of (a/λ) as,

$$E_s = \eta(a/\lambda)^2 \lambda a \tag{2.24}$$

in which η is effective viscosity of the matrix.

And next, surface energy increase (E_i) by depression becomes,

$$E_i = \gamma\lambda\left(1 + (a/\lambda)^2\right)^{1/2}. \tag{2.25}$$

Therefore, the total increase of energy U_t of the height of dent a and n of number of dent holds,

$$U_t = n\left(\eta(a/\lambda)^2 \lambda a + \gamma\lambda\left(1 + (a/\lambda)^2\right)^{1/2}\right). \tag{2.26}$$

The critical conditions of the periodic depression can be represented by the minimum of the above total energy increase in the decompression.

Accordingly, the critical conditions can be evaluated by,

$$U_t = -an\left(\eta ax + \gamma\left(1 + x^{-2}\right)^{1/2}\right) \tag{2.27}$$

with $x = a/\lambda$

Then, we have,

$$(1/an)\mathrm{d}U_t/\,\mathrm{d}x = \eta a - \gamma\left(1/x^3\right)\left(1 + \left(1/x^2\right)\right)^{-1/2} = 0 \tag{2.28}$$

Thus, we obtain,

$$a^2 = Kx^{-6}\left(x^{-2} + 1\right)^{-1} \tag{2.29}$$

and,

$$K = (\gamma/\eta)^2. \tag{2.30}$$

Here, the relation between a and y under $K = 1$ will be obtained by simulation as shown in Fig. 2.26, and then it holds,

$$1/a = 1.4(a/\lambda)^4 \tag{2.31}$$

Thus, we obtain,

$$\lambda = 1.08\, a^{5/4} \tag{2.32}$$

Therefore, it holds,

$$\log\lambda - (5/4)\log a + C \tag{2.33}$$

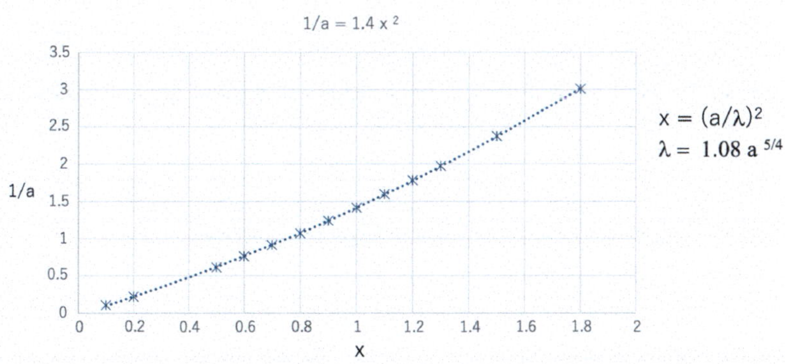

Fig. 2.26 Theoretical relation of amplitude, a and wavelength, λ of the decompression dent on jog surface

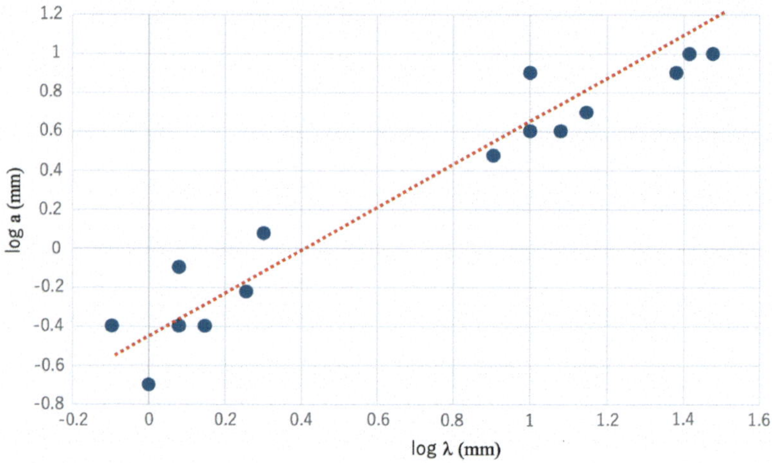

Fig. 2.27 Diagram of logarithmic amplitude and wavelength of the decompression dent of the jog surface from Sanbagawa low grade greenschists in Nagatoro, Kanto Mts

In this equation, C is the constant term involving K.

The observed relationship in Fig. 2.27 between the wavelength and amplitude of the depression in the shear crack jog is in accordance with the theoretical critical depression geometry obtained above.

The microstructure of the decompression dent displays characteristic feature that the outline of the dent commonly appears babble-like composite circular shapes, and irregular cloud-like ones, and their composite shapes as seen in Fig. 2.28, thereby suggesting the depression collapse into the albite sealing jog volume. Even under the thin section parallel to stretching axis and shortening one, it is obvious that there is common feature showing the rounded dent boundary into the albite sealing jog, and the outline seemingly looks like the frame structure in the gravitationally unstable soft sediment layers (Fig. 2.28).

It is important that the depression zone of the surrounding matrix regions is composed of fine-grained chlorite and epidote, and often small amount of fine-grained albite and quartz, as is seen in Fig. 2.29. It seems that there are some varieties of dendrite-like aggregates of fine-grained albite and quartz.

2.3 Microcrack Clusters in the Plate Boundary Metamorphic Rocks

In the previous book, the author proposed the dissolution–precipitation–fluid flow model of the metamorphic banding in the simple system that contains chlorite, albite, and pore in the basic schist analogue (Toriumi 2022). It is suggested that the regional fluid flow of aqueous solution is assumed to be vertical to the schistosity plane

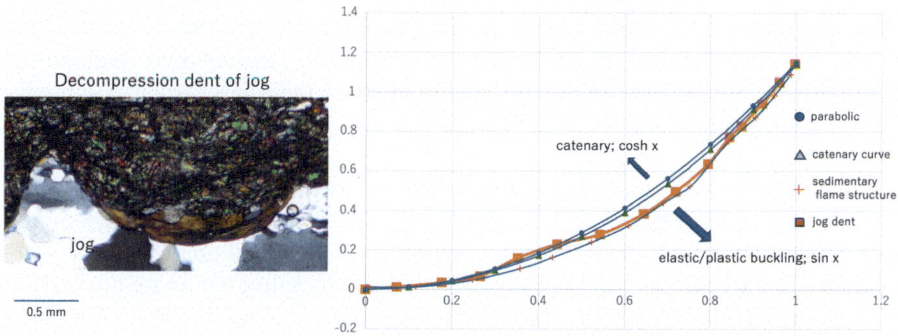

Fig. 2.28 Outline curve of decompression dent of jog surface (left photograph) and various curves of catenary case (closed triangle), sedimentary flame (cross), decompression case (closed square), parabolic (closed circle), and buckling case (sin *x*)

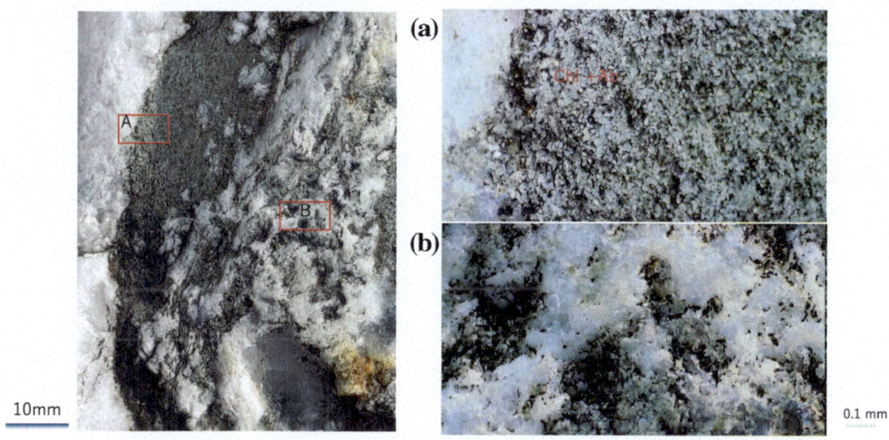

Fig. 2.29 Photographs from the top of the jog showing fine-grained chlorite and albite in the jog exterior and albite in the jog interior. Left figure displays overview of the jog and dent, and in the right figure, **a** is the close-up view near the dent boundary, and **b** is the close-up of the diffusive boundary of the dent composed of fine-grained albite and chlorite

because of common occurrence of fluid filling microlens elongated vertically to the schistosity plane by TEM study (Toriumi et al. 1986). The critical texture suggesting the orientation of fluid flow during the metamorphism is that the clusters of the type I microcrack elongated normal to the schistosity sealed by chlorite, actinolite, albite, and quartz are very frequent, as discussed in the previous book (Toriumi 2022).

In this section, the low-grade metamorphic rocks in the Sanbagawa metamorphic belt of Kanto Mountains and central Shikoku were studied. The samples are of greenschists in the chlorite and garnet zones and are composed of mainly amphiboles, epidote, chlorite, albite, quartz, sphene, and garnet (Toriumi 1990). Greenschists show commonly a banding texture composed of albite band, chlorite band, epidote band, and rarely calcite band. Albite bands contain small amounts of chlorite and quartz, chlorite band does small amounts of amphiboles, and epidote, and epidote band contains chlorite and amphiboles. Amphiboles in the bands display commonly fibrous shape with aspect ratio with over 10, showing elongation parallel to the schistosity.

Microstructures of bands vary along the schistosity plane from the single layer to bifurcated layer, and array of blobs as seen in Fig. 2.30. In addition, there is common occurrence of thick rectangular block composed of albite and quartz which partly continued to the bands as shown in Fig. 2.31. These block-like bands are not derived from brecciated thick layer, but may be a variety of the band, because of no flow texture around the block and sudden change from block to bands.

The sealed open crack and shear cracks are very common in the banded schists of the studied Sanbagawa metamorphic belt as studied in Toriumi and Hara (1995) and

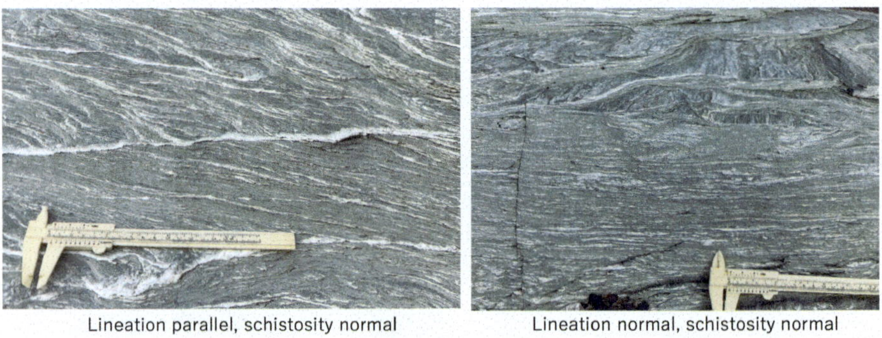

Fig. 2.30 Photographs showing multiple banding structure of albite and quartz in low grade Sanbagawa greenschist in Nagatoro, Kanto Mts

Fig. 2.31 Photograph of rectangular blocks continuing to schistosity parallel band of albite and quartz in low grade Sanbagawa greenschists of Nagatoro, Kanto Mts

Okamoto and Sekine (2011). These cracks are mainly sealed by albite, chlorite, and quartz, and minorly of actinolite and calcite. Fibrous actinolite elongates commonly parallel to the direction of the displacement of the crack. In the case of the shear crack, the jog volume formed by shear displacement is filled with albite and actinolite.

The microcracks in the banded basic schists appear as regular alignment vertical and subvertical to the banding plane, and nearly normal to the lineation of the matrix. The microcracks almost always terminate at the schistosity parallel bands, but sometimes continues across the bands (Fig. 2.32). There are two types of sealed microcracks: one is the albite-filling microcrack and another is chlorite-filling one. The thickness of both microcracks ranges from 0.1 to 1 mm, and the distance between microcracks is almost always about 10 mm. Clusters of both microcracks extend in the whole schists having banding structure, but somewhere, they show sporadic in space. It is noteworthy that there are three types in the microcrack feature between the albite bands: the one is folded, the second is bulged, and the third is straight as shown in Fig. 2.33. The former two features display plastic deformation of microcracks sealed with albite and chlorite associated with plastic deformation of the matrix between the albite bands. On the other hand, the sealing of straight type should be coeval with deformation of the matrix, and it seems that the sealing mineral components of the open microcracks should be supplied from the matrix just at their plastic deformation, being controlled by the precipitation and dissolution mechanism of the plastic deformation. This type of microcrack-sealing plastic deformation of metamorphic schists has been proposed in the brittle deformation zone of the low-grade Sanbagawa metamorphic belt by the author (Toriumi 1990; Toriumi and Hara 1995).

There are irregularly shaped microcracks sealed by albite and chlorite between the folded and displaced albite bands in the basic schists, suggesting that the sealed microcracks deformed by both of ductile and brittle modes as like as the band and

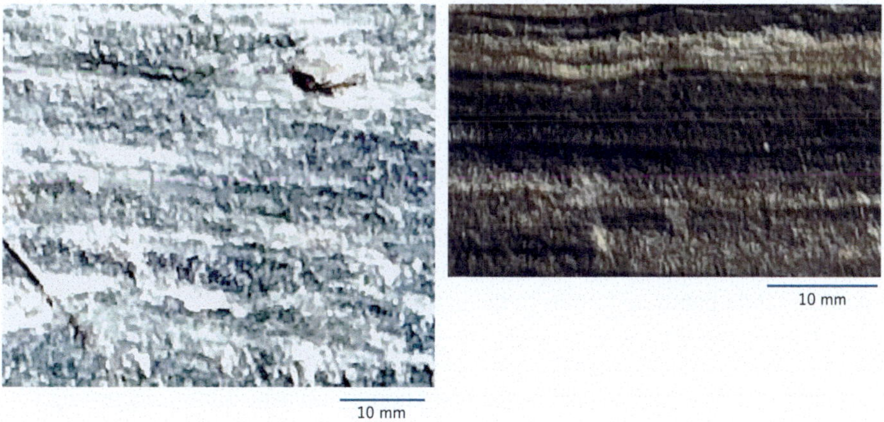

Fig. 2.32 Microcrack clusters sealed by albite which is normal to the schistosity parallel bands in low grade greenschist (left) and stilpnomelane schist (right), Nagatoro, Kanto Mts

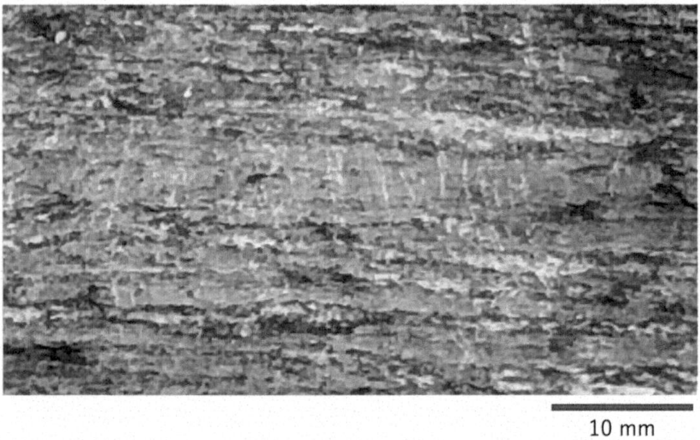

Fig. 2.33 Photograph showing partly folded and fractured microcrack cluster filled with albite and quartz in low grade greenschist, Nagatoro, Kanto Mts

matrix deformation. Some of them appear as small blobs composed of albite and/or chlorite with small amounts of fibrous actinolite.

Under the optical microscope, the sealed microcracks are constituted mainly with albite and chlorite and minorly with fibrous actinolite and calcite. The fibrous actinolite occurs comb-like feature across the microcrack (Fig. 2.34). It is noteworthy that the albite and chlorite sealing the microcracks with 0.1–1 mm of thickness occur alternatively with each other at the distance of about 0.1–1 mm. The occurrence of albite and chlorite in the microcracks appears commonly alternative with 0.1–1 mm width, showing the repeated precipitation of albite and chlorite from solution in abundant microcracks.

On the other hand, it shows that the albite bands normal to the microcracks sealed by albite and chlorite change their texture from the continuous layer to the layer-forming cluster of microcracks with short length of which orientation is nearly parallel to the microcracks described earlier as shown in Fig. 2.34. It is, therefore, suggested that the albite bands may be formed by the densely aligned microcrack cluster sealed with albite. As described earlier, there are abundant the albite-sealed microcracks of the bulge microstructure with coherent side by side, suggesting that the periodic and coherent precipitation of albite within the aligned microcracks. Thus, it seems that the coherent precipitation in the mutually neighboring microcracks makes the albite bands and blobs in the low-grade basic schists.

Even in the pelitic schists, there are abundant chlorite bands replaced after muscovite matrix and microcrack cluster sealed by chlorite: both of chlorite bands and matrix of pelitic schist contain abundant porphyroblastic albite, suggesting that the chlorite band is formed by precipitation of chlorite after muscovite. These chlorite bands crosscut the perpendicular microcracks filled with chlorite and sometimes albite. The spacing of the sealed microcracks ranges from 1 to 10 mm and their length

Fig. 2.34 Photographs of microcrack cluster filled with albite, chlorite, actinolite, and epidote, normal to the schistosity parallel albite band (left) and microcrack filled with fibrous actinolite, chlorite, calcite, and albite under optical microscope in Sanbagawa low grade greenschists, Asemigawa. Shikoku

are also from 1 to 10 mm. These microcracks terminate the shear slip plane/zone, judging from the offset of the high-angle quartz/albite-sealed open cracks.

Microcracks formed during the period of the metamorphism have been sealed by metamorphic minerals such as albite, chlorite, actinolite, and quartz through abundant permeable solution saturated by their minerals. Furthermore, it is sure that the crack opening and related dissolution and precipitation are the important contributions to the relaxation of bulk stress acted around the large shear crack as like as the viscous relaxation around the crack tips. In this case, the fracture toughness model involving microcrack formation should be employed as follows.

Let us consider the microcrack formation in the volume around the large shear crack as shown in Fig. 2.35.

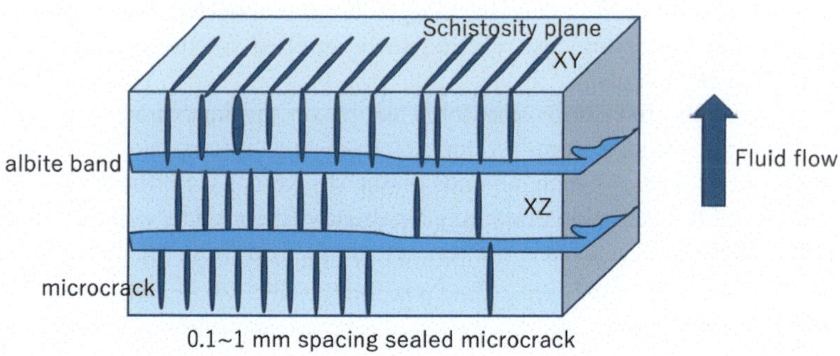

Fig. 2.35 A model of microcrack cluster vertical to the schistosity parallel albite bands which is formed by shear crack jog migration. Microcrack is of open crack (type I) and thus is fast path of fluid flow during metamorphism

In this case, the Griffith failure by the crack propagation is attributed to the balance of increasing rates of dissipation energy and elastic strain energy by incremental main shear crack length c,

$$\mathrm{d}W/\mathrm{dc}{-}\mathrm{d}U/\mathrm{dc} = 0 \tag{2.34}$$

$$dU/dc \ = \pi c\sigma^2/E \tag{2.35}$$

$$\begin{aligned}\mathrm{d}W/\mathrm{dc} &= 4\gamma + 4\pi\rho_\mathrm{m}c_\mathrm{m}\gamma \\ &= 4\gamma(1 + \pi\rho_\mathrm{m}c_\mathrm{m})\end{aligned} \tag{2.36}$$

in which W and U are total surface energy and strain energy by the main shear crack, respectively, γ is the surface energy, and c_m and ρ_m are length of surrounding open microcrack and its density, respectively.

Then, we obtain,

$$\begin{aligned}c &= \left(2\gamma E/\pi\sigma^2\right)(1 + \pi\rho_\mathrm{m}c_\mathrm{m}) \\ &= c_g(1 + \pi\rho_m c_m)\end{aligned} \tag{2.37}$$

Therefore, it follows that the velocity of large shear crack should be controlled by the crack growth mode instead of propagation mode in the case of microcrack cluster formation associated with shear slip growth. Then, the slow shear crack velocity V_s becomes (Wei 2010),

$$V_\mathrm{s} = a\Delta\sigma^n + b(n = 1 \sim 3) \tag{2.38}$$

This type of shear crack growth law is commonly applied for the small-scale cracks of many materials, but in this study, possibly for large-scale shear crack of the deep crust and mantle in the plate boundary zone, it is required that the mechanics of large-scale landslides may be referred to be applied for the mechanics of the large-scale slow slip events.

In the case of the landslides, the force acting on the landslide mass M is the gravity g, and the resistive force is derived from the viscous one within the slip zone with finite thickness, which depends on the slip velocity. Furthermore, for both terminal edges of the landslide mass may be already fractured by coupled shear and open cracks, these terminal back stresses are small, and these are independent on the major slip velocity. Therefore, the Newtonian equation of the landslide mass movement becomes,

$$M\ \mathrm{d}v/\mathrm{d}t = Mg\cos\theta - \eta Sv/d \tag{2.39}$$

in which v is velocity of the mass, S is the surface area of the mass, d is the thickness of the boundary zone, and η is its viscosity.

The initial slip velocity should be zero, and thus we can obtain the steady-state velocity v as,

$$Mg\cos\theta - \eta Sv/d = 0 \tag{2.40}$$

where θ is the angle of the shear surface under the mass.

Then,

$$v = Mg\cos\theta/S\eta \tag{2.41}$$

On the other hand, if it is in the case within the crust and mantle along the plate boundary zone, the mechanical system is constituted with continuous subduction of the oceanic plate, and the overriding wedge mass of the crust and mantle should be dragged by the moving plate. The effective viscosity of the thin slip boundary zone should strongly depend on the volume content of fluid phase which occupies the open cracks, jogs of shear cracks, and voids. Furthermore, the boundary condition of the moving mass is satisfied to be free at the top but at the bottom the boundary slip velocity is just slip velocity. Therefore, the governing equation of the overriding mass holds,

$$M\ \mathrm{d}v/\mathrm{d}t = F - Kv \tag{2.42}$$

$$F = \tau S - Mg\cos\theta \tag{2.43}$$

in which τ is the shear stress on the shear plane under the mass,and then, we have,

$$\begin{aligned} K &= \eta S/d \\ M &- \rho S\,D \end{aligned} \tag{2.44}$$

Therefore, considering the terminal state at dv/dt = 0, the velocity becomes,

$$\begin{aligned} v &= (\tau - A)d/\eta \\ A &= \rho Dg\cos\theta \end{aligned} \tag{2.45}$$

in which D and ρ are thickness of the overriding mass and its density, respectively.

It yields logarithmic relation between the slip velocity, slip area, and effective viscosity as,

$$\log v = \log(\tau - A) - \log(\eta/d). \tag{2.46}$$

Therefore, it concludes that the slip velocity of the shear crack in the low stress region varies with changing effective viscosity of the slip boundary zone, which depends on porosity filled with aqueous fluid. This type of the slip motion of the shear cracks should correspond to the various slow slip event in the plate boundary zone, and it is suggested that the velocity of the slow slip events ranging from 10^{-3} to 10^{0} mm/s (Fukao et al. 2021) may be derived from the various porosity of the boundary layer.

By the way, the microcrack cluster is the controlling factor of the permeable flow of solution in the plate boundary zones as suggested by the author (Toriumi 2022). There are several important characters of the microcrack cluster in the deep-seated metamorphic rocks of the plate boundary zone: one is the crack-opening orientation parallel to stretching lineation and perpendicular to the schistosity and monomineralic banding structure as described previously, and the other is the periodic and ordinary spacing between microcracks sandwiched by the shear plane. By these characteristic features, the permeability of the schist with microcrack clusters is strongly anisotropic, and the large permeability is just perpendicular to the schistosity and banding plane which is parallel to shortening axis of the deformed schists. The third point of the microcrack clusters is that the density of the microcracks governs the volume fraction of the fluid, and thus it controls the seismic velocities and their ratio. In addition, Poisson's ratio of elastic moduli depends strongly on the V_s/V_p ratio, and thus the large shear crack growth associated with microcrack cluster formation should be checked by the seismic observation around the large slow slip event regions. Furthermore, the model presented here should give the critical failure size of the large shear crack from the microcrack cluster density proposed previously.

By the way, the repeated earthquakes along on the plate boundary have been found commonly around and in the region of the large asperity region (Matsuzawa et al. 2004). This type of earthquakes is probably derived from the shear crack propagation that is the high speed and velocity softening mode of slip velocity, and thus it belongs to the regular earthquake. On the other hand, the asperity region is just the large seismic moment emission zone. Repeating earthquakes and giant earthquakes are both derived from the high-speed shear crack propagation over the critical shear crack size which is defined by the microcrack density proposed in the previous section. Apparently, it seems that the repeated earthquakes (so that it may be called as the repeated shear crack) are also the shear cracks associated with large shear crack growth with slow slip velocity. It may be probable that the slow slip events (SSEs) may take place during the formation of repeated and associated shear cracks around/near the growing large shear crack region until reaching the critical size determined by associated shear cracks and open microcracks. It may be expected that there are low velocity and high V_s/V_p regions around the repeated and large asperity zone, as shown by Igarashi et al. (2003).

As described in the preceding sections, the open microcracks in the cluster show the ordinary and periodic spacing and sealing by metamorphic minerals such as albite, quartz, chlorite, epidote, actinolite and sometimes calcite, as like as those in the matrix in basic schists. It seems that the sealing of the open microcracks was in the period of the metamorphism. In this section, the porosity and related permeability will be estimated from the occurrence of the sealed microcracks in the basic schists.

In the previous descriptions, the microcrack cluster appears very commonly in the basic and pelitic schists of the plate boundary rocks, and they are filled with common metamorphic minerals. Thus, it shows that the apparent microstructures of

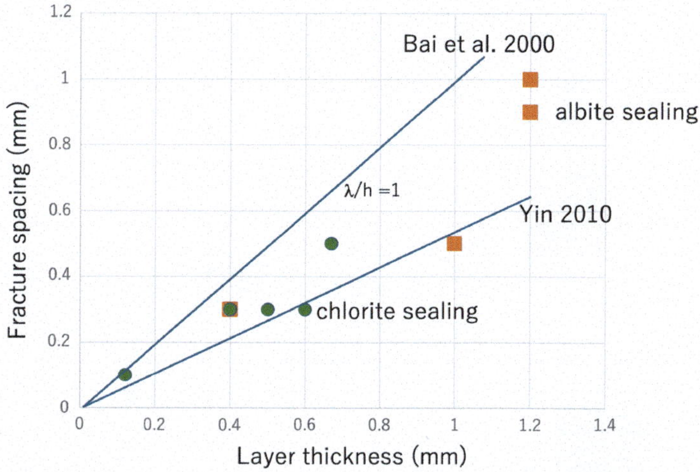

Fig. 2.36 Diagram of fracture spacing, and width of the micro-open crack sealed by albite and chlorite in low grade greenschists of Nagatoro, Kanto Mts. Theoretical relations by Bai et al. (2000) and Yin (2010) are shown in the same diagram

them display often sharp contrast between the microcracks and surrounding matrix, but sometimes, there is no contrast between them because of the same mineral assemblages. This is probably due to the reason why the microcracks were continuously formed and sealed during the metamorphism, and the early microcracks should become mechanically mixing with surrounding matrix not to be identified clearly. Accordingly, in this study, the spacing and length of the distinctive sealed microcracks were measured as shown in Fig. 2.36.

It is obvious that the spacing of the microcracks in the matrix is nearly proportional to the spacing of the monomineralic bands composed of albite and carbonate, which is approximated to be the spacing of the shear slip as discussed in the previous sections. From Fig. 2.37, it provides that both width and spacing of the microcracks are dominantly about 0.3 mm, and the spacing of the band-forming shear slip is about 0.5 mm, thereby showing the ratio of fracture spacing by slip spacing to be about 0.6, which is nearly in the range of the simulation by Bai et al. (2000) and Yin (2010). On the other hand, the length of the sealed microcrack is possibly measured on the schistosity plane as shown in Fig. 2.38, to be about 10 mm.

The porosity defined by the ratio of microcrack volume to the reference rock volume should be estimated from the following relation.

$$\phi = kwl\big/ (d + w)l = 0.5k \tag{2.47}$$

in which k is the probability of microcrack toward the y-axis of Fig. 2.38, to be around 0.1. Therefore, the porosity ϕ may be around 0.05 or so.

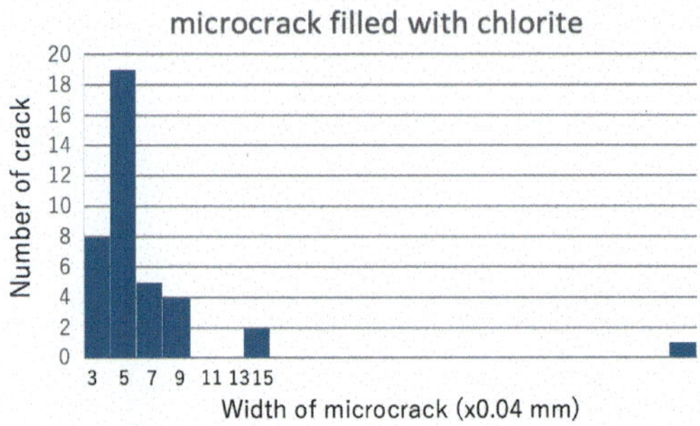

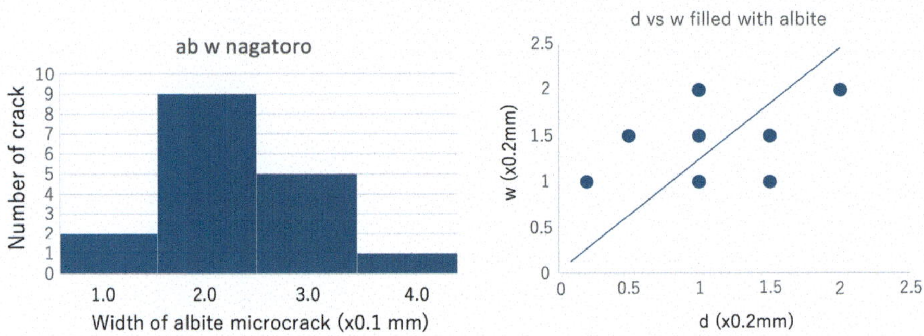

Fig. 2.37 Frequency distribution (left) of width (w), and relation (right) between width and spacing (d) of albite sealing microcrack in the low grade greenschists of Nagatoro

Fig. 2.38 Photograph showing albite sealing microcrack on the schistosity plane extending normal to the lineation (tectonic axis x) in the Nagatoro greenschist

2.4 Shear Crack Jog Wave and Mineral Banding of Plate Boundary Rocks

There is common occurrence of the mineral bands in the high-pressure metamorphic rocks of the subduction boundary zone, and they show parallel alignment with schistosity plane and shear slip one as shown in Fig. 2.39. It suggests that the mineral bands were formed by the precipitation process in the solution-filled open voids, that is the jog, aligned along the shear crack plane as suggested by Toriumi (2022). The jog along with the shear crack plane increases with shear crack growth during the deformation of rocks, and thus, it shows clearly that the permeability of the sheared rocks increases with increasing the density of jog, and so does with shear crack growth.

The author suggested that the porosity of the plate boundary metamorphic rocks should change as a waveform curve of its density during the precipitation and dissolution process in the plate boundary metamorphism, even without consideration of shear deformation (Toriumi 2022). In the same book, it showed that the mineral banding is responsible for the preferential precipitation of some kinds of metamorphic mineral in the growing shear crack jog, thereby suggesting coeval process of mineral banding and shear crack growth, that is the shear deformation rate.

By the way, there is wide variation of the band width and length, but the mineralogy of various mineral bands shows monotonous assemblage that core of the band is composed mainly of albite and minorly of quartz, epidote, actinolite, and the top and bottom is of chlorite. These minerals filling the bands are the same as the constituent ones of the matrix, but not the prograde minerals of amphiboles and garnet in the lower grade zones of the Sanbagawa metamorphic belt. It should be noticed that the fine and dense band structure is characteristic in the lower grade zones of the

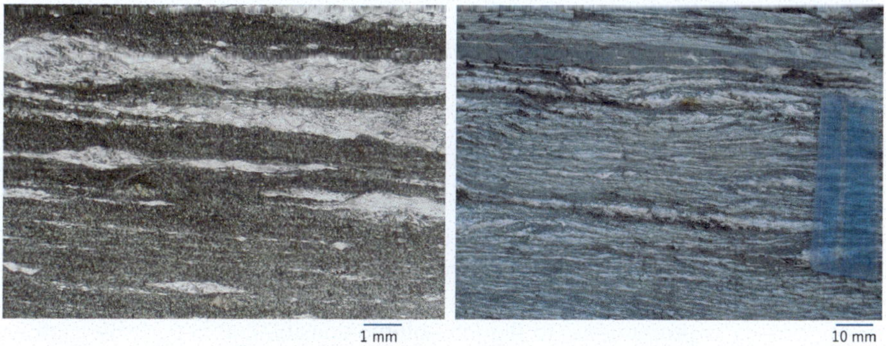

Fig. 2.39 Photographs showing parallel alignment of albite and quartz bands with schistosity plane, formed by shear crack jog growth in low grade greenschists of Nagatoro. Left; microstructure of bands and lens composed of albite, quartz, chlorite, and epidote under optical microscope with open polar, right; periodic bands composed of albite and quartz with various thicknesses

plate boundary metamorphic belt, but in the higher grade zones such as epidote–amphibolite and eclogite zones, there is common occurrence of thick bands and porphyroblastic layers of constituent minerals such as plagioclase, garnet, omphacite, and hornblende. Judging from the diffusional zoning of Mn, Fe, and Mg in these minerals in the high-grade metamorphic rocks, it seems that the mineral banding structures above should be derived from the annealing and recrystallization of fine and dense monomineralic bands in the lower grade metamorphic rocks. In addition, it may be obvious that the monomineralic bands with various thicknesses and mineralogy, being found in the low-grade metamorphic rocks, are responsible mostly for the shear crack growth and propagation with various slip velocity in the plate boundary zone.

In the previous book (Toriumi 2022), the author suggested somewhat enigmatic nature that there are various thicknesses of the monomineralic bands ranging from 0.1 mm to several meters, and the microstructure of the constituent minerals shows the waveform outline that manifests the unstable growth (but not plastic deformation) by rapid growth from aqueous solution. Then, it seemingly means that the contemporaneous shear crack jog growth and mineral precipitation with entrainment of aqueous water have occurred, and their magnitude should be inhomogeneous in space, thereby resulting in the concentration of shear slip motion of rocks.

Thus, in this section, the author intends to propose a model of shear slip concentration and resulting shear crack jog concentration forming the thick monomineralic bands in the low-grade plate boundary rocks by the interaction between the jog, flow-in of aqueous fluid, and mineral precipitation.

As described in the previous section, it should be apparent that the shear crack jog is important for the carrier of the fluid flow, which enhance the shear crack growth and propagation. The interaction between the shear crack motion and the fluid flow is essential for modeling the mineral band formation and related shear crack excitation through jog formation and jog sealing by mineral. The basic equations are the same as those proposed in the previous book (Toriumi 2022), and these are constituted with conservation law, continuity equation, momentum equation, and geometrical relation of jog and shear crack under deformation of rock.

First, let us assume the volume of rock that schistosity and shear crack planes are parallel to tectonic xy plane, shear crack jog is randomly aligned on shear cracks having number density per unit area of shear plane, and fluid flows parallel to the z-axis as shown in Fig. 2.40.

Porosity f is defined as $\phi = V_p/V$ ratio in which V_p is the pore volume and V is the rock volume, and in this equation, the pore volume is approximated to be the jog volume in this study. Then, the porosity ϕ is approximated to the jog volume density. For jog volume increases with increasing shear crack slip, the increasing rate of porosity is proportional to the job volume density, that is porosity itself as seen in Fig. 2.40.

On the other hand, the fluid flux is governed by the Darcy law to be,

$$q = k\phi^3 \nabla P_f \tag{2.48}$$

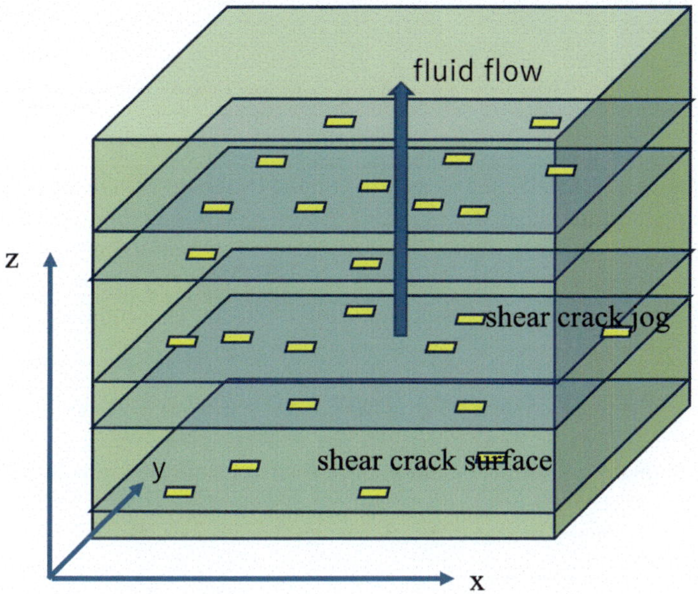

Fig. 2.40 Model of mechano-chemical development of jog, shear crack, and porosity during fluid flow of aqueous solution across the shear plane (schistosity plane). In this model, shear crack interface is parallel to *yz* plane, shear crack slip orientation is toward *y*-axis, and aqueous fluid flows along *x*-axis

in which P_f is fluid pressure of the jog because of fluid flow through jog and matrix. Considering that the fluid pressure in the jog is enough lower than pressure in the matrix because of mechanical volume increase of jog by shear slip as discussed in the previous section, the gradient of fluid pressure becomes using equation of state of fluid $P_f\phi$ = constant,

$$\begin{aligned} \text{grad } P_f &= \text{constant} * \nabla\phi^{-1} \\ &= -\text{constant} * \phi^{-2} \end{aligned} \tag{2.49}$$

Thus, we obtain fluid flux q as,

$$q = -K_1\phi \tag{2.50}$$

in which K_1 is constant.

On the other hand, as the fluid pressure in jog is lower than that of that of the matrix, the mineral should precipitate in the jog because of oversaturation of ionic species concentration product against the matrix where the present mineral must be saturated with aqueous solution. Therefore, the jog volume should decrease corresponding to the volume of precipitated mineral. The degree of oversaturation of mineral becomes constant x $(P - P_f)/P_s$, and then, the decrease of porosity by newly precipitated mineral becomes,

$$\delta\phi = K_2\left(1 - \phi^{-1}\right) \tag{2.51}$$

in which K_2 is constant

Furthermore, it is available that the jog volume increases with increasing amount of shear slip γ, and the shear slip rate $d\gamma/dt$ is assumed to be proportional to the jog porosity filled with aqueous fluid. Then, the generation rate of the jog porosity due to shear slip increment becomes $K_0\phi$.

Accordingly, adding the diffusion term of fluid, the conservation law of fluid (aqueous solution) becomes,

$$\partial\phi/\partial t + q\nabla\phi + \phi\nabla q = K_0\phi - K_2(1 - B/\phi) + D\Delta\phi. \tag{2.52}$$

Thus, we obtain the simple equation of porosity change (= jog volume change),

$$\begin{aligned} \partial\phi/\partial t &= K_o\phi + K_1\phi\nabla\phi - K_2(1 - B/\phi) + D\Delta\phi \\ V_s &= K\phi \end{aligned} \tag{2.53}$$

in which V_s is the shear slip velocity, and K_i, D, B, and K are the constant terms.

In this equation, the system is such that albite dissolves into solution in the matrix, and precipitates some amounts in the shear crack jog, and shear crack growth velocity is proportional to the shear crack jog porosity. In addition, the fluid disperses into jogs on the single shear crack plane by diffusion. Therefore, the above porosity equations display the spatial–temporal variation of the jog density and shear crack velocity with fluid flow. Consequently, it seems that the fluid flow and mineral precipitation process should govern the shear crack excitation (activation) to mineral band formation and development.

The above equation is the nonlinear partial differential equation and thus the simulation study is needed for obtaining the characteristic behavior of the evolving shear crack jog density. In this study, the author intends to simulate the evolution of the jog in the space and time, which means at the same time the shear crack growth velocity and localization of shear slip in the rocks. The simulation of this equation has been performed by R-language on RStudio of the platform installed in the MacBook Pro, using simple coding of time and space increments with their discretized adjustment method.

The simulation was performed by available parameterization of the jog porosity equation mentioned above, using the $\delta r/\delta t$ value about 1 for available parameters under $\delta t = 0.01$–0.001. Simulation code is made on RStudio in the one-dimensional system with simple time increment method. The initial condition of the jog porosity is given by homogeneous random function, and boundary condition is periodic on the influx and outflux of every time step.

Results are shown in Fig. 2.41. The jog porosity distribution changes gradually forming the waveform with various spacing and wavelength, and the porosity wave

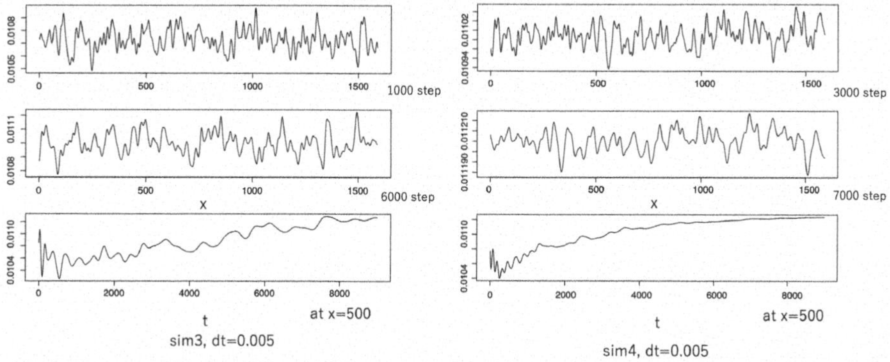

Fig. 2.41 Results of simulation for jog porosity development and its distribution along *x*-axis, showing the migration of jog porosity wave (thus shear crack slip rate too) upward stream, and its wavelength becomes enlarge with time. The jog porosity gradually increases with time in the bottom two figures

migrates slowly upstream direction. This behavior of the porosity wave seems to be mainly derived from coupling of advection of fluid and precipitation reaction of albite from fluid in the equation, but diffusion of fluid along on the shear plane gives rise to smoothing the waveform, as seen in the simulation results. After enough development of the jog porosity wave, the jog porosity increases and forms large peaks, which manifests accumulation of the fluid in the jog, and then the shear slip rate increases accompanied with increase of jog porosity. It implies that the mode of shear crack motion changes from slow growth to propagation one, because the increase of jog porosity is obviously due to the fluid accumulation by spatial–temporal variation of the permeability change, and then the length of the shear crack with many jogs may enlarge over the critical size, and it gives rise to shear crack propagation. It should be noticed that large and wide peak of jog porosity is made after enough time steps as shown in Fig. 2.41, thereby suggesting the shear crack propagation starts at the late stage of jog development and mineral precipitation process.

In the early stage of simulation experiments, the jog porosity wave develops slowly and migrates upstream direction as previously mentioned, and it suggests that the mineral precipitation takes place frequently in the increasing jog density, thereby suggesting the monomineralic band formed in the present simulation experiments. In addition, it is noteworthy that the monomineralic bands migrate to the upstream direction, and then the band width may increase with advancing the time step.

On the other hand, it seems that the jog porosity wave gradually becomes wider noted previously. In addition, the spacing of peaks of jog porosity wave is apparently proportional to the wavelength of them both in the early to late stages of the simulation, and it shows that the peaks move from the downstream to upstream of the general flow of aqueous fluid in the simulation. These results are clearly common in the simulation experiments with various parameters, thereby suggesting the stable pattern evolution of the jog porosity in the studied nonlinear dynamical system.

By the way, it is interesting that how the distribution pattern of the wide and gentle peaks of the jog porosity wave changes in the process of the present shear crack growth. To obtain the progressive pattern change of the jog density distribution, it should be necessary that the width and spacing of the jog porosity peaks are to be measured in the simulation results. The results are shown in Fig. 2.42 in the diagrams of *w*/*d* ratio and *d*, and *w*/*d* versus *w*. It seems that the general trend of *w*/*d*—*d* relation appears a gradual decrease with increasing the spacing of the high porosity peaks, and inversely, the trend of *w*/*d*—*w* relation does a gradual increase. For the time experiments of the simulation, it is obvious that the above relations between *w*/*d* versus *d* and *w* display the universal relation even in the various time steps. Furthermore, the *w*/*d* and *w* values are not dependent on the spatial coordinate *x* as shown in Figs. 2.43 and 2.44, thereby manifesting that the above-mentioned relations are invariant in the space and time.

As described previously, there are abundant shear crack jog filled with albite and the same metamorphic minerals as those of the matrix, thereby suggesting the coeval formation of sealed shear crack jog with the metamorphism. It shows that

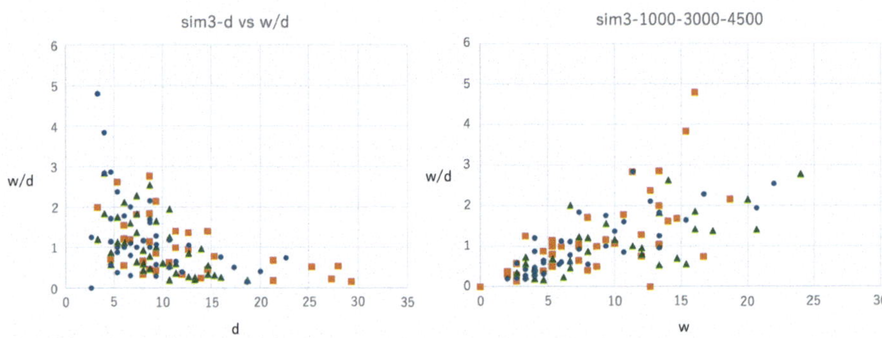

Fig. 2.42 Diagrams of width (*w*) and spacing (*d*) of the jog porosity against d and w of simulations, showing the invariant relations for advancing time steps from 1000 (triangle), 3000 (circle), and 4500 (square)

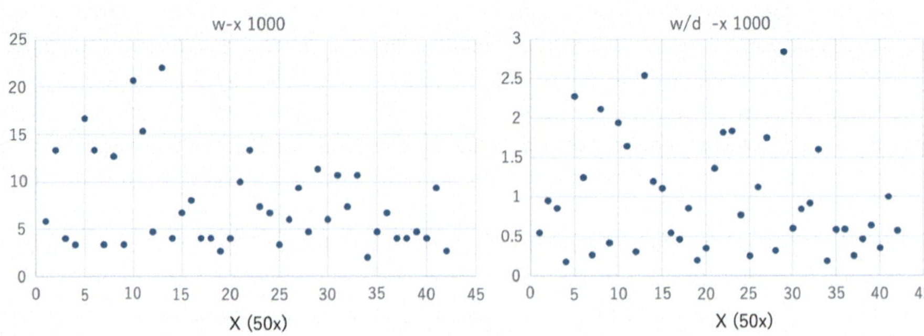

Fig. 2.43 Diagrams of variations of width (*w*; left) and ratio (*w*/*d*; right) against distance *x* at the early time step of simulation showing sporadic distribution of high *w* and *w*/*d* layers

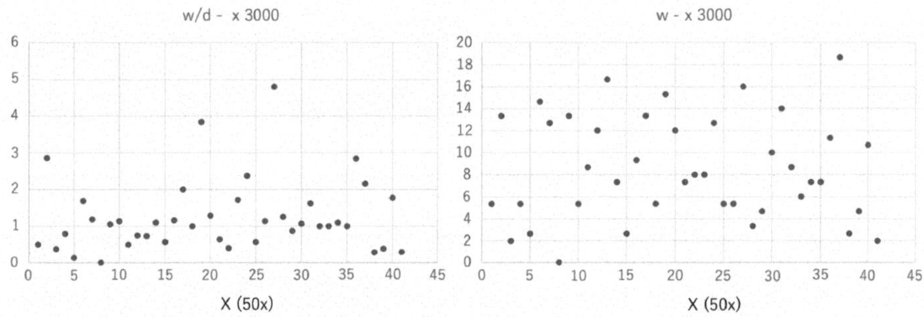

Fig. 2.44 Diagrams of variations of width, w and ratio of w/d at the enough time step of simulation, showing the sporadic pattern of w and w/d like to those at the early time step

the wide variation of the length and width is responsible for the active jog migration along on the shear crack with its growth, but not the boudinage deformation of the existing veins. It seems to be justified that the present monomineralic bands are formed by the mechanism constituent with open space generation of shear crack jog and rapid precipitation of metamorphic minerals from aqueous solution entrained into the jog, because the geometry of strain of these schistose rocks is the maximum shortening axis nearly perpendicular to schistosity plane and elongation axis parallel to the mineral lineation, and main shear deformation is performed by schistosity planes, and the banding is nearly parallel to the schistosity plane. Furthermore, it is obvious that the sealed jog in non-schistose rocks is filled with tabular grains of calcite and albite elongated toward the shear displacement, as like as the short length jog in the schistose rocks (Fig. 2.45). In addition, it manifests the above model that monomineralic bands commonly are accompanied with shear displacement of albite-sealed open crack at their crossing. Therefore, in this work, it should be sure that the monomineralic bands nearly parallel to the schistosity plane have been formed by the continuous precipitation of minerals in the shear crack jog.

The width and spacing of the sealed jog band were measured on the photographs of the field and of the thin section parallel to the shear direction (mineral lineation) and normal to the shear plane defined by schistosity plane by means of image analyzer software, ImageJ, installed on MacBook Pro. The samples are collected from the basic schists from the chlorite zone (350–400 °C) in the Sanbagawa metamorphic belt in Asemigawa and Nagatoro regions. The detailed descriptions of this metamorphic belt are referred to Itaya et al. (2011) and Miyazaki et al. (2024).

The diagrams of the *w*/*d* versus *d* and versus *w* from these schistose rocks are shown in Fig. 2.46. It is obvious that the data points of the *w*/*d* and *d* are widely scattered in the diagram, but they show the trend of inverse relations of them. On the other hand, the ratios of *w*/*d* are surely proportional to width, *w*, and the range of dispersion of the data increases with increasing *w*, thereby suggesting the consistency with the simulation results. There is no difference in the relationships between the Nagatoro and Asemigawa samples, thereby showing the universal relationship

Fig. 2.45 Photographs of shear crack jog array of non-schistose gabbroic rock of Shimanto belt in Muroto, and jog containing parallel-elongated albite grains under optical microscope with crossed polar in the low grade greenschist of Asemigawa, Shikoku

between *w*/*d* and *w* of the banding structures in the schistose rocks in the plate boundary metamorphic belt.

On the other hand, the distribution patterns of the *w* and *d* appear random in every spatial range, as shown in Fig. 2.47, and it may conclude that the distribution patterns of the sealed jog bands display the invariance against the parallel shift normal to the schistosity plane, suggesting the available model proposed here for the metamorphic band.

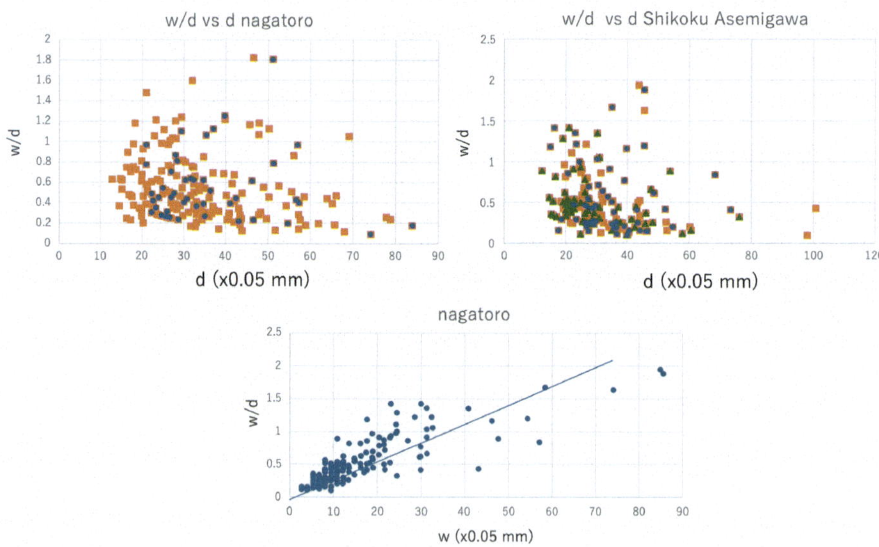

Fig. 2.46 Diagram of *w*/*d* versus *d* (top two) and *w*/*d* versus *w* (bottom) of albite bands in the low grade greenschists of Nagatoro and Asemigawa of the Sanbagawa metamorphic belt, showing wide scattering of *w*/*d* versus *d*, but rather sharp pattern of *w*/*d* versus *w*

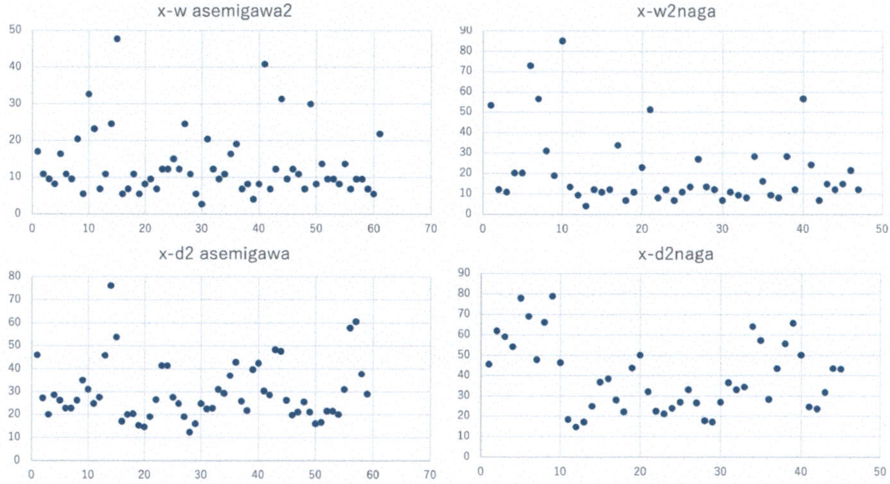

Fig. 2.47 Spatial variations of width (w) and spacing (d) of albite bands in the Nagatoro and Asemigawa low grade greenschists, showing the sporadic distribution of thick and wide spacing of the bands likely to the simulation

It seems to be important that the large shear crack jog bands are commonly found in the studied schists. The large bands filled with albite and quartz together with small amounts of chlorite and carbonates are derived from the mechanism that whether it is controlled by multiple jog formation, or single large jog formation associated with large shear crack motion. Considering that the multiple structure of the grain size layering of albite and carbonates in the large sealed-jog as seen in Fig. 2.48, it seems that once the shear crack jog was formed and filled with metamorphic minerals, the sealed jog behaves as an obstacle against shear slip of schistose rocks and the jog will be formed newly over the present sealed jog during the successive shear slip motion. Therefore, it likes that the sporadic occurrence of thick bands as shown in Fig. 2.47 is responsible for the concentrated jog formation along the shear slip plane, as discussed in the simulation section.

In this section, it has been clarified that the shear crack jog is formed by shear slip motion and jog volume is immediately filled with fluid which flows across the shear crack in the surrounding rocks. The fast path of the fluid flow is obviously the cluster of many micro-open cracks that are nearly perpendicular to the shear crack interface, as shown in the previous sections. On the other hand, the monomineralic bands are frequently found in the schists, and they are nearly parallel to the schistosity plane, thereby suggesting that those are associated with the shear slip motion. This is surely supported by the shear displacement of the sealed open cracks which show the high-angle crosscut with the schistosity plane, that is the shear crack interface (Fig. 2.48). Therefore, it concludes that the monomineralic bands in the schistose rocks of the low-grade metamorphic belt are formed by the successive migration of shear crack jogs along the shear slip interface, and thus, they appear the schistosity parallel banding structure.

Fig. 2.48 Photograph of the large shear crack jog sealed with albite and quartz, showing the periodic banding of coarse-grained layer and fine-grained one in the medium grade basic schist outcrop of the Dozangawa, Shikoku

The simulation experiments, therefore, are based on the physical model involving shear crack motion, jog formation, fluid flow, mineral precipitation, and resulting the spatial–temporal distribution of jog porosity. It concludes that the jog density on the shear crack interface gradually develops to form the waveform distribution in the space and time, and accordingly, the high intensity wave of the jog density gives rise to the high shear slip rate because of hydrolytic weakening of the crack tip and crack interface (Wei 2010). In this model, the jog density is governed by the shear slip rate and fluid entrainment rate, and thus, inversely it shows that the shear slip seems to be localized during the present mechanics with geochemical fluid flow in the plate boundary rocks.

Therefore, it seems that the largely developed sealed jogs and monomineralic bands in the schists of the plate boundary rocks manifest the large amount of shear displacement along just neighboring shear crack. Even in the higher grade schists of the garnet zone and biotite zones of the Sanbagawa metamorphic belt, there are many of the large monomineralic bands with meter-scale thickness, thereby suggesting the sealed and annealed large jog formed by large shear slip fault movement in the basic schists. These largely sealed and annealed jogs appear the lens-like blocks, but they are not associated with pressure shadow structures in both sides of their tail, thereby showing not ductile pulled-apart texture around the sealed jog block.

Furthermore, it has been clarified that the relations of the width and spacing of the multiple monomineralic bands nearly parallel to the schistosity plane, that is the shear slip plane, in the natural schists of the plate boundary metamorphic rocks are very consistent with those of the simulation experiments. The results show that the width w and spacing d of the high jog density layers show the sporadic distribution, and the ratio of w/d is nearly proportional to the width w, but it is inversely proportional to d.

2.5 Kelvin–Helmholtz Instability of Shear Slip with Ductile Layer

The shear slip plane in the metamorphic and basement rocks is commonly associated with ductile shear zone of apparent viscous materials, that are composed of fine brecciated aggregates and low-viscosity crystalline materials such as mylonite. In the central part of this ductile band, the shear slip plane displays curvilinear shape, and its wavelengths and amplitudes are something like the characteristic features. These ductile shears slip structure seems to be scale-free, for the very large waveform in the plate boundary zone and microscale shear structure under the optical microscope in the metamorphic schists, suggesting that the above-mentioned wavy shear plane is derived from the hydrodynamics of the shear slip motion with surrounding ductile materials.

The neutral plane of the ductile shear is classified into two types: one is finite continuous gradient of shear velocity at the midplane, and the other is the very large to infinite gradient of shear velocity at that point. The former is the homogeneous shear zone, and the latter is shear slip with neutral plane. It is an interesting pattern that the shear slip with neutral plane appears looking like boundary flow after the obstacle in the fluid system, and it is hydrodynamically unstable flow with finite vortex even under the low Reynold's number of the flow. This flowage has a neutral plane of the bilateral flow and instability of the neutral plane is called as Hadamard instability (Kelvin–Helmholtz instability).

In this section, the author intends to propose a hydrodynamical model of the instability of shear slip plane with neutral plane in the middle layer by means of Kelvin–Helmholtz instability.

General theory of Kelvin–Helmholtz instability (Hadamard instability) has been developed by Funada and Joseph (2001), based on the experimental results of instability of air and water boundary interface. In this study, it requires that the system to be investigated is moved from air–water system to the system which consists of two viscous layers having the same viscosity and density, and at their boundary surface the velocity of viscous layers changes its direction, that is the shear deformation gradient is maximum at the boundary as shown in Fig. 2.49.

First, we must consider the continuity equations of velocity $v(u, w)$, and potential function $\phi(u, w, t)$ in the two dimensions (x, z) as,

$$\frac{\partial u}{\partial x} + \frac{\partial w}{\partial z} = 0 \tag{2.54}$$

$$v = (u, w) \tag{2.55}$$

$$\phi = \phi(u, w, t), \ \nabla\phi = (u, w) \tag{2.56}$$

$$\frac{\partial^2 \phi}{\partial x^2} + \frac{\partial^2 \phi}{\partial z^2} = 0 \text{ for } a \text{ and } b \text{ layer.} \tag{2.57}$$

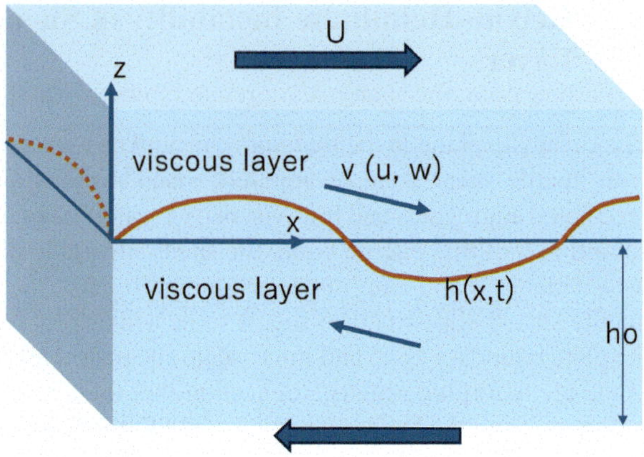

Fig. 2.49 Model of the Kelvin–Helmholtz instability of the ductile shear zone near the shear neutral interface. Initial planer shear plane (blue flat interface) changes waveform interface (red curve) due to the sigmoidal cross section of the shear velocity in the ductile layer around the neutral plane

The boundary is defined by $h(x, y, t)$ as shown in Fig. 2.49,

$$Wa = 0 \text{ at } z = ho$$
$$Wb = 0 \text{ at } z = -ho. \tag{2.58}$$

Then, the boundary conditions at the interface are the interface elevation as,

$$\frac{\partial h}{\partial t} + \frac{U_a \partial h}{\partial x} = Wa$$
$$\frac{\partial h}{\partial t} + \frac{U_b \partial h}{\partial x} = Wb \tag{2.59}$$

Let us consider the following solutions,

$$\phi a = Aa \cosh[k(z - ha)] \exp(\sigma t + ikx) + \text{cc},$$
$$\phi b = Ab \cosh[k(z + hb)] \exp(\sigma t + ikx) + \text{cc}. \tag{2.60}$$

In which k is wave number, σ is rate of increase (decrease) of amplitude, and cc is constant terms. The position of interface is $h(x, z)$ as shown in Fig. 2.49.

Then, we obtain,

$$(\sigma + ikUa)A = -kAa \sinh(kha)$$
$$(\sigma + ikUb)A = -kAb \sinh(khb) \tag{2.61}$$

Other boundary condition is the normal stress balance at the interface.

Then, we obtain,

$$-p + \frac{2\mu \partial Wa}{\partial z} + \rho gh - \left(-p + \frac{2\mu \partial Wb}{\partial z} + \rho gh\right) = -\gamma \frac{\partial^2 h}{\partial x^2} \tag{2.62}$$

in which p is the pressure at the interface, μ is the viscosity, ρ is the density, and γ is the surface tension.

Using Laplace equation,

$$\rho\left(\frac{\partial h}{\partial t}+\frac{Ua\partial h}{\partial x}\right)=\frac{\partial pa}{\partial x} \tag{2.63}$$

And,

$$\begin{aligned}\rho\left(\frac{\partial^2 Wa}{\partial t\partial z}+\frac{Ua\partial^2 Wa}{\partial x\partial z}\right)&=\frac{\partial^2 pa}{\partial x^2}\\ \rho\left(\frac{\partial^2 Wb}{\partial t\partial z}+\frac{U_b\partial^2 Wb}{\partial x\partial z}\right)&=\frac{\partial^2 pb}{\partial x^2}\end{aligned} \tag{2.64}$$

Then, the normal stress balance holds,

$$\begin{aligned}&\rho\left(\frac{\partial^2 Wa}{\partial t\partial z}+\frac{U_a\partial^2 Wa}{\partial x\partial z}\right)+\frac{2\mu\partial^3 Wa}{\partial x^2\partial z}+\rho\left(\frac{\partial^2 Wb}{\partial t\partial z}+\frac{U_b\partial^2 Wb}{\partial x\partial z}\right)\\ &-\frac{2\mu\partial^3 Wb}{\partial x^2\partial z}=-\gamma-\partial^4 h/\partial x^4\end{aligned} \tag{2.65}$$

$$h=A\exp(\sigma t+ikx)+\text{cc}. \tag{2.66}$$

$$V=Ua-Ub$$

The neutral stability condition holds,

$$\sigma_R=0$$

Then, we have,

$$\begin{aligned}V^2&=2\delta\rho g(\alpha k+1/k)\tanh(kh)/\rho\\ \alpha&=\gamma/\delta\rho g\end{aligned} \tag{2.67}$$

As the minimum neutral stability condition can be obtained at the minimum of V^2 (Funada and Joseph 2001), it becomes from Fig. 2.50,

$$hk_c=2\pi h/\lambda c=1. \tag{2.68}$$

Thus, the preferential wavelength of KH instability of viscous shear layer holds,

$$\lambda c=2\pi h. \tag{2.69}$$

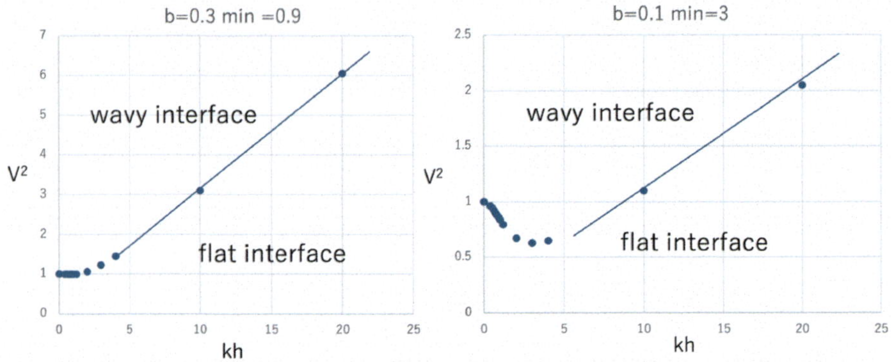

Fig. 2.50 Diagram of square of shear velocity difference U and kh of the perturbation of the interface at the neutral stability condition of the interface. The minimum condition of this relations gives preferential wavelength of the shear slip interface in the ductile shear zone k; wave number, $2h$; thickness of ductile layer. The regime over the neutral curve shows planar interface, and that under the neutral curve does waveform interface. The critical waveform is given on the line given by shear velocity

And it holds,

$$V^2 = 1.5\delta\rho g(\alpha k + 1/k)/\rho. \tag{2.70}$$

On the other hand, the amplitude of the waveform of the boundary surface should be stopped at the top of the viscous layer, h, and thus the maximum amplitude of the growing wave of the surface is surely near the thickness of the viscous layer.

Therefore, it is possibly approximated as,

$$\log\lambda = \log h + \text{cc}. \tag{2.71}$$

And, in this condition, the shear slip velocity V becomes,

$$V^2 = 2\delta\rho g(\alpha k + 1/k)/\rho. \tag{2.72}$$

If $1/k < 1$, we have,

$$V^2 = 2\gamma k/\rho, \tag{2.73}$$

and,

$$\log\lambda = -2\log V + \text{cc}. \tag{2.74}$$

If the case is $1/k > 1$, it holds,

$$\log\lambda = 2\log V + \text{cc}. \tag{2.75}$$

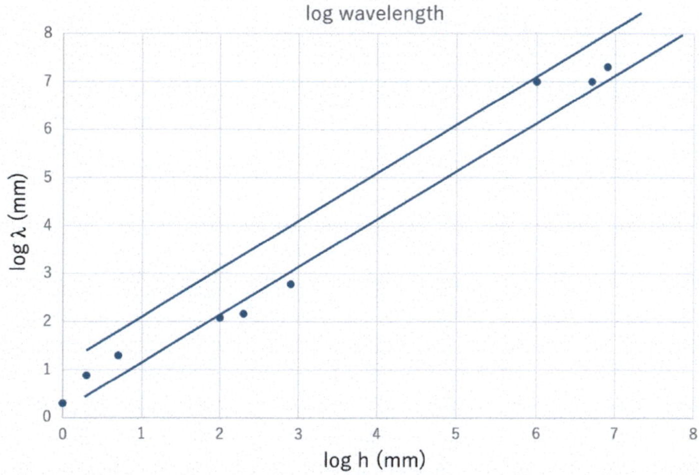

Fig. 2.51 Inferred relationship between the wavelength (l in mm) and effective thickness of ductile shear zone (~ amplitude of wavy shear interface h in mm) in logarithmic diagram. Points represent natural examples shown in late figure

From the above equations, it should be important that the instability is independent on the viscosity between two blocks, showing that the above relation seems to be universal of the viscous shear fault having scale-free relation. It is very important that the two segments of the wavelength–shear velocity relation: one is that it decreases with increasing shear velocity, and the other is that it increases with increasing shear velocity as shown in Fig. 2.51.

In this formulation, the above universal relation between wavelength and thickness of the viscous layer around the shear slip plane is applicable for the ductile shear deformation in the wide range of the scale from 1 mm to several ten km. On the other hand, it seems that in the case of different viscosity of the upper and lower layers sandwiching the shear slip plane, the modified relation between wavelength and thickness of the viscous layer can be obtained as,

$$V^2(k) = 1.5\delta\rho g(\alpha k + 1/k)(1 + r)^2 / \rho\left(1 + r^2\right)$$
$$r = \eta_1/\eta_2 \tag{2.76}$$

In this equation, r is the viscosity ratio between 1 (η_1) and 2 (η_2) layers, and density of the layer is ρ. Therefore, the neutral condition of KH instability becomes the same as,

$$\log\lambda = -2\log V + C(r, \delta\rho, \gamma). \tag{2.77}$$

Therefore, it is noteworthy that the wavelength of the wavy interface of shear slip plane is governed by the shear flow velocity together with viscosity and density contrast between top and bottom layers. However, the wavelength is inversely proportional to the square of shear velocity as well as the case of no viscosity contrast.

The occurrence of waving shear slip plane (wavy fault plane) varies in the wide range of tectonic situation from small to large ductile shear faults in plate boundary metamorphic rocks and overriding accretion sedimentary rocks to the subduction boundary of the oceanic plate, and the oceanic to on-land transform fault such as San Andreas Fault. It is very important that the range of the wavelength is much wide from 1 mm to 10 km in the above-mentioned geologic situations.

First, the small-scale wavy shear slip plane will be described in the low-grade metamorphic schists of the northeastern Kyushu in the Shimanto accretion along the Nankai Trough. The metamorphic conditions of the present basic schists are 250–300 °C in temperature and 0.2 to 0.5 GPa in pressure (Toriumi and Teruya 1988). As shown in Fig. 2.52, it is obvious that the waveform of the thin albite layer is derived from the shear slip motion, and the wavelength and amplitude of wavy shear slip interface appear very periodic likely to sinusoidal curve, and ranges from 0.1 to several mm in length, judging from the displacement of pre-cut high-angle sealed cracks at the crossing with wavy thin albite bands. In Fig. 2.52, the thickness of the albite layer appears asymmetric between the left- and right-hand sides of the peak so that the right-hand side layer shows thinner than that of the left one. It seems that the above imbalance of the thickness is derived from the left-hand shear movement along the wavy shear plane, so that the left-hand side becomes dilatant likely to the shear crack jog. This explanation is supported by the evidence that the Riedel and penetrative shear microstructures are found around the peaks of the wavy shear plane.

On the other hand, there is abundant folded and wavy boundary interface of schists with different mineral assemblages in the Sanbagawa metamorphic schists (Fig. 2.53). As shown in Fig. 2.53, it is obvious that the boundary shear plane between stilpnomelane schist (brown) and greenschist layers are clearly folded tightly. It

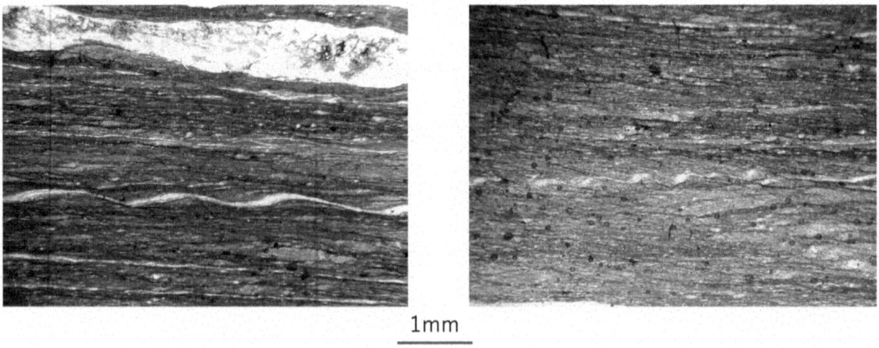

Fig. 2.52 Photographs of ductile microshear slip interface showing periodic waveform in the low grade pumpellyite-actinolite schists of Shimanto belt, eastern Kyushu under optical microscope with open polar. The photographs are parallel to the stretching lineation and normal to the schistosity plane. Wavy shear interface is filled with albite and quartz by precipitation in the volume of migration jog. The layer amplitude of the large wavelength is larger than that of the small one

seems that these wavy shear interface in the schistose rocks are surely derived from the shear plane instability of the neutral plane of the ductile shear, that is due to the KH instability. The clear example of the periodically wavy interface of the ductile shear fault is shown in the boring-core recovery of the Nobi fault zone in the central Japan by the Japanese Center of Nuclear Power Control (Fig. 2.54). The wavelength of these periodic wavy shear plane varies from 100 to 150 mm.

On the other hand, it is quite interesting that the plate boundary between the subducting slab of the Philippine Sea plate and overriding accretionary prism of the Nankai Trough of the SW Japan displays clearly the periodic wavy interface as shown in the multichannel seismic survey (MCS) image of Fig. 2.55. In this case of the plate boundary, the surrounding unconsolidated rocks of the plate boundary consist mainly of accretionary sediments and trench fill sediments composed mainly of sand, and mud and their alternation, and thus the plate boundary is surrounded

Fig. 2.53 Photographs showing wavy shear slip interface of greenschist (**a**; yellow arrow, Satamisaki, Shikoku) and tightly folded boundary of stilpnomelane schist and greenschist (**b**; yellow arrow, Nagatoro, Kanto Mts.)

Fig. 2.54 Photograph of periodic waveform fault boundary appeared in the boring core of Neodani Fault, showing the ductile layers around the fault interface. Data provided by Japanese Center of Nuclear Power Control

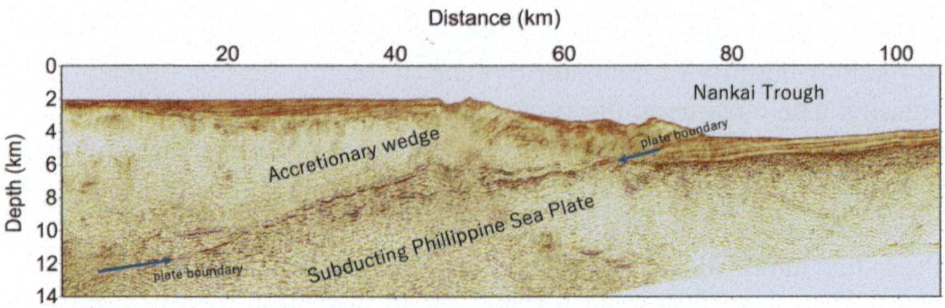

Fig. 2.55 Multi-channel Seismic Survey (MCS) image of the subducting boundary of the Philippine Sea Plate and Island Arc of SW Japan along the Off-Muroto transect, showing waveform subduction boundary. Data provided by JAMSTEC seismic survey database site (https://doi.org/10.17596/0002069)

by somewhat ductile materials, thereby suggesting the appearance of KH instability even in the very low plate velocity. The wavelength of above plate boundary ranges from 1 to 3 km in length. The thickness of the ductile layer around the boundary is estimated to be about 1–5 km.

These data are plotted in the diagram of logarithmic wavelength and estimated thickness of the apparent ductile layer as shown in Fig. 2.56, suggesting the relationship between them in accordance with the inferred relation from the KH instability of the shear interface. It seems to be surprised that there is a simple relation of the wavelength and thickness of the ductile layer in the very wide scale from 1 to 10^7 mm. Furthermore, it seems that the above relation should be applied for inferring the shear slip velocity as formulated previously. In Fig. 2.56, the range of the constant parameters of the relationship between the logarithmic wavelength and thickness is derived from the shear slip (fault slip) velocity, thereby suggesting the higher velocity in Nobi fault and Miura fore arc sediment faults than that in shear slip of the Nobeoka shear zone and plate boundary slip in Nankai Trough.

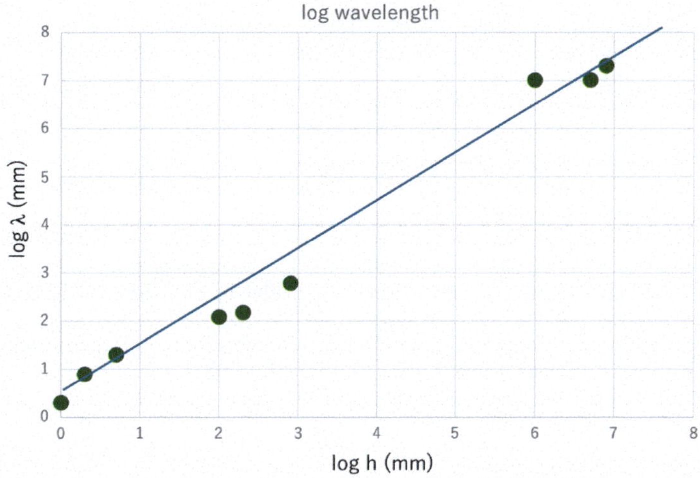

Fig. 2.56 Logarithmic diagram of wavelength (*l*) and thickness (*h*) of the ductile layer (amplitude) of the wavy shear slip interface. Full line shows the theoretical relation by KH instability studied here

References

Aki K, Richards PG (1980) Quantitative seismology: theory and methods. W. H. Freeman, San Francisco, California

Atkinson BK (1984) Subcritical crack growth in geological materials. J Geophys Res 89:4077–4114

Bai T, Pollard DD, Gao H (2000) Explanation for fracture spacing in layered materials. Nature 403:753–756

Fukao Y, Kubota T, Sugioka H, Ito A, Tonegawa T, Shiobara H, Yamashita M, Saito T (2021) Detection of "rapid" aseismic slip a the Izu-Bonin Trench. J Geophys Res Solid Earth 126:e2021JB022132. https://doi.org/10.1029/2021JB022132.

Funada T, Joseph DD (2001) Viscous potential flow analysis of Kelvin-Helmholtz instability in a channel. J Fluid Mech 445:263–283. https://doi.org/10.1017/S0022112001005572

Guvendiren M, Yang S, Burdick JA (2009) Swelling-induced surface patterns in hydrogels with gradient crosslinking density. Adv Funct Mater 19:3038–3045

Holland TJB, Powell R (1998) An internally consistent thermodynamic data set for phases of petrological interest. J Metamorp Geol 16:309–343

Igarashi T, Matsuzawa T, Hasegawa A (2003) Repeating earthquakes and interpolate aseismic slip in the northeastern Japan subduction zone. J Geophys Res 108(B5):2249. https://doi.org/10.1029/2002JB001920

Itaya T, Tsujimori T, Liou JG (2011) Evolution of the Sanbagwa and Shimanto high-pressure belts in SW Japan: insight from K-Ar (Ar-Ar) geochronology. J Asian Earth Sci 42:1075–1090. https://doi.org/10.1016/j.jseaes.2011.06.012

Kodaira S, Hidaka T, Kato A, Park J-D, Iwasaki T, Kaneda Y (2004) High pore fluid pressure may cause silent slip in the Nankai trough. Science 28(May):1295–1298

Kurz W, Fisher DJ (1986) Fundamentals of solidification. Trans Tech Publications

Langer JS (1980) Instabilities and pattern formation in crystal growth. Rev Mod Phys 52:1–30

Matsuzawa T, Uchida N, Igarashi T, Okada T, Hasegawa A (2004) Repeating earthquakes and quasi-static slip on the plate boundary east off northern Honshu, Japan. Earth Planets Space, 56:803–811

Miyazaki K, Wallis S, Ishii K, Annen C (2024) Thermal modeling of the Sanbagawa and Ryoke belts. Elements 20(2):110–116. https://doi.org/10.2138/gselements.20.2.110
Okamoto A, Sekine K (2011) Textures of syntaxial quartz veins synthesized by hydrothermal experiments. J Structural Geol 33:1764–1775
Sakai I, Hayakawa Y (2006) Shape selection of Kompeitoh. J Phys Soc Japan 75(10):104802
Spry A (1969) Metamorphic texture. Pergamon, Oxford
Toriumi M (1990) The transition from brittle to ductile deformation in the Sambagawa metamorphic belt. J Metamorphic Geol 8(4):457–466
Toriumi M (2022) Geochemical mechanics of the earth and deep neural network modelling. Springer
Toriumi M, Teruya J (1988) Tectono-metamorphism of the Shimanto belt. Mod Geol 12:303–324
Toriumi M, Hara E (1995) Crack geometries and deformation by the crack-seal mechanism in the Sambagawa metamorphic belt. Tectonophysics 245:249–261
Toriumi M, Teruya J, Masui M, Kuwahara J (1986) Microstructures and flow mechanisms in regional metamorphic rocks in Japan. Contrib Mineral Petrol 94:54–62
Wei RP (2010) Fracture mechanics: integration of mechanics, materials science, and chemistry. Cambridge University Press
Yamamura T, Tawara K, Tazaki K, Chiba H, Ishibashi J (1997) Needle—like minerals at Iheya Ridge. J Mineral Japan 26:113–116
Yin HM (2010) Fracture saturation and critical thickness in layered materials. Int J Solids Struct 47:1007–1015. https://doi.org/10.1016/j.ijsolstr.12.016

Chapter 3
Characteristic Microstructure Related to Multiphase Shear Flow

Abstract Metamorphic rocks of the plate boundary zones contain commonly large grains of albite and their clusters. The characteristic features of clusters are investigated by pairwise correlation and grain density maps with Gaussian mixed modeling (GMM). It shows that the compact cluster size is fitted well by Poisson's distribution function. The simulation of mechanical coagulation of grains in the shear flow deformation clarified the periodically discrete GMM patterns of the grain density and pairwise correlation. The albite band formed by jog migration on the shear crack slip surface gives rise to make chlorite band, epidote band, and sometimes actinolite band in the sequential order outside the albite band, thereby suggesting symbolic dynamics of multi-mineral reactions with diffusive component ions in grain boundary solution. Crystal size distribution of albite and quartz in jog manifests decompression crystallization at jog formation during shear slip. In addition, the decompression brecciation of metamorphic rocks is possibly investigated from grain size distribution of breccia. Quartz grains in jog often have fluid inclusion array formed by healing and annealing of micro-open crack filled with aqueous fluid. Various time scales of annealing are inferred from size distribution of fluid inclusion to be in the range from 100 to 10,000 years after healing.

Keywords Shear flow coagulation of grains · Symbolic dynamics of fluid mineral reaction · Annealing of fluid inclusion

3.1 Distribution Patterns of Grains of Albite, Cluster Size, and Shear Flow Clustering Simulation

In the high-pressure type metamorphic rocks of the plate boundary, albite is essential minerals with amphiboles, sodic pyroxenes, garnets, epidote, chlorite, and micas. The grains of albite often display anhedral and amoeboid shapes, and the size of them rapidly changes from several ten microns to millimeters with increasing metamorphic temperature across the garnet zone in the Sanbagawa and Alpine metamorphic belts. In the Sanbagawa metamorphic rocks, the porphyroblastic albite grain is commonly

M. Toriumi, *Physics of Geochemical Mechanics and Deep Neural Network Modeling with Diffusion Augmentation*, Advances in Geological Science,
https://doi.org/10.1007/978-981-97-9376-1_3

Fig. 3.1 Photograph showing mode of occurrence of albite porphyroblast (albite spot) in medium grade basic schist in the Sanbagawa belt of Asemigawa, Shikoku. Albite spot aggregate shows layer and rounded cluster. Photograph is lineation parallel and schistosity normal

called as albite spot, and it has been believed as the index texture of the metamorphic grade as shown in Fig. 3.1.

On the other hand, the author studied the microstructures surrounding the albite spots of the Sanbagawa metamorphic rocks, and concluded that the albite spots are formed by the mechanical coarsening process during the plastic deformation of the metamorphic rocks (Toriumi 1975a, b), and later he proposed the same mechanism of the garnet clusters of the Sanbagawa pelitic schists, being compared with the spatial distribution of the garnet grains in the undeformed contact metamorphosed pelitic rocks (Toriumi 1986). In this paper, he suggested coagulation coarsening of garnet for common occurrence of multiple cores in the euhedral grains of garnet. The mechanical segregation of these plastically deformed plate boundary rocks seems to be basic process in the plate boundary metamorphic rocks, and it may induce the progressive change of mineral distribution patterns during the plastic deformation of the rocks and suggests the deformation mechanism of the polymineralic metamorphic rocks in the plate boundary zone.

Besides, it has been clarified that the spatial distribution patterns of like-mineral grains and unlike-mineral grains should be investigated sharply by the pairwise correlation map (or two-point correlation map) on the schistosity parallel and normal planes to show the stochastic cluster structure of the mineral grain dependent on the environmental conditions as shown by Toriumi (1986, 1994). In the case of cumulative dunite containing chromium-rich spinel, it is shown that there is a systematics of chromian spinel cluster between the mass-correlation index and point correlation one (Toriumi 1994). On the other hand, the correlation of unlike-mineral grains (epidote vs magnetite) also makes a spatial pattern that nearest neighboring grain of epidote is magnetite in the low-grade metamorphic rocks, suggesting the reaction relation between them.

Therefore, the reason why albite spots appear as various cluster and collision coarsening in the metamorphic rocks should be clarified by data-driven scientific method including correlation matrix and Gaussian mixed cluster analysis methods. In this paper, the author intends to study the spatial distribution pattern and cluster structure of albite grains in the plate boundary metamorphic rocks.

The studied rock samples are from Dozangawa and Asemigawa routes in the Sanbagawa metamorphic belt of Shikoku, and photographs of the schistosity—parallel plane and schistosity and lineation—normal plane are taken at the field. The coordinates of albite spots on these photographs are acquired by means of IMAGE J applications installed on OS15 of Mac PowerBook Pro. The data of number of albite grains are ranged from 200 to 500. The pairwise correlation matrix in the orthogonal spatial coordinates was calculated using in-house software on R 4.0 platform. Furthermore, the Gaussian Mixture Model was obtained by using the application package of 'mcluster.R' on R-platform. In this application, the mixture model of Gaussian distribution function is estimated using expectation maximalization (EM) algorism with finite number of cluster model under Bayesian parameter fitting. The selection of the finite number of clusters was made by comparison of Bayesian Information Criterion.

In this study, the albite spot is defined as the single crystal/polycrystal surrounded by matrix aggregate that is constituted with amphiboles, epidote, chlorite, titanites, and garnet. The studied samples are photographs taken on the outcrops in the field, and the orientation of the photographic samples are parallel to schistosity plane and normal to schistosity plane with parallel to lineation. The measured albite spots are over 200 grains for imaging the number density patterns of rocks. The two-dimensional position data of the center of albite spots are obtained by means of ImageJ software on wide scale photographs of the planer outcrops. At the same samples, the number of individual grains of albite in the albite clusters was measured for all albite spots in the studied area.

The partial pairwise correlation patterns were investigated because whole pairwise correlation pattern displays the broad peak at the center position which should involves several peaks arranged near the center. The partial pairwise correlation diagram is obtained by the spatial correlation function between several ten albite spots in the small target square and all albite spots of the whole studied area. It is available that the partial pairwise correlation function represents the characteristic distribution patterns of albite spots in the rocks because the partial correlation show the same patterns in every target square within the studied area, suggesting that the partial correlation patterns is invariant feature as shown in Fig. 3.2. In addition, it is obvious that the simulated distribution pattern constituted with lattice-like random distribution gives also the same partial correlation pattern even though the whole pairwise correlation function has the broad peak at the center.

The results of distribution patterns by partial pairwise correlation patterns of albite spots are shown in Fig. 3.3, suggesting the lattice-like and layered structure. The layered structure of albite spots can be surely identified in the thin section and

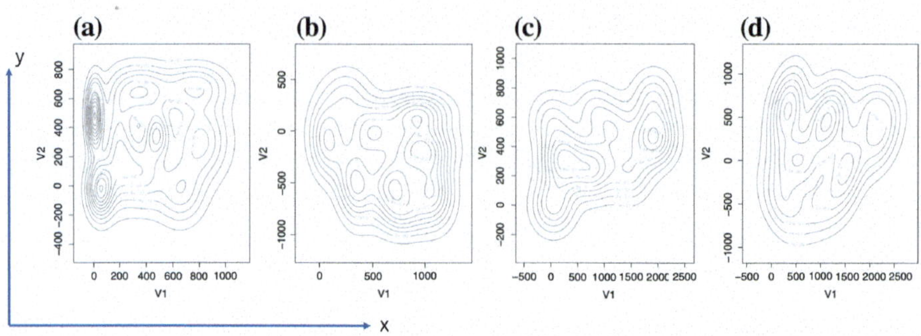

Fig. 3.2 Pairwise correlation maps of positions of albite spots in the medium grade basic schists (four samples) on the plane of lineation (x) parallel and schistosity parallel (xy) obtained from Gaussian mixed model with bootstrap data, showing the various clusters periodic to x- and y-axis

naked eyes, but the lattice-like pattern appears sharply in the pairwise correlation pattern. Furthermore, the typical distance between apparent clusters in the correlated pattern varies in the samples showing the difference of plastic strain by means of shape change of albite spots.

The simulation of collision coarsening during plastic flow of the metamorphic rocks was carried out by using MacBook Pro on the R-platform. The plastic flow deformation of the system is classified into three types: one is the simple shear flow, second is the shear flow with constriction, and third is shear flow with flattening. The deformation gradient matrices are exhibited as,

$$D_1 = \begin{pmatrix} 0 & 0 & \gamma_1 \\ 0 & k_1 & 0 \\ 0 & 0 & k_2 \end{pmatrix} \tag{3.1}$$

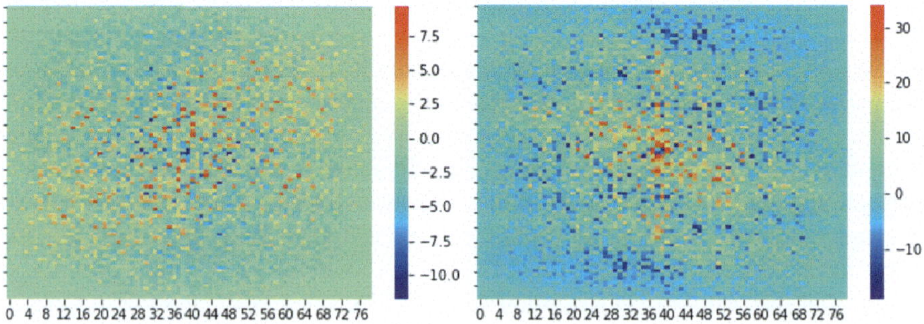

Fig. 3.3 Heat maps of the pairwise correlation density (xz) of albite spot positions in the medium grade basic schists (two samples) with filtered central density, showing various discrete layering and clustering microstructures

where γ_1 is shear velocity gradient. In the simple shear regime, k_1 and k_2 are zero (simulation 1), but, in the case of constrictive shear, the parameters are $k_1 < 0$ and $k_2 < 0$ (simulation 2). On the other hand, the case of $k_1 > 0$ and $k_2 < 0$ is the flattening shear (simulation 3). The model system is assumed to be $100 \times 100 \times 30$ volume and number of grains of 1000. The three-dimensional positions of them are given by the uniform random number in the above volume. The grain size is to be 1 + 0.3 x (uniform random distribution between 0 and 1).

The simulation was performed by the following:

1. Grain is moved by the above deformation gradient matrix.
2. The distance between neighboring grains is estimated for all pairs.
3. If this distance is less than the total grain size of both grains, the smaller one is deleted. by replacing coordinates to be (0, 0, 0), and if not, the next pairs will be processed.
4. Above step will be repeated until given time steps.

The results of simulation are represented by two-dimensional frequency diagram of the pairwise correlation density and GM model of the grain density in Fig. 3.4. It shows that the pattern of the pairwise correlation changes from random one with high frequency peaks to periodic broad peaks with low frequency. But it seems that the grain density distribution pattern displays composite patterns of periodic multiple peaks and layers parallel to the major shear flow as shown in Figs. 3.5 and 3.6. It seems that the progressive shear flow deformation of the matrix induces formation of layered and periodic distribution of albite spots by shear flow coagulation from the random distribution. Furthermore, it looks like that the lattice structures of grain clusters in the layers seems to be developed in the large shear deformation.

By the way, the change in the multiple Gaussian distribution patterns of pairwise correlation and spatial grain density are basically derived from the coagulation growth of the albite compact clusters, and therefore it should be satisfied that the size of the grain clusters increases with developing the shear flow deformation of the matrix. It seems to be seen that the frequency distribution of the grain cluster size displays the change from sharp peak at the single grain cluster to broad and multiple peaks at the large size of grain clusters. In this context, the simulation was performed in the condition that the single grain distinguishes in the case of coagulation of cluster and smaller grains, and the cluster changes the compact cluster of the grains (Fig. 3.7).

By the way, metamorphic pelitic schists contain abundant garnet grains in the higher grade metamorphic rocks, and often display the cluster of them in their highly deformed cases. The author has proposed a mechanical segregation model of such grains of garnet during solid-state shear flow of these schists, for the pairwise correlation functions of garnet grains clearly appear the change from the random type in the non-deformed contact metamorphosed rocks to the clustered type in plastically deformed schists (Toriumi 1986). The mechanical model is based on the mass balance equation without newly formation of albite as

$$dN/dt = q_{in} - q_{out} \quad (3.2)$$

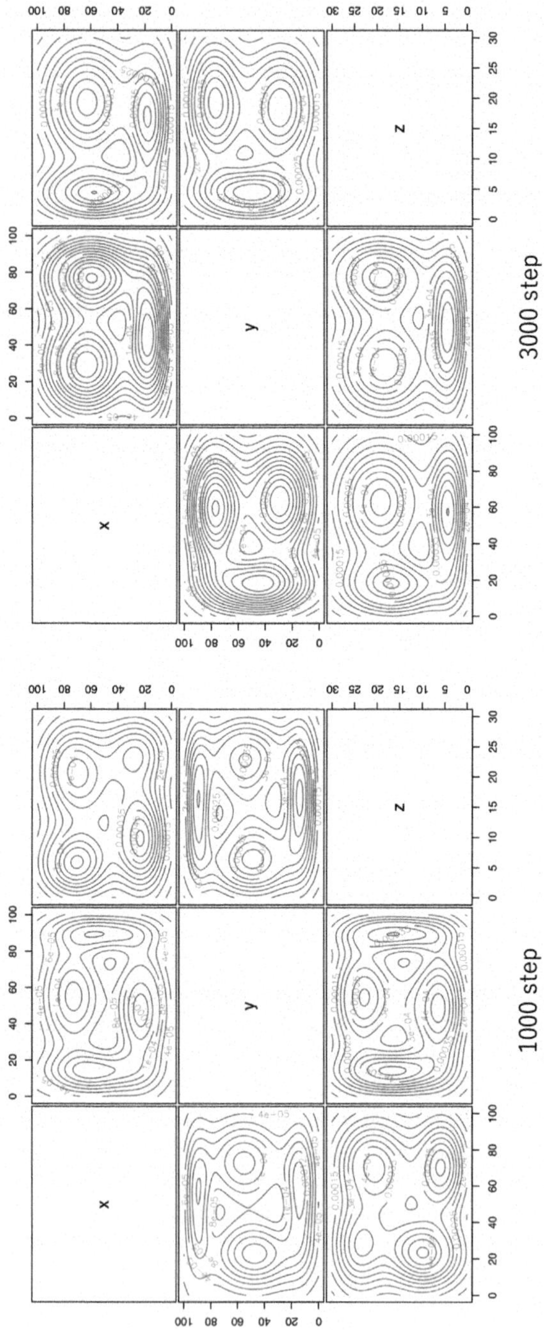

Fig. 3.4 Gaussian mixed model of simulated grain density in the three dimensions (*x*, *y*, *z*) under simple shear along x *at* 1000 and 3000 step, showing the lattice-like distribution of cluster in both time step simulations

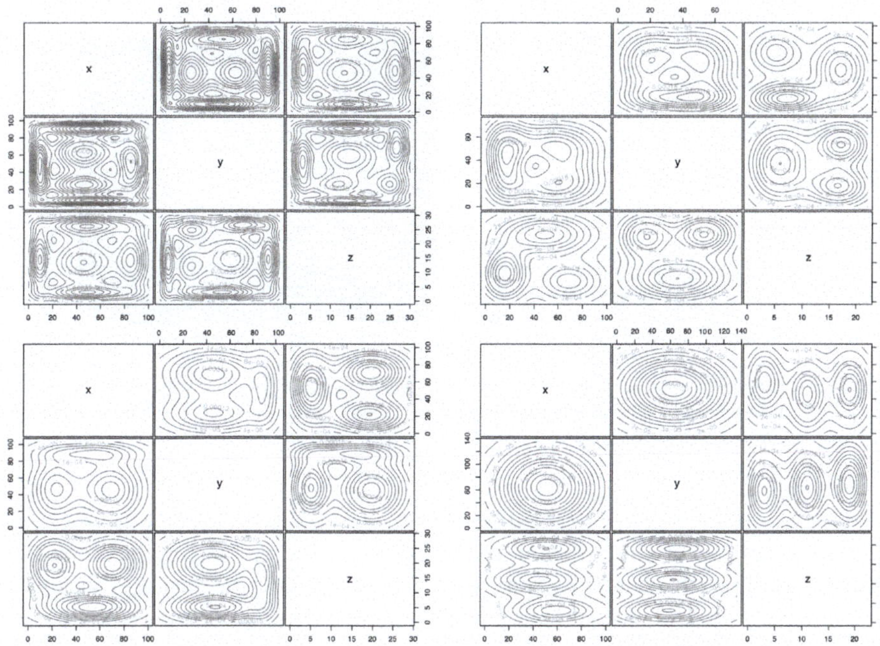

Fig. 3.5 Gaussian mixed model of simulated grain density in the three dimensions (x, y, z) in the cases of simple x-shear (**a**), x-shear and yz-constriction (**b**), x-shear and y-constriction, (**c**) and x-shear and z-constriction (**d**) experiments

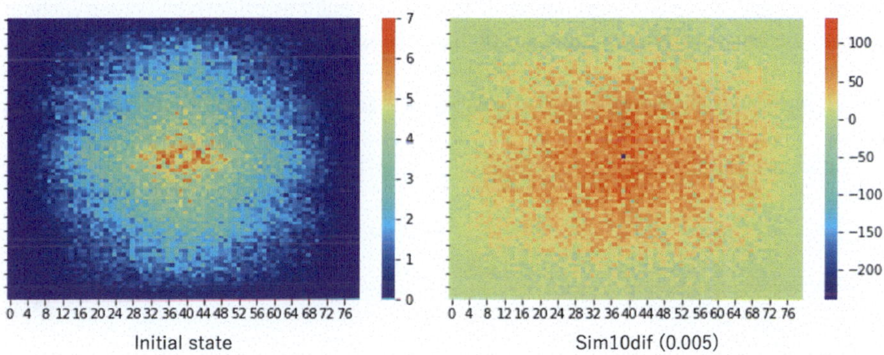

Fig. 3.6 Heat map of simulated pairwise correlation density in the case of simple shear, projected on xz plane that filtered by center density of initial state (left) and low shear strain (right), showing weak clustering after initial state

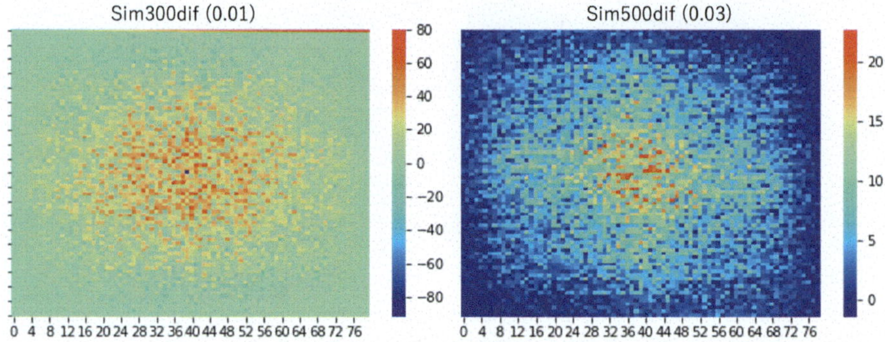

Fig. 3.7 Heat map of simulated pairwise correlation density in the case of simple shear projected on *xz* plane at middle shear strain (left) and large shear strain (right), showing the change of cluster structure during shear deformation

in which q_{in} means number of input grains into the compact cluster with N grains across the cross section of collision of the matrix grains, and q_{out} does number of output grains from the cluster by viscous shear stress around the cluster. The author (Toriumi 1986) studied the geometry of collision-clustering system and obtained the following logistic equation,

$$dN/dt = AN(B-N) \tag{3.3}$$

in which A and B are the parameters of the system such as effective viscosity, shear strength of cluster, number density of the grains, and effective system size. It is obvious that this type of logistic equation has typical bifurcation and successive chaos solution with increasing control parameter of AB value (Beck and Schlogl 1993).

In the case mentioned above, the parameter AB becomes,

$$\begin{aligned} dx/dt &= rx(1-x) \\ x &= N/B \\ r &= AB \end{aligned} \tag{3.4}$$

in which B is the possible maximum cluster size in the system, and if r is smaller than 3, the solution appears a unique value. In the range from 3 to 3.5 of r, bifurcation of cluster size should appear, and if r is in the range from 3.5 to 4.0, the chaotic change of x does. It is sure that garnet grains show commonly the single core, but some of them appears double cores and triple cores in the progressively zoned garnet grains as reported in Toriumi (1986).

On the other hand, the albite spots in the Sanbagawa metamorphic rocks appear as the cluster that constituted with smaller albite grains as shown in Fig. 3.8, as suggested by the author (Toriumi 1975a, b). He clarified the mechanical coagulation of albite spots in the basic schists of the Sanbagawa metamorphic rocks and the mineral separation process in the closing interstitial space between two albite spots. The separation process is mathematically represented by the logistic equation of mineral separation by grain boundary slip of the matrix. In this model, the fine-grained aggregates filling the interstice between closing two albite spots are plastically deformed like Poiseuille flow by grain boundary slip mechanism. At this case, the like-mineral boundary is rather strong coupled than unlike-mineral boundary. In the system of bi-mineral fine grains such as chlorite and epidote/amphiboles, volume fraction of the epidote/amphiboles in the closing interstice described above may obey the Rayleigh equation as

$$d(\ln x)/d(\ln t) = K \tag{3.5}$$

in which x is the volume fraction of epidote and amphibole in the interstice between albite spots, and K is the system parameters. In natural metamorphic rocks, the volume fraction approaches to unity with increasing time, being as uniform sigmoidal pattern of x versus time.

By the way, it is possible that the coagulation growth mechanism of albite spots may operate by several times during prograde metamorphism. If this is the case, the single and multiple grains of albite spots should display the texture that the albite spots are constituted with domains having different sized epidote/amphibole inclusions. In this process, it may be satisfied that the coagulated albite spots should involves remnant epidote inclusions in their inherited grain boundary of albite spots that now show the uniform crystallographic same orientation likely to a single grain as shown in Fig. 3.9. In this figure, it seems that there are three stages of the coagulation clustering and growth during metamorphism, and the coagulation clustering

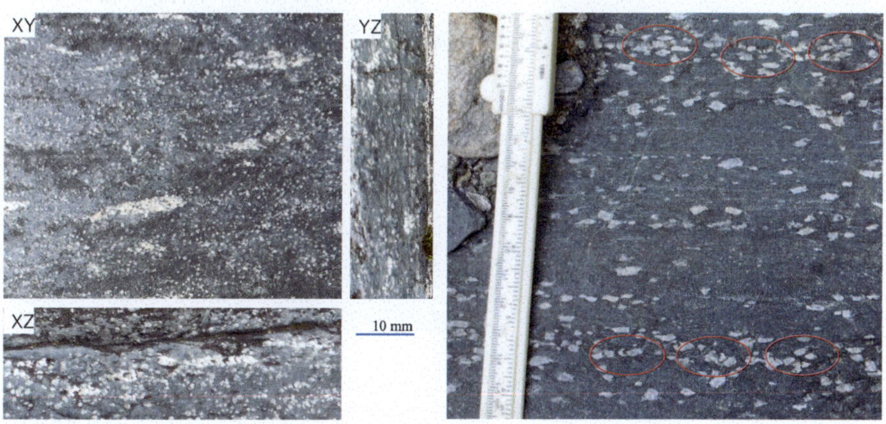

Fig. 3.8 Photographs of large clusters of albite spots on *xy*, *xz*, and *yz* planes (**a**), and lattice-like clusters (**b**; encircled by red) of Dozangawa and Asemigawa basic schists

may appear intermittent because of discontinuous increase of grain size of epidote/amphibole in a single grains of albite spots. The first stage of coagulation is very fine-grained inclusion-rich domain, and the second is the rather medium- grained inclusion domain beside fine-grained inclusion one. The third stage is manifested by the rim of contacting two or three albite spots including coarse -grained amphiboles/epidote as like as the matrix amphiboles and epidote (Fig. 3.9). The grain size of inclusion minerals increases discontinuously from several microns to several ten microns, and to several hundred, suggesting the shear flow coagulation of albite was intermittent during metamorphism with increasing matrix grain size.

The coagulation growth process of albite spots should be identified to be two periods during prograde metamorphism, and the other one process took place at the retrograde and thus exhumation stage of the metamorphic rocks, because of inclusion size likely to the grain sizes of matrix amphiboles and epidote. Therefore, it seems that the present configuration of albite compact clusters in the schists has taken place during the main process of the exhumation of the metamorphic rocks (Fig. 3.9).

Thus, in this study, the number of the constituent grain of albite in the compact clusters was measured on the photographs of tectonic *xy* and *xz* planes in which *x* is parallel to stretching lineation and normal to schistosity plane, *y* is parallel to schistosity plane and normal to lineation, and *z* is normal to both axis, number of total clusters counted attains about 100 to 150. The frequency diagrams of the results are shown in Figs. 3.10 and 3.11, thereby showing there are two types of distribution pattern: one is simple decreasing pattern from single cluster, and the other is multiple cluster dominant type as shown in Fig. 3.11. The latter type occurs very rarely in the

Fig. 3.9 Photograph of compact cluster of albite spots under optical microscope with crossed polar, showing various sized inclusions of amphibole and epidote. Center of single albite spot includes fine-grained amphibole, epidote and sphene, but boundary regions of the spots contain coarse-grained amphibole and epidote

higher grade basic schists. On the other hand, the former type is found commonly in the basic schists even in the higher grade zones of the studied metamorphic belt.

The multiple clusters of albite spots under the microscope appear various shapes as shown in Fig. 3.12, showing the boundaries including grains of amphiboles, epidote, and sphene. The clusters display the alignment along the *y*-axis on the *xy* plane, and along the *z*-axis and *x*-axis on the *xz* cut plane. Large cluster over 4 grains of albite spots displays two-dimensional aggregate as shown in Fig. 3.12, suggesting the various coagulation stages during the clustering process. In addition, the arrays of mineral inclusions within the cluster show the discrete change of their grains size from the core region to outer one of the clusters as described earlier, thereby suggesting the intermittent coagulation stages with increasing metamorphic temperature. Furthermore, it seems that the coagulation frequency was small in the second stage, compared with the latest and first stages of the coagulation. For the coagulation frequency may increase with increasing shear flow deformation rate, it is

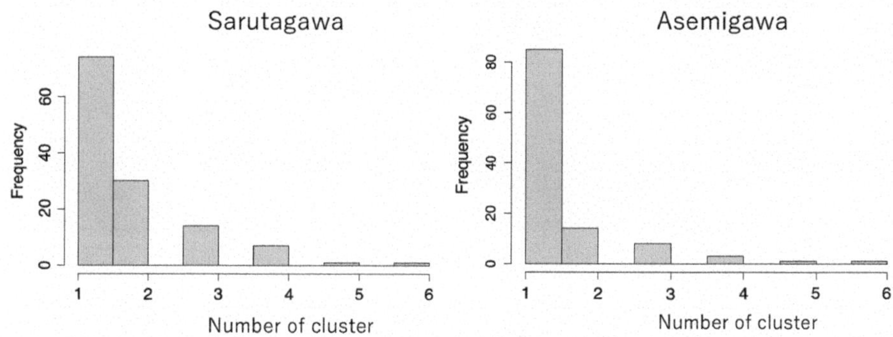

Fig. 3.10 Frequency distribution of compact cluster size of albite spot from Sarutagawa and Asemigawa of the Sanbagwa basic schists, showing the gradual decrease of cluster size

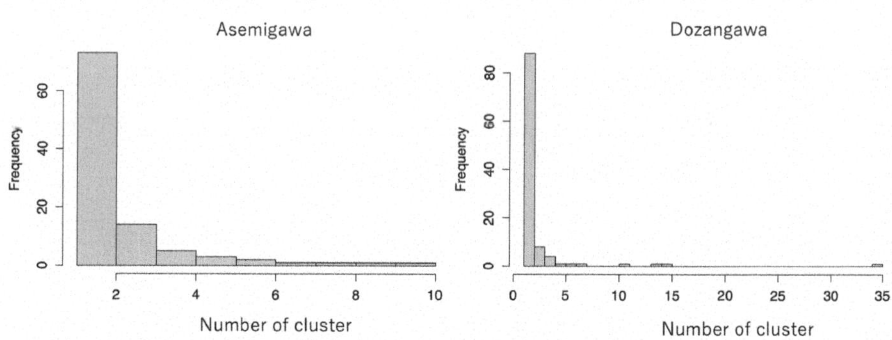

Fig. 3.11 Frequency distribution of compact cluster size of albite spot from Asemigawa and Dozangawa basic schists, showing the extending the frequency to the large cluster (left) and the high peak of the very large size (right)

probable that the latest stage deformation rate should be very high after the peak of metamorphism, that is the exhumation stage of the metamorphic belt.

On the other hand, the size distribution P of the compact clusters of grains in the simulation experiments appears to be of the Poisson's process with unidirectional collision as shown in Fig. 3.13, thereby suggesting that the ratio of cluster size $n =$ 1 and $n = 3$ can be expressed by average coagulation frequency, q as,

$$\log\left(P(3)/P(1)\right) = 2\log q - 0.778, \tag{3.6}$$

because of the Poisson' s distribution function as,

$$P(n) = q^n e^{-q}/n! \tag{3.7}$$

Fig. 3.12 Photographs of the compact clusters of albite spots on yz (left) and xy (right) plane of schistosity under the optical microscope with crossed polar

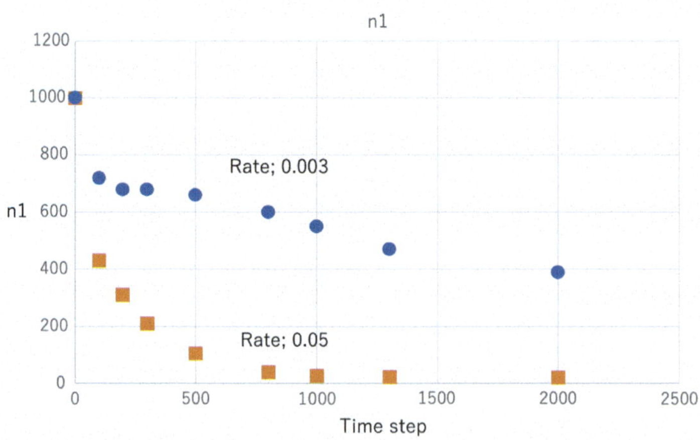

Fig. 3.13 Evolution of n_1 in the simulation of two deformation rates, showing the high decreasing rate in the high deformation rate

For q is the average coagulation frequency, it is obvious that the value of q increases with increasing shear flow rate of the matrix and the local cluster size, and thus with increasing the metamorphic temperature, for shear flow rate of the matrix should increase with temperature rise. However, in Fig. 3.14, it is possible to consider that the total strain of the matrix controls the Poisson's coagulation frequency in the simple simulation experiments. It seems surely that the coagulation frequency of the simulation experiments shows the sigmoidal change from the low value to the maximum one as displayed in Fig. 3.14, suggesting that there are two regions of the low coagulation frequency regime in the small strain magnitude and the high frequency one in the large strain. The intermediate magnitude of coagulation frequency appears the transient change depending on the total volume fraction and grain size of coagulating particles in the system.

To clarify the progressive change of the n_3/n_1 ratio in the low and high shear flow rates, the simulation experiments with time study from 100 to 2000 time steps have been carried out as shown in Fig. 3.15, and it shows that in the high deformation rate, the logarithmic n_3/n_1 ratio displays up and down and reaches to about 0.8, but in the case of low deformation rate, it shows asymptotic change to 0.8. Thereby, it seems that the logarithmic ratio becomes approaching the value of about 0.8 in the large strain magnitude. On the other hand, in the low strain magnitude, it appears that there keeps low value of the logarithmic n_3/n_1 ratio as shown in Fig. 3.15.

So far, the coagulation frequency rate of the Poisson's process is the collision number per time step, and thus it depends on the cross section area of the two clusters of collision that is governed by square of cluster sizes of them and shear velocity gradient (shear flow rate). Therefore, as seen in Fig. 3.14, the q increases with increasing the shear flow rate. On the other hand, the cluster size distribution should change with increasing shear flow rate because of the increase in coagulation rate previously mentioned. Next, it will be shown that the cluster size frequency distribution changes with increasing the shear flow rate.

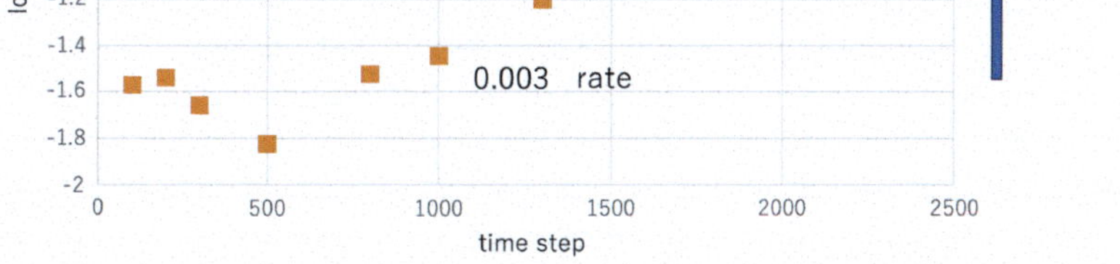

Fig. 3.14 Evolution of the coagulation of grains in simple shear simulation with increasing deformation rate from 0.003 to 0.05, showing the increase of coagulation frequency, q

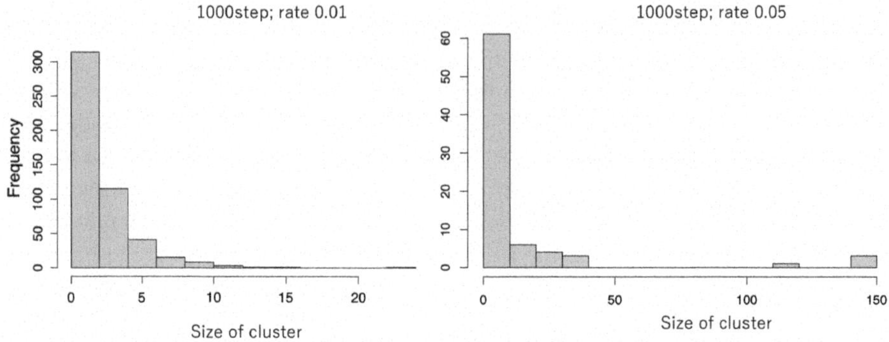

Fig. 3.15 Frequency diagrams of cluster size with increasing shear strain rate (displayed in number), showing the long-term tail of large cluster size with increasing deformation rate

In Fig. 3.15, it is shown that the cluster size frequency distribution patterns change drastically from the simple Poisson's pattern without long-term tails to that with very long-term tails reaching about several hundred cluster. It is obvious that such size distribution pattern of rapid decrease of small number of clusters having very long-term tails does not display the Poisson's process, but the process should be governed by the preferential absorption into large sized cluster due to large collisional cross section. As shown in Fig. 3.15, the drastic change appearing the large cluster tails takes place in the of the higher coagulation frequency rate regime.

As shown in the previous sections, there are two types of cluster size distribution patterns in the plate boundary rocks of Sanbagawa metamorphic belt: one is simple Poisson's distribution of small coagulation frequency, and another is composite type that shows simple Poisson's type in the small cluster size range and long-term tails of very large cluster sizes. The latter type occurs commonly in the basic schists of higher grade zones, but the former type appears frequently in the lower grade zones. It should be noted that the albite grains in the lowest grade zones appear fine to medium grained aggregates with actinolite, chlorite, epidote, and quartz, but not show the porphyroblastic texture (Toriumi 1975a, b).

The compact clusters of albite spots (porphyroblastic grain of albite) display frequently the composite structure that is constituted with albite spots containing many fine-grained epidotes and sphene inclusions and those containing medium-grained epidote and hornblende inclusions, as shown in Fig. 3.12. These medium-grained inclusion minerals are smaller than the grain sizes of hornblende and epidote in the matrix surrounding albite spots. Considering that the grain sizes of hornblende and epidote increase with increasing the metamorphic grade, it is obvious that clustering of albite spots with fine inclusions was earlier than that with medium inclusions, and that late clustering of albite spots was earlier than overgrowth of matrix amphiboles and epidote. According to Okamoto and Toriumi (2004), matrix amphiboles of the higher grade basic schists of Sanbagawa metamorphic belt show the growth zoning of the retrograde metamorphism, and thus it is concluded that the clustering of albite spots with medium amphibole inclusions should take place during the retrograde stage, that is the period of extrusion process of the deep-seated plate boundary rocks. Therefore, the shear flow deformation of these spotted schists seems to manifest the extrusion-related deformation of the deep-seated plate boundary masses.

On the other hand, it is concluded that the present simulation results in this study sharply clarify the long-term tails of the cluster size distribution appearing in the case of large shear flow rate and large shear deformation magnitude, extending over 10. Therefore, it may be concluded that the natural occurrence of very large clusters of albite spots in the higher grade zones indicates the large shear deformation rate and large deformation intensity in the higher grade rocks.

By the way, the very large clusters of albite spots appear in the amphibolite in the vicinity of the boundary zones between oligoclase-biotite zone and albite-biotite zone, thereby suggesting that the boundary zone between them is just the high shear deformation zone in the exhumation stage. Furthermore, the present large deformation zone should be derived from very high shear deformation rate. This is consistent with the conclusion that the highest-grade zone of the Sanbagawa metamorphic belt has been experienced by the very rapid exhumation process by the boundary-friction-controlled extrusion mechanism (Toriumi 2022). In that book, he proposed the frictional extrusion model of the rapid and extremely rapid exhumation of the metamorphic core sheets that is based on the velocity-independent very low frictional drag against negative buoyancy force acting on the metamorphic mass. In this case, the model may be needed for the massive flow of aqueous fluid which makes great lowering the frictional coefficient and does enhancement of the hydration reaction of the matrix minerals in rocks along the boundary zone, as was clarified by Uno et al. (2015).

On the other hand, it is to be discussed that the distribution pattern of albite spots manifests lattice-like texture development accompanied with shear deformation intensity in the simulation study. The physical mechanism of this pattern evolution may be governed by the collisional cross section that is mainly dependent on the size of large clusters. Accordingly, the distribution patterns of the albite spots containing various clusters should be controlled by the spatial distribution of large clusters which may appear the topologically equal-distance configuration in the isotropic random motion case of albite spots. But, for the present case is simple shear flow deformation of matrix, then it should be constrained that the differential velocity of each albite spots arises normal to the flow velocity. Thus, the collision of them may take place normal to the flow velocity, thereby suggesting that layers with large cluster may develop and those with small one extinct on the contrary. Then, it may easily be considered that the large cluster layers appear intermittently and the clusters themselves occur separately by the distance within the layer. As a result, it seems that there appears a lattice-like texture of albite spot clusters.

In this paper, the author considers that the albite grains should collide with each other during the shear flow of the matrix constituted by amphiboles, epidote, chlorite, and quartz with minor amount of garnet. These matrix grains appear much smaller than porphyroblastic albite and their clusters. The collision and successive coagulation of albite grains and clusters should be responsible for their behavior like suspension particles in the viscous fluid flow, and thus it seems that the albite grains and their clusters probably look like rigid particles, being compared with deformable matrix.

Metamorphic rocks in the plate boundary commonly show the texture that displays periodic bands rich in large clusters and single large grains of albite both in the basic and pelitic schists. The formation of these clusters and porphyroblastic grains is responsible for mechanical coagulation of smaller grains of albite by shear flow deformation of the matrix, judging from the Poisson's distribution patterns of the cluster size and microstructures of compact cluster containing massive inclusion-rich parts. It is still strange that there are common size frequency distribution patterns showing the Poisson's pattern in the small cluster size but having the isolated peaks of very large cluster. The three-dimensional simulations of collision and clustering of albite grains during the shear flow were investigated, and their results show that the size distribution patterns of the cluster display clear Poisson's pattern in the low deformation/deformation rate conditions, but do the complex patterns composed of Poisson's pattern in the small cluster size region and multiple peaks in the very large cluster sizes, being like the natural occurrence mentioned above.

Furthermore, it was shown by the simulation experiments that the distribution patterns of clusters and single grains of albite appear as the periodic cluster-rich bands and discrete occurrence of cluster in the bands, thereby suggesting that the natural occurrence of albite clusters in the highly deformed metamorphic schists has been derived from the collision and coagulation growth of the smaller grains of albite during shear deformation process during the retrograde metamorphism. Therefore, the common occurrence of albite cluster and porphyroblastic albite grain in the high-grade zones in the high-pressure metamorphic belt should be responsible for the highly shear deformation rates of the matrix in the exhumation stage of the metamorphic belt.

3.2 Phase Petrology and Symbolic Dynamics of the Metamorphic-Metasomatic Serial Reaction

Magma genesis and metamorphic process have been investigated by means of thermodynamical framework and phase diagram based on the theoretical and experimental works together with geochemical and petrographic studies. The basic concepts of these disciplinary works have been phase equilibrium of minerals recrystallized in the metamorphism. In this framework, equilibrium mineral assemblages are governed by the recrystallization temperature, pressure, fluid pressure, and chemical compositions of the related bulk rock, and then the phase rule holds,

$$f = c + 2 - p \tag{3.8}$$

in which f is the degree of freedom, c is the number of chemical components, and p is the number of mineral phases (number of mineral species).

On the other hand, the solid-state mineral reactions in the deep-seated metamorphic rocks take place in the limited rates of diffusion of components and advection by fluid flow rate through grain boundary, and therefore it shows that the bulk equilibrium among whole minerals is not available even if at the grain boundary of growing minerals and interstitial fluid, minerals and fluid keep the equilibrium conditions at the changing environmental parameters. Furthermore, it seems that many ionic components dissolved from minerals into the grain boundary and pore fluid migrate by advection on fluid flow, and they contribute the replacement of minerals by metasomatic reactions. Uno et al. (2015) and many authors have clarified the bulk chemistry of metamorphic rocks changing largely during the prograde and retrograde metamorphism in the plate boundary zone.

By the way, the fluid flow is controlled by the porosity distribution of rocks, and in the deformed rocks it should be very anisotropic because of anisotropic permeability derived from fluid-filled open crack and shear crack jogs as shown in the previous book (Toriumi 2022). It shows in the book that the mineral precipitation and dissolution in the flowing fluid takes place repeatedly along the fluid path, and then the mineralogical banding may appear vertical to the fluid flow orientation. This is why the chemical composition of the dissolve species in the flowing fluid changes repeatedly and saturates cyclic with different minerals. Therefore, the mineral banding of the metamorphic and metasomatic rocks in the anisotropic fluid flow should manifest the characteristic order of the mineral bands for the fluid flow rate and reaction rate. In this section, the author intends to study this type of banding structures as the mineralogical symbolic dynamics in the nonlinear reaction—fluid flow dynamics.

There are abundant mineral bands that develop nearly parallel to the schistosity plane in the plate boundary metamorphic rocks. In the low-grade metamorphic rocks, bands appear commonly mono- to bi-mineral ones with thickness of about 0.1 to 10 mm, and these are constituted with albite, albite + quartz, albite + chlorite, chlorite, chlorite + epidote, epidote, epidote + calcite, actinolite, actinolite + chlorite, and actinolite + epidote + chlorite + albite + quartz + alkali amphiboles as shown in Fig. 3.16. The last assemblage is composed of matrix, and other bands appear often repeatedly with some order of succession.

The bands change to arrays of lens and blob horizontally. They look like boudin structure deformed from bands by plastic deformation of the bulk rock, but the bands often display termination with other bands, forming the Y-shape junction. Furthermore, the bands appear intralayer tight folds between the bands. Therefore, lens and blob are not derived from the extension breakout as a pinch and swell structure forming boudin. It seems that the bands, thin lens, and blobs should be formed simultaneously in various mode of occurrence as mono- or bi-mineral aggregates.

The microstructures parallel to schistosity display island-like texture of bands which look like the cross section of sand dune as shown in Fig. 3.17, suggesting that the unidirectional origin of formation. This type of microstructures on schistosity plane is favorable for the jog model of the band origin on the shear crack surface as modeled by the previous book (Toriumi 2022).

Fig. 3.16 Photographs showing metamorphic banding structure composed of albite band (white), chlorite band (deep green), and epidote band (yellowish green). Left; lineation parallel and schistosity normal, and right; lineation and schistosity normal

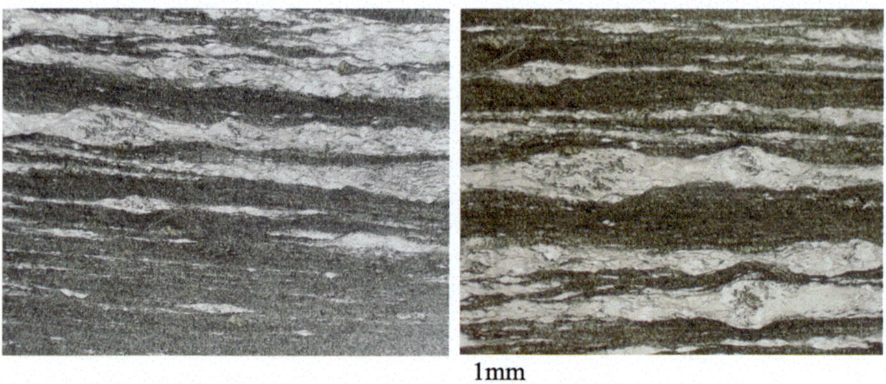

Fig. 3.17 Photographs showing albite band and lens composed of albite, fine-grained epidote, and chlorite under optical microscope with open polar

The repeated patterns of order of different mineralogical bands in the low-grade metamorphic schists observed in the basic schists are,

1. ACETM
2. ACEM
3. ACM
4. AM.

in which T is the actinolite dominant band, E is the epidote dominant one, C is the chlorite dominant one, A is the albite dominant one, and M is the matrix composed of actinolite, epidote, chlorite, albite, and quartz. Above order from 1 to 4 is the increasing order of frequency in basic schists.

The mode of occurrence of the bands are (1) A-band often contains chlorite, C-band does epidote, E-band does chlorite. It means that the aqueous solution saturated in these minerals is not multi-component system with one mineral phase but with two mineral phases, thereby showing that the system of solution + two minerals has freedom $f = c - 2$. For c is nearly 8, the freedom of each band system becomes 6. It is important that the solution flowing the grain boundary pores has a variable chemical composition relevant with solubility of minerals in bands and controlled by the phase rule.

The chemical reactions of the grain boundary solution in bands A and C should be formulated as,

$$4NaAlSi_3O_8 + 3Mg^{2+} + H_2O = Mg_3Al_4Si_3O_{10}(OH)_2 + 4Na^+ \\ + 9SiO_2 + 2H^+$$

On the other hand, the reaction between C and E holds,

$$3Mg_3Al_4Si_3O_{10}(OH)_2 + 8Ca^{2+} + 3SiO_2 = 4Ca_2Al_3Si_3O_{12}(OH) \\ + 9Mg^{2+} + H_2O$$

Furthermore, the E and A are related to be,

$$Ca_2Al_3Si_3O_{12}(OH) + 3Na^+ + H^+ + 6SiO_2 = 3NaAlSi_3O_8 + 2Ca^{2+} + H_2O$$

In the case of T and E bands, it holds,

$$Ca_2Mg_5Si_8O_{22}(OH)_2 + 3Al^{3+} + H^+ = Ca_2Al_3Si_3O_{12}(OH) + 5Mg^{2+} \\ + 3SiO_2 + H_2O$$

These reactions should take place along the boundaries between relevant mineral bands, so that for example, in the case of growth of albite consuming chlorite along their band boundary, Na^+ will be consumed from solution but Mg^{2+} dissolves into solution. Furthermore, it will take place nearby that the second reaction advances consuming Mg^{2+} and dissolving Ca^{2+} in solution with growth of chlorite instead of epidote. In addition, the third reaction will take place by replacement of epidote after albite, consuming Ca^{2+} in solution. It seems that above sequential reactions proceed between the nearest neighboring bands by flowing aqueous solution.

In these sequential reactions associated with flowing aqueous solution, the chemical gradients of these ionic species should be represented by the phase diagram as shown in Fig. 3.18. Then, the systematic relations of two bands will be discussed in the phase diagram of Na^+–Mg^{2+}–Ca^{2+} activities in aqueous solution. The center is the invariant point among epidote, chlorite, albite, and aqueous solution, which represents the chemical composition of matrix solution, and the boundary between

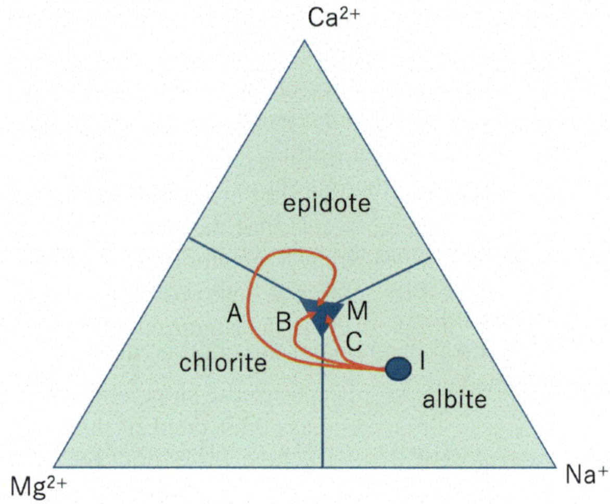

Fig. 3.18 Illustration of various compositional trends of aqueous solution in the grain boundary from the initially banded albite I to the matrix of albite, epidote, and chlorite assemblage M in basic schists in the triangular diagram of activity of ionic components in solution. The various trends of chemical compositions in solution are controlled by the invariant point M in the triangular diagram as discussed in the text and later figure

two phases such as albite and chlorite is the di-variant composition of solution at the band boundary, and the volume of single phase is to the tri-variant composition of solution in the band. Considering that the system contains SiO_2, FeO, Fe_2O_3, Al_2O_3, TiO_2, K_2O, and other minor components in solution and constituent minerals, it should be required for the actual banding process that the di-variant boundary of ionic concentration ratio in Ca^{2+}–Mg^{2+}–Na^+ diagram must represent the somewhat extended boundary region and area around point, respectively in the diagram.

Consequently, it may conclude that the chemical composition of the aqueous solution in the grain boundary should be traced from the multiple phase matrix region to the single-phase mineral band through various band boundaries as shown in Fig. 3.19. It is interesting that the sequences of bands in basic schists mentioned above manifest the various type of cycloid—like traces (shown in Fig. 3.18) from matrix to every monomineralic bands, thereby suggesting the evolutionary track of the chemical composition and mineral bands with progressive banding process. This is interpreted by the symbolic dynamics of the nonlinear dynamics (Beck and Schlogl 1993) of the dissolution–precipitation-diffusion process in the directional solution flow in the metamorphic rocks.

In the symbolic dynamics model for various banding, it should be required that the mineral bands such as albite band may be formed by precipitation in the shear crack jog volume in the sheared metamorphic rock, that is a schistose rock, as proposed by the author in the previous book (Toriumi 2022). The successive banding process after jog precipitation should make the chlorite, epidote, and actinolite bands in the above reaction–diffusion dynamics in the fluid flow system mentioned previously.

On the other hand, the directional flow of fluid (aqueous solution) is probably performed by the periodic open crack cluster between the shear slip planes shown in the previous chapter. It shows that the microcracks occur perpendicular to the schistosity plane represented by albite and chlorite bands, and they appear to be the periodic clusters as shown in Fig. 3.20. Microcracks are of open crack type and

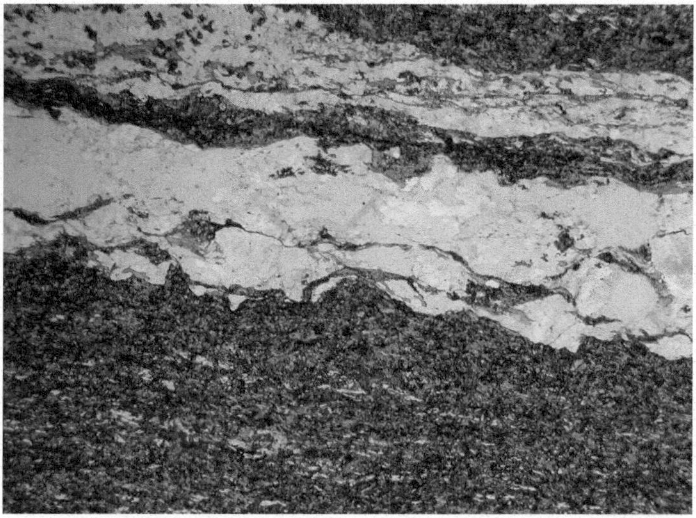

Fig. 3.19 Photograph showing the migrating boundary between albite band and matrix of the greenschist under optical microscope with open polar in the Sanbgawa metamorphic belt of Asemigawa, Shikoku. The migration of the boundary toward the matrix is apparently derived from replacement of albite after matrix chlorite and epidote, which is shown by enclosing of matrix in the grain boundary of albite

filled with commonly albite and chlorite. As described in the preceding chapter, the microcrack walls appear subvertical to the mineral lineation, that is the stretching orientation of the rocks, and thus the fluid flow suggests nearly vertical to the schistosity and mineral banding planes. The microcracks filled with albite terminate to the albite band, and those filled with chlorite to the chlorite band, thereby suggesting the banding minerals precipitated from the aqueous fluid flowing through microcrack clusters.

By the way, it has been shown in the previous section that the large shear crack slowly grows in the situation of flow-in of aqueous fluid along the shear crack itself and being associated with formation of the microcrack cluster beside the main shear crack by dissipation of the concentrated stresses around the shear crack tips. Thus, it seems that the shear crack growth continues even over the critical length of Griffith crack in the multiple banding schists. Inversely, it can be said that the metamorphic schists containing abundant multiple banding of albite and chlorite record large amount of shear slip by shear-crack growth and perhaps propagation.

In addition, it must be important that the shear crack growth and propagation is surely associated with the rapid decompression and successive recovery of fluid pressure in the arrayed jog along the shear crack, as being simulated in the previous chapter. The various temporal variation of the decompression path in the shear crack jog are shown in the previous section, thereby suggesting that the pressure drop of

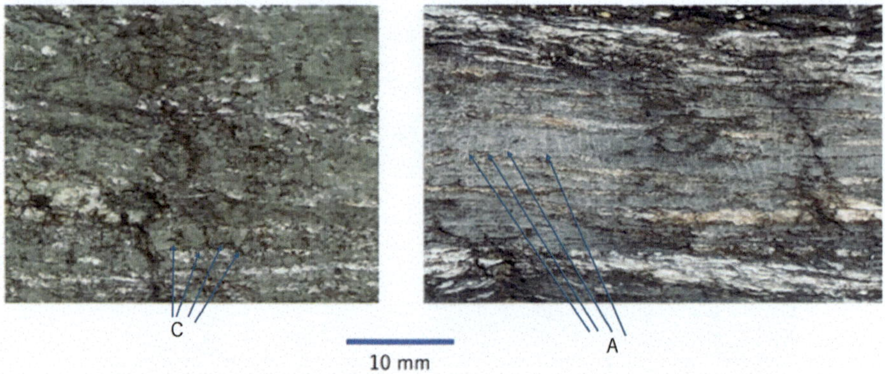

Fig. 3.20 Photographs showing the cluster of micro-open cracks sealed by chlorite (C, left) and albite (A, right) of low grade greenschists of Nagatoro, Kanto Mts

decompression varies in the wide range according to the slip velocity of the shear crack. It follows that the phase relation in the shear crack jog probably varies in response with the pressure drop of decompression, being compared with the phase relation of the matrix surrounding jog.

Considering the nearly invariant phase relation of the matrix greenschists showing the system of actinolite-chlorite-epidote-albite-quartz-magnetite in the given T and P, when the fluid pressure rapidly drops down, the invariant point of the jog system with aqueous fluid and mineral particles should move to the chlorite apex in the phase diagram (Fig. 3.21), and the chemistry of the aqueous solution which has the invariant composition of matrix seems to be plotted in the single-phase region of albite. Then, the pressure of the jog gradually increases to that of the matrix, and consequently, the invariant point of the phase relation in the jog returns possibly to that of the matrix. As a result, the stable mineral changes from chlorite to epidote in the jog as shown in Fig. 3.21.

The metamorphic minerals represent the stable assemblages in the aqueous solution of the shear crack jog at the process of rapid pressure drop and successive recovery by shear crack slippage as noted in the previous section. However, the detailed process above should involve the basic processes of fluid flow, chemical reactions with precipitation and dissolution in solution, and diffusion of ionic and molecular species in solution. The simulation model of dissolution and precipitation of minerals and jog formation in the multiple shear crack system with flowing of aqueous solution is presented in the previous chapter, and it is concluded that the layering of high jog density is formed by the coupling of the above basic processes, and it enhances the shear crack motion and the pressure drop in jogs. The ionic and molecular concentrations of aqueous solution are governed by the kinetics of mineral reactions and diffusion of them, together with changes of pressure and temperature which control the velocity of mineral reactions. These processes may contain the non-linearity, and therefore, there seems to be various stable and unstable layers of mineral precipitation perpendicular to the fluid flow of aqueous solution. Therefore,

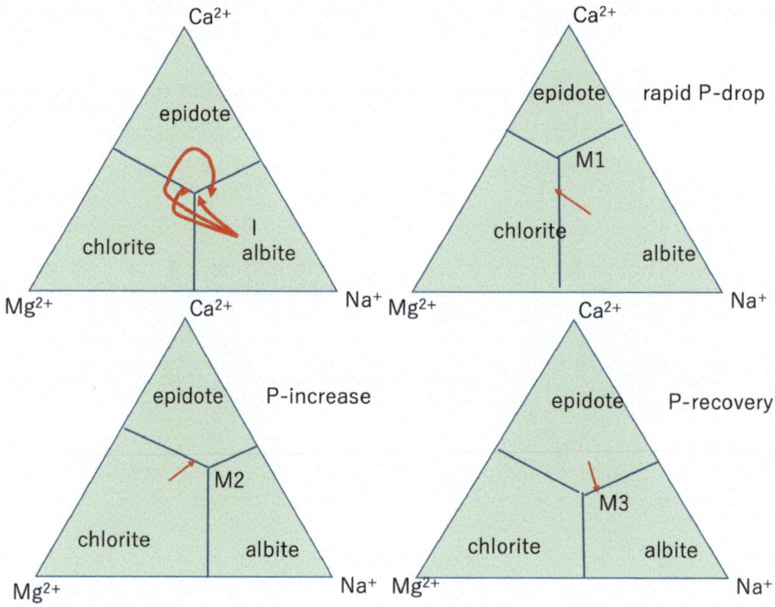

Fig. 3.21 Illustration of various processes producing the three types of compositional trends of the aqueous solution by changing the invariant point M dependent on the effective pressure around the albite band formed by shear crack jog motion. **a**; possible compositional trends shown in Fig. 3.18, **b**; Initial condition of M around albite band, **c**; in changing M2, chlorite stable to epidote stable field, and **d**; in M3, epidote stable to matrix assemblage field

in this section, it may be available that simple repeating patterns of minimal nonlinear differential equation is discussed later.

Considering the symbolic dynamics associated with the one-dimensional logistic differential equation.

The differential equation having two variables u and v in this study will be considered,

$$\begin{aligned} \mathrm{d}u(t)/\mathrm{d}t &= au - k_1uv^3 \\ \mathrm{d}v(t)/\mathrm{d}t &= bv - k_2uv^3 \end{aligned} \tag{3.9}$$

in which a, b, k_1 and k_2 are constant parameters in these nonlinear equations.

Above simple NLD can be simulated as shown in Fig. 3.22, suggesting complicated timeseries of u and v which are the function of ionic concentrations controlled by effective pressure in the moving shear crack jog with advancing time. In the phase map of u and v for various P (Fig. 3.23), it is available that the set of u and v can be classified into three domains of A, B, and C by means of their ranges. These domains are mapped possibly in the column of the test region of the (u, v) variations as displayed in Fig. 3.23, thereby indicating clearly the repeating structure of the domains from the left hand side as like as,

A B C A B A C A C A B C B A B

This is the symbolic sequence of the chaotic dynamics in the one-dimensional space. This is also available for the symbolic sequence of the temporal variation.

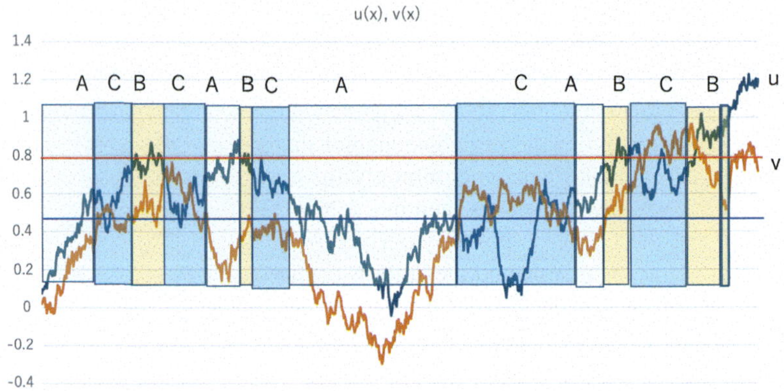

Fig. 3.22 Evolutionary tracks of u and v of binary nonlinear dynamics in the text and symbolic sequence corresponding the assemblages of u and v value denoted in the phase map of u and v shown in Fig. 3.23

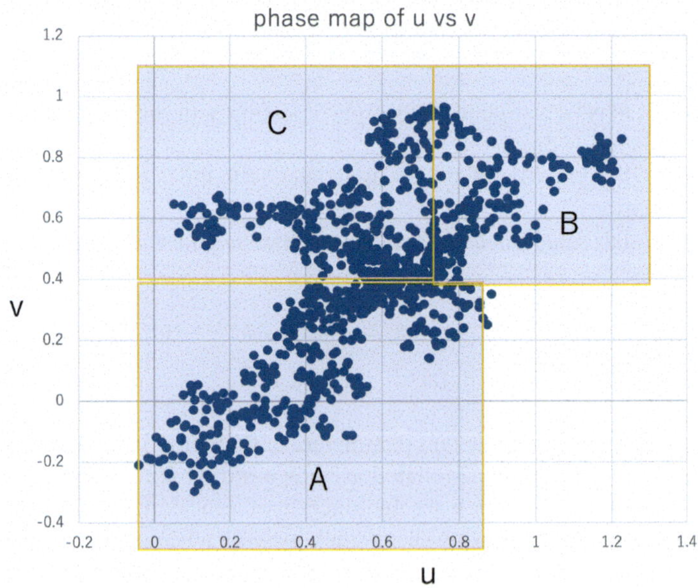

Fig. 3.23 Phase map of u and v of binary nonlinear dynamics in the text and regimes of symbols (A, B, and C) corresponding to domains defined by u and v ranges

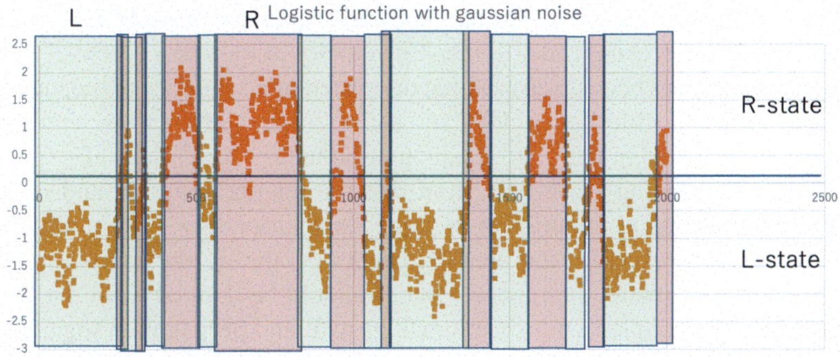

Fig. 3.24 Evolution of logistic dynamics and the symbolic (A and B) sequence defined by the parameter level (horizontal bar)

On the other hand, more simple logistic dynamics has also various temporal sequence of variables in terms of the symbolic dynamics as,

$$dx/dt = 1 - ax^2 \tag{3.10}$$

In the logistic dynamics, the symbolic sequence varies depending on the intensity of parameter, a, but taking the labels of L ($-1 < x < 0$) and R ($0 < x < 1$), the repeating of their labels appears in the temporal sequence, such as LRLRLRLR …, and the length of the duration of them varies randomly as shown in Fig. 3.24.

By the way, the mineral banding sequences of the metamorphic greenschists are represented by three repeaters embedding in the matrix as,

ACET, ACE, AC, and A.

These systematic sequences may be modeled by the successive precipitation of the albite, chlorite, epidote, and actinolite in the process of decompression and pressure recovery of the shear crack jog growth as described in the previous chapter. It seems that the sudden change of invariant point of aqueous fluid in the growth jog is derived from the rapid decompression at first, and the stable assemblage of the initial jog changes from that of matrix to albite monophase as displayed in Fig. 3.21. Consequently, during the pressure recovery in the jog, it holds that the invariant point moves toward albite and chlorite apexes, and thus the aqueous fluid enters the chlorite stable field, precipitating the chlorite. Furthermore, becoming the pressure of matrix, the invariant point comes back to the matrix system, and thus, aqueous fluid enters the epidote stability field where epidote becomes precipitating.

If the shear crack velocity is not so high, the degree of decompression is not large, and then the change of invariant point does not move so much, thereby showing the precipitation path likely to different one (Fig. 3.21): one path in the very low slip velocity of the shear crack may exhibits in the albite stability region, and the other path of the medium slip velocity enters in the chlorite region from albite one as shown in Fig. 3.21, thereby suggesting that the repeated sequences from ACET, ACE, AC to A manifest the increasing order in the slip velocity of the shear cracks.

In this section, it has been discussed that the systematic sequence of the monomineralic bands of albite, chlorite, epidote, and actinolite in the rather low-grade metamorphic schists formed in the plate subduction boundary zone. Basically speaking, the metasomatic banding is derived from dissolution–precipitation and diffusion kinetics of the constituent minerals in the aqueous solution flow system through the schistose rocks. However, for the banding occurs parallel to the schistosity plane, that is the shear slip plane, the repeated bands should be derived from the dissolution—precipitation coupled with shear crack growth and propagation together with aqueous fluid flow as proposed in the previous sections. Moreover, in the previous model of the mineral band parallel to the shear slip, it is concluded that the precipitation of albite and quartz took place in the shear crack jog which migrates along with shear crack motion.

Otherwise, it is also possible that repeated systematic sequences of the mineral bands parallel to the shear slip is derived from the metasomatic reaction of matrix mineral assemblage with flowing aqueous solution in the decompression zone formed by shear crack jog motion as proposed in this section. In this model, the author did not treat the detailed mechanism of nonlinear dynamics involving decompression reaction of minerals with aqueous solution and diffusion, but as discussed in the section of the symbolic dynamics of the nonlinear dynamics, the various chemical concentration paths of the decompressed aqueous solution can be identified by the sequential systematic banding of stable minerals. In addition, it should be concluded even in this model that the different symbolic paths of the mineral bands seem to be responsible for the magnitude of the decompression pressure, and thus the shear crack velocity in the schistose rocks of the plate boundary metamorphic zone.

3.3 Grain Size Distribution in Decompression Growth of Minerals in Shear Crack Jog

There is common occurrence of monomineralic layers parallel to the schistosity plane (shear slip plane) of the subduction plate boundary metamorphic schists, and they are mainly composed of albite in basic schist and quartz in quartzose schists. In the previous chapter, it is concluded that these monomineralic layers were surely formed by mineral precipitation in moving shear crack jog in sudden decompression and successive pressure recovery process. In these monomineralic layers, it is noteworthy that there are abundant very fine grains of albite which occur as mineral inclusion in large grains of quartz, and on the contrary, fine-grained quartz in large albite grains.

It seems that the fine grains of albite in large quartz grains showing roughly euhedral shape display almost always platy euhedral shape as shown in Fig. 3.25, thereby suggesting growth of albite grains in the aqueous solution and/or gaseous fluid in the shear crack jog space. It is still strange that the euhedral-platy fine grains of

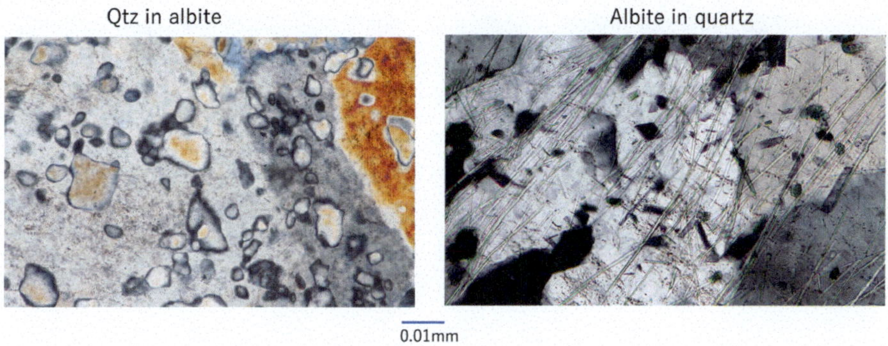

Fig. 3.25 Photographs showing vermicular quartz inclusion in large albite grain (left) and euhedral albite inclusion in large quartz grain in shear crack jogs of low grade greenschists of Nagatoro, Kanto Mts

albite occur showing the random orientation in a single quartz grain. If the nucleation and growth of albite took place in the fluid environment of very low density, it should be sure that the grains of albite sediment on the floor of the jog space with alignment of them. However, random distribution of grains and their orientations manifests keeping the random motion in the jog, suggesting the turbulent fluid in the jog, and very rapid growth of host quartz. Moreover, it suggests that host grains of quartz grew rapidly including the random situation of fine-grained albite.

On the other hand, in the above jog, the very long grains of acicular actinolite occur as clusters of parallel alignment and perpendicular to the wall of the jog. The acicular actinolite freely elongates without fine-grained albite and host grains of quartz, suggesting its rapid growth in the jog associated with its enlargement by shear crack growth and propagation. It is noteworthy that some of long acicular grains of actinolite contain several segments of alkali amphibole, which is assigned by its blue–violet color and sharp boundary under optical microscope. Judging from the coexisting actinolite, alkali amphibole, chlorite, albite, and quartz, the mineral precipitation and growth was in the conditions of 300–400 °C of temperature.

The occurrence of quartz inclusions in albite host is different from above-mentioned euhedral grains of albite inclusion in the jog: fine grains of quartz appear rounded shape and often occurs as random cluster of equant grains under several micrometers size. Many of quartz grains in host albite and calcite occur near the grain boundary region of the albite. It is noteworthy that the very large grains of quartz show the concave feature of the boundary against the large grains of albite, in which the fine grains of quartz are commonly aligned near the boundary as shown in Fig. 3.26. It seems that the concave boundary is derived from the enlargement of large grain of quartz, being likely to absorb the fine-grained quartz inclusion in host albite. In this figure, it is very impressive that there is a large discrepancy of the grain size of quartz between the coarse grains and fine grains of inclusion quartz in albite, and the grain size distribution of the quartz in the jog space may display the lognormal type with very large long-term tail.

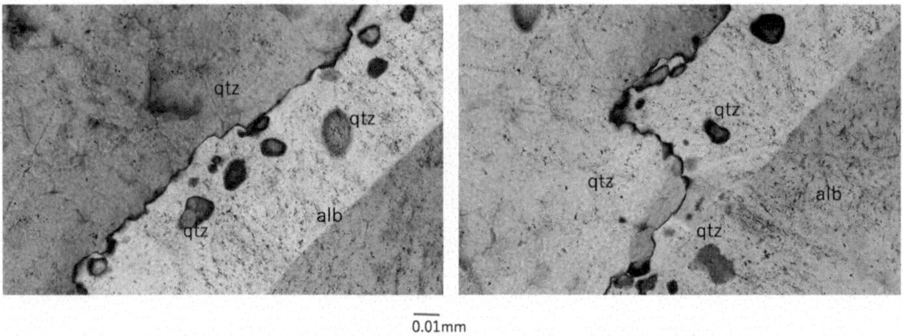

Fig. 3.26 Photographs showing large quartz grain and neighboring vermicular quartz inclusions in large albite grains in shear crack jogs of low grade greenschists of Nagatoro, Kanto Mts

Decompression modeling of crystal size distribution patterns in the volcanic rocks has been investigated by Cashman and Marsh (1988), and they obtained the exponential type CSD which characterizes the decreasing rate of pressure under nearly constant temperature model of the volcanic eruption. In this model, the physical process of the eruption of magma is represented by the rapid pressure drop during the uprising in the conduit that is the increasing nucleation rate and constant growth rate of crystal with ascending magma.

On the other hand, decompression crystallization of metamorphic minerals in the shear crack jog is different from the magmatic process even in the decompression crystallization in the magma. The crystallization of metamorphic minerals in the shear crack jog is represented by first decompression nucleation in the large degree of supersaturation, and later gradual crystal growth in the recovery process of pressure in the high-pressure condition. Therefore, it is obvious that the initially formed nuclei should experience the successive crystal growth, and resultant CSD must show the low frequency of the small size mineral grains. On the other hand, the CSD pattern in the large size region surely displays the exponential distribution likely to the CSD by Cashman and Marsh. In this section, the author would like to propose the CSD model of the decompression growth during the rapid formation of shear crack jog.

First, let us consider the conservation law of grain size distribution function as follows (Toramaru 2019):

$$\partial f(R,t)/\partial t + \partial f(R,t)G(R)/\partial R = 0 \tag{3.11}$$

in which $f(R, t)$ is crystal size frequency distribution, and $G(R)$ is the growth velocity of crystal. R is the grain size of crystal.

Assuming the variables separation about size and time in the frequency distribution function as,

$$f(R) = f_1(R)f_2(t) \tag{3.12}$$

Then, the conservation equation becomes,

$$(1/f_2)\partial f_2/\partial t = -(1/f_1)\partial(Gf_1)/\partial R \tag{3.13}$$

Let us take transformation as,

$$g = Gf_1 \tag{3.14}$$

$$f_1\partial f_2/\partial t + f_2\partial g/\partial R = 0 \tag{3.15}$$

$$\begin{aligned}(1/f_2)\partial f_2/\partial t &= (-1/f_1)\partial g(R)/\partial R\\ &= C(C;+\text{if } R > Ro,\ C;-\text{ if } R < Ro)\end{aligned} \tag{3.16}$$

Furthermore, it holds,

$$\begin{aligned}-(G/g)\partial g/\partial R &= C\\ (1/f_2)\partial f_2/\partial t &= C\end{aligned} \tag{3.17}$$

Therefore, we obtain,

$$f_2 = f_o \exp(Ct), \tag{3.18}$$

$$g = g_o \exp\left(-C\int \frac{1}{G}\mathrm{d}R\right)$$

It concludes,

$$\begin{aligned}\log f(R,t) &= -CR/Go + B(t)(R > Ro; C > 0, R < Ro; C < 0)\\ G &= Go(\text{constant})\end{aligned} \tag{3.19}$$

in which B is a function of the time.

Therefore, the frequency distribution function of crystal size in the decompression jog is represented by Fig. 3.27, thereby showing the characteristic feature that in the small R region, the positive exponential curve appears, and in the large region of R it does negative one. The critical size between them should depend on the sampling time and constant grain growth rate Go. It is probable that the frequency distribution function may appear smooth curve around the peak of the frequency curve.

The actual frequency distribution of the vermicular quartz inclusion in albite and euhedral albite inclusion in quartz were measured under the photographs taken by optical microscope. The euhedral albite inclusion measured the length of its long axis of prismatic shape because euhedral albite grain appears thin platy shape and its random orientation. On the other hand, vermicular quartz inclusion appears rounded

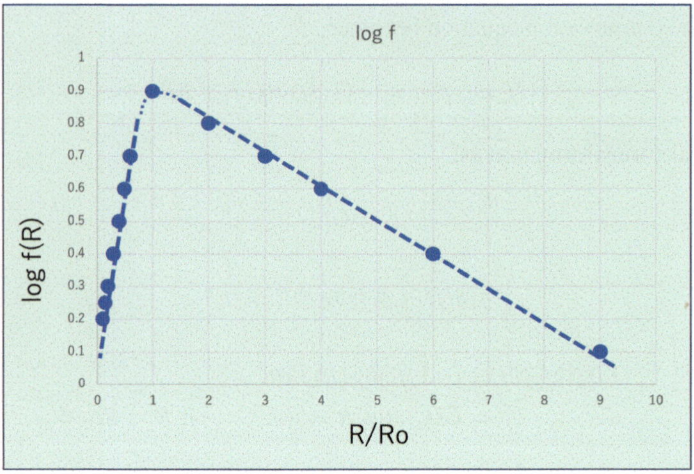

Fig. 3.27 Logarithmic diagram of normalized crystal size distribution (CSD) in simulation model in the case of sudden decompression and gradual compression process in the precipitation from solution

and sometimes irregular shape and thus the grain size is measured the length of its long axis.

The histograms of their crystal size distributions are shown in Fig. 3.28, thereby suggesting the lognormal distribution of grain size. In addition, the logarithmic frequency distribution patterns of quartz and albite inclusion are shown in Fig. 3.29, and it suggests that the pattern displays the exponential type CSD in large size, but in the small range it shows rapid decrease, suggesting the model CSD pattern mentioned previously. Furthermore, it seems that $\log R$ and $\log f(R)$ diagram of the model CSD looks very likely to that of the quartz inclusion in albite, thereby suggesting that the present CSD model of the fast large nucleation rate in the rapid decompression, and later continuous growth but minor nucleation rate in the slow recovery of pressure in the shear crack jog.

It remains what is the reason that fine and euhedral grains of albite are distributed in the single large host quartz in the shear crack jog and rounded fine grains of quartz are included in the large albite grains. As stated previously, the rapid decompression in the shear crack jog surely makes the turbulent state of fluid (aqueous solution and/or gas), and rapidly nucleated fine grains should rotate and move randomly in the jog space. Later, it may be happened that at that turbulent state, the very fast growth of quartz controlled by the jog extension took place and include fine-grained euhedral albite. It seems to be supported by the fact that the very long acicular actinolite occurs as a parallel cluster and as inclusion mineral of large tabular quartz grains, as like as euhedral albite grains. Furthermore, above acicular actinolite stings appear perpendicular to the jog wall together with the tabular quartz host grains, thereby suggesting that the growth of acicular actinolite and tabular quartz is coeval and guided by the jog opening and thus the shear slip motion.

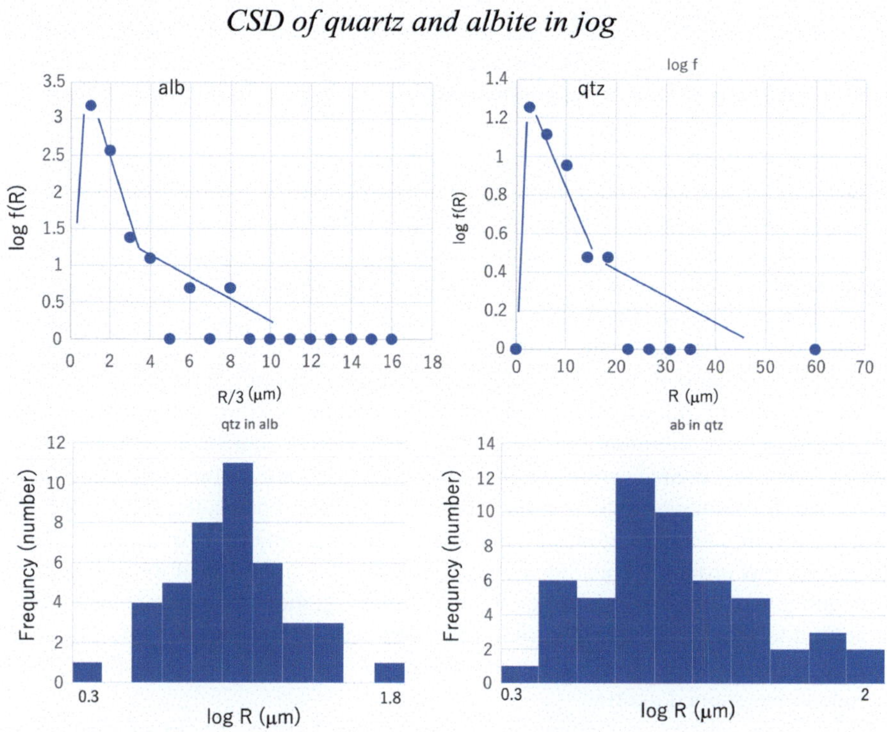

Fig. 3.28 Crystal size distribution patterns of albite grains (left, top) and quartz grains (right, top), and quartz inclusion in large albite grain (left, bottom) and albite inclusion in large quartz grain (right, bottom)

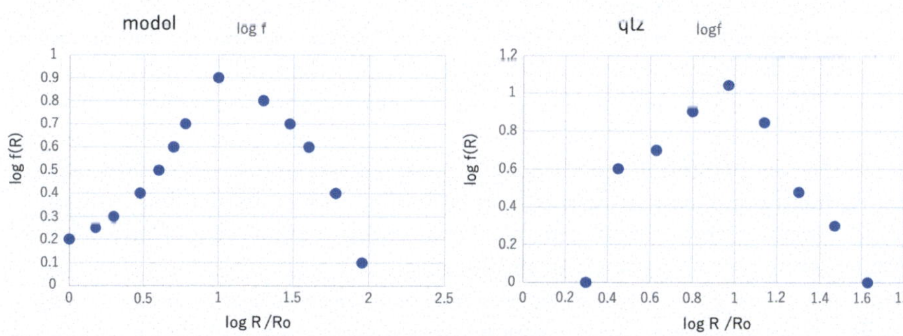

Fig. 3.29 Logarithmic diagrams of normalized CSD of model shown in Fig. 3.27 and quartz grains in natural shear crack jog of low grade greenschist in Nagatoro, Kanto Mts

On the other hand, it seems that the vermicular quartz inclusions are derived from the rapid coarsening of the host albite which contains abundant fluid inclusion arrays. The arrays in host albite grain display the closed loop with flat surface likely to the extinct grain boundary, suggesting that the single host grain of albite is derived from the polycrystalline aggregate containing polygonal grains of quartz which are included by host at that stage. Judging from keeping the very small fluid inclusion arrays, the change from polycrystalline aggregate to the single crystal of albite is very fast, even in the rather low temperature condition but very low-pressure condition of jog opening.

In addition, it seems that polycrystalline aggregate is composed with fine grains of albite and quartz, and grains of albite contain commonly very fine-grained quartz as inclusion, suggesting that very fine-grained aggregates of quartz and albite with minor amounts of chlorite and calcite were the primary texture at the jog opening. Furthermore, it is common that there is heterogeneous occurrence of fine-grained aggregates in the coarse -grained aggregates, and these are mutually gradational in grain size. It is noteworthy that in the coarse-grained aggregates, the euhedral shape of albite is found and surrounded by the very fine-grained matrix of quartz and albite. Therefore, it is probably concluded that the primary precipitation of the jog space at its opening was very fine-grained quartz and albite together with very acicular stings of actinolite and vermicular chlorite. Then, the rapid coarsening of fine grains of albite and quartz took place, and during the rapid coarsening process, albite grains included fine-grained quartz.

Prior to the above coarsening process, it seems that fine grains of albite may adsorb randomly to form a diffusion-limited-aggregate (DLA) as shown in Fig. 3.26, suggesting that rapid decompression in the jog space made aqueous fluid containing fine particles of albite and quartz to be turbulent.

It is somewhat strange that the above-mentioned rapid coarsening may take place in the rather low temperature conditions, for rapid thermal annealing technique (RTA) in ceramic industry is performed under the high temperature conditions with decompression. This process needs the lattice rotation and grain boundary migration of fine grains in the aggregates within the short time, considering that there are abundant arrays of the micro fluid inclusions along the early grain boundary of constituent particles. Furthermore, the microstructures of layered grain size distribution as shown in Fig. 3.30, manifest probably that the rapid coarsening of grains should have occurred in the pressure recovery process of the jog space.

In this case of coarsening in the pressure recovery process, it can be imagined that the coagulation and lattice rotation coarsening of like grains of albite and/or quartz should occur successively side by side in various portions in the fine-grained aggregate of them, as like as the albite grains include the grains of quartz as remnant grains. It seems that the rate of coarsening is governed by the grain rotation rate and the number density of rotation boundary sites in the fine-grained aggregate. However, it should be not the case that the grain rotation is controlled by boundary diffusion of constituent elements, but by the dissolution and precipitation together with interfacial energy dissipation, thereby suggesting the appearance of micro fluid inclusion array along the primary grain boundary as shown in Fig. 3.31. As a result, it is sure that

Fig. 3.30 Photograph showing the grain size layering of albite and quartz grains sealing in the shear crack jog of low grade greenschist in Nagatoro, Kanto Mts. Crossed polar

the primary boundary displays the low-angle sub-boundary in the large single grain of albite.

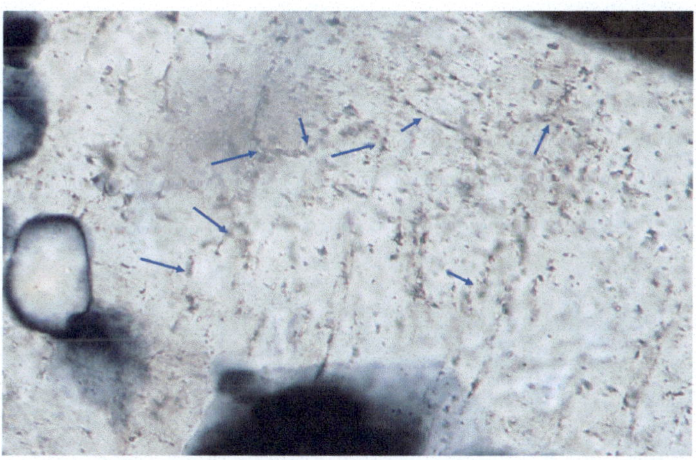

Fig. 3.31 Photograph showing very fine-grained mineral inclusion arrays in large albite grain in the shear crack jog of low grade greenschist in Nagatoro, Kanto Mts. The very fine inclusion arrays shown in arrows appears trace of outline of fine euhedral albite grains formed at the rapid coarsening of albite in shear crack jog

3.4 Annealing and Growth of Fluid Inclusion of Healed Microcrack

Fluid inclusions in the mineral grains are commonly formed by various geological processes such as metasomatism, metamorphism, and recrystallization in the mylonitic and cata-clastic process. Fluid inclusion is simply composed of fluid with tiny crystals and bubble, though there are some varieties of related microstructure; for examples, dislocation loop formed by collapse of vacancy cluster, nanopore, grain boundary lens, and dislocation core. These fluid inclusions have a wide range in size from several nano meter to millimeter.

For fluid inclusion itself in the single crystalline host should be isotropic, the boundary between fluid and host must be anisotropic depending upon the lattice structure of the host crystal, thereby showing that the interfacial energy of the boundary may be estimated by surface integral of the interfacial energy. Therefore, if the fluid inclusion attains enough equilibrium state, its shape should appear euhedral (negative crystal) and it is surrounded by singular surfaces of host crystal. However, many fluid inclusions display rounded to irregular outline (Kerkhof and Hein 2001). Thus, it seems that many fluid inclusions may record a shape transformation process.

By the way, the author has proposed the shape transformation model of mineral inclusion in single crystalline host (Toriumi 1979, 1981) in the cases of the quartz in albite, chromian spinel in olivine, orthopyroxene in olivine, and their inverses. These shape transformation process reveals the kinetics of rounding of inclusion minerals from polyhedral shape during various duration of annealing under high temperature and pressure conditions. The present type of shape transformation kinetics is very available for estimation of the time scale of the relevant physics of the geological process. In practice, the author has pointed out the very rapid process of exhumation of the metamorphic rocks, being based on the annealing time after peak metamorphism (Toriumi 2022).

In this section, the author intends to propose a new growth model of fluid inclusion as the intermediate state between rounded and irregular shape to euhedral-negative crystal shape.

Let us start the configuration of microcrack formation in the single crystal as shown in Fig. 3.32. In this figure, the microcrack is open type I circular crack filled with aqueous fluid, and aperture is a, and radius is L. After an open crack is formed by applied stress in the single crystal of the rock, and immediately it is filled with fluid, and then the healing process starts to close the open crack associated with formation of many micro fluid inclusions array along on the crack surface. It seems that the size of the primary fluid inclusion is enough smaller than 100 nm order, for there are abundant very small fluid inclusions along the healed microcracks in the natural metamorphic rocks as seen in Fig. 3.33. The healed host crystal may be rich in dislocations and vacancies in the narrow zone along the open crack because of initial stress intensification at the growth of microcrack tip and around the micro fluid inclusions.

The healing process should be controlled by the following kinetics.

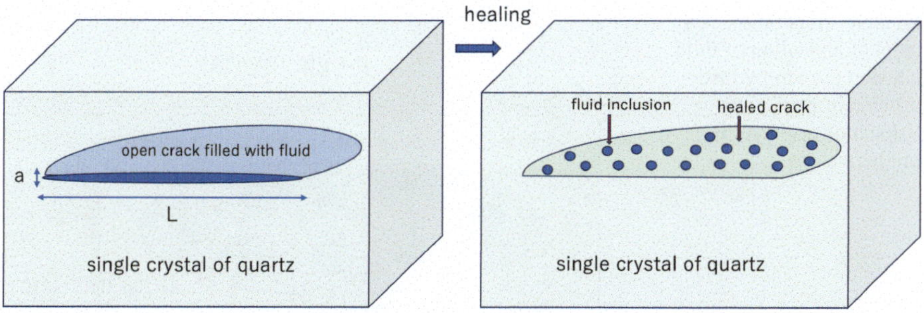

Fig. 3.32 Illustration of healing of open crack filled with aqueous fluid (left) in the single grain of quartz to form array of nanoscale fluid inclusions (right). L; length of open crack filled with fluid, a; aperture of open crack

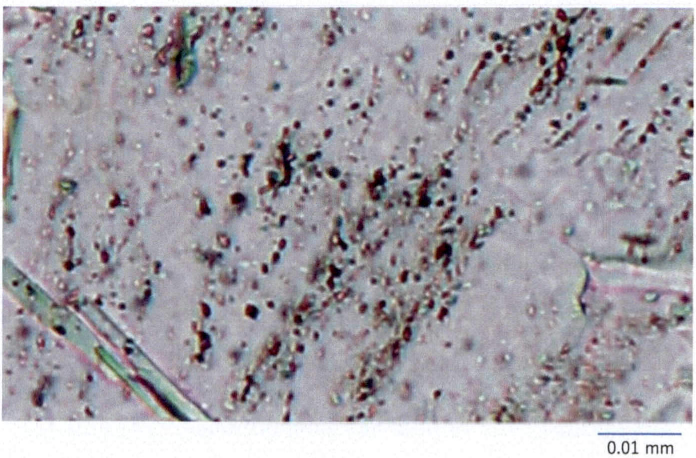

Fig. 3.33 Photograph showing occurrence of arrays of micro-nano fluid inclusions in quartz single crystal in shear crack jog of low grade greenschist of Nagatoro, Kanto Mts. Green acicular grain; actinolite, low index grain in right bottom; albite

First, let us compare the internal free energy between the open crack state filled with fluid and healed crack with arrayed fluid inclusion along the crack surface as illustrated in Fig. 3.34.

The free energy U in the open crack state becomes,

$$U_1 = 2\pi L^2 \gamma + mg \tag{3.20}$$

in which γ is interfacial energy between fluid and host crystal, and m and g are mole number of aqueous water in open crack and molar free energy of fluid, respectively.

On the other hand, after healing of open crack and formation of micro fluid inclusions along the crack surface, and the free energy of this situation becomes,

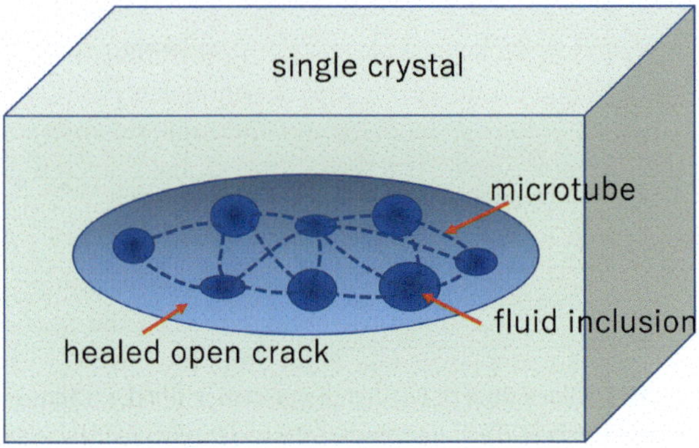

Fig. 3.34 Illustration of a model of annealing of fluid inclusion growth by fast diffusion of microtube (dislocation) connecting fluid inclusions

$$U_2 = 4\pi n r^2 \gamma + mg \tag{3.21}$$

and

$$aL^2 = (4/3)nr^3 \tag{3.22}$$

in which r is the size of fluid inclusion sphere, a is the thickness of aperture of the primary open crack, and n is the total number of fluid inclusion.

Therefore, we obtain the difference between these energy as,

$$\begin{aligned}\Delta U &= 2\pi L^2 \lambda (1 - 3a/2r) \\ &= 0 (\text{at equilibrium condition}).\end{aligned} \tag{3.23}$$

Therefore, the primary size of fluid inclusion becomes,

$$r = 2a/3. \tag{3.24}$$

Considering the length and aperture of open crack about 100 μm and 0.1 μm, respectively (corresponding to aspect ratio of open crack is about 0.001), then the size of fluid inclusion holds,

$$r = 0.067\,\mu\text{m}. \tag{3.25}$$

The average spacing of fluid inclusions becomes,

$$\begin{aligned}s &= \left(4r^3/3a\right)^{1/2} \\ &= 0.16\,\mu\text{m}\end{aligned} \tag{3.26}$$

where s is the average spacing of inclusions.

It is obvious that the set of the size and spacing of the micro fluid inclusion shows a primary structure after healing of the microcrack filled with fluid.

In this study, the author intends to investigate the annealing process after crack healing, and thus he must discuss the morphology of the fluid inclusion. From this point of view, it should be important that there are two types of shape of the fluid inclusions in the metamorphic and metasomatic rocks: one is rounded and irregular shape, and the other is euhedral and polyhedral shape surrounded almost always by singular low index surface of host crystal (called as negative crystal) in the high temperature metamorphic rocks (Kerkhof and Hein 2001). Considering the equilibrium temperature over 500 °C and enough long duration of metamorphic event, it is probable that the euhedral fluid inclusion (negative crystal) may be changed from the rounded and irregular shaped inclusion by reduction of interfacial free energy of the fluid-host boundary. On the other hand, it seems that the rounded fluid inclusion did not attain the long-term equilibrium shape even in the duration of the rather low temperature conditions.

It seems a little strange that the state of the annealing process of inclusion shape change appears still rounded, for the enough annealed shape is surely negative crystal state in the higher temperature conditions. It is necessary to be discussed that whether the stable shape of fluid inclusion in single crystalline host is rounded or negative euhedral state under enough annealing.

Considering the energy difference between the cubic and rounded fluid inclusions in crystalline host, and the crystalline host is cubic, and inclusion boundary is (100), (010), and (001), total interfacial energy is contributed by eight edges, twelve corners, and six surfaces. Assuming the corner and edge to be tube and sphere shape of radius δ the surface energy of them are $6\pi r\delta\gamma$ and $\pi\delta^2\gamma$, respectively, and the flat surface has $6r^2\gamma_s$, in which γ_s is surface energy of singular surface. On the other hand, the surface energy of the sphere of the iso-volume with cubic inclusion becomes $4\pi(0.62r)^2$.

Therefore, the condition, that the spherical inclusion is stable against the cubic one, holds,

$$4\pi(0.62r)^2\gamma \leq 6r^2\gamma_s + 6\pi r\delta\gamma + \pi\delta^2\gamma \tag{3.27}$$

Neglecting the second order small number, it takes,

$$\begin{aligned}&(4.8-6k)6\delta/r\\&k = \gamma_s/\gamma\end{aligned} \tag{3.28}$$

Now, if k is about 0.7, the critical size of spherical inclusion is 10δ, which is about 10 μm.

If k is over 0.8, the critical size becomes infinite, so that actual size of spherical fluid inclusion seems to be 10–100 μm. Therefore, the fluid inclusion of the small size does not change its spherical shape during annealing process, though rather large fluid inclusion may change it shape to polyhedral having facet, as is seen in the high temperature large fluid inclusions (Kerkhof and Hein 2001).

By the way, the growth kinetics of fluid inclusion along the healed microcracks is governed by the following Langevin type equation,

$$dr/dt = Md\Delta U/dr + \text{Noise} \tag{3.29}$$

in which M is the growth mobility of the fluid inclusion, and it has a thermally activation term.

From previous equations, the potential gradient term holds,

$$\begin{aligned} d\Delta U/dr &= (3\pi a L^2 \gamma)/r^2 \\ &= k/r^2 \end{aligned} \tag{3.30}$$

Thus, we obtain,

$$\begin{aligned} dr/dt &= K/r^2 \\ K = Mk &= 3\pi a L^2 \gamma M \exp(-Q_f/RT) \end{aligned} \tag{3.31}$$

In which Q_f is the activation energy for fluid inclusion growth, which should involve the processes of diffusion of H_2O and host components along the healed open crack.

Therefore, we obtain the relation between the annealing time after healing and the average size of fluid inclusion as,

$$r^3 - r_o^3 = 3Kt \tag{3.32}$$

in which r_o is the initial size of fluid inclusion after healing of crack, and it has been obtained to be 0.067 μm.

Assuming $r >> r_o$, it holds,

$$r^3 = 3Kt \tag{3.33}$$

Therefore, the above equation becomes simple linear relation between logarithmic annealing time and fluid inclusion size as,

$$\log t = 3 \log r - \log K(T) - 0.48 \tag{3.34}$$

and at the isothermal annealing,

$$\log t/t_1 = 3 \log r/r_1 \tag{3.35}$$

in which t_1 is the reference inclusion size at the r_1. The latter relation is very convenient for comparison of time scale of the annealing after open crack healing. The diagram showing the above relationship is shown in Fig. 3.35.

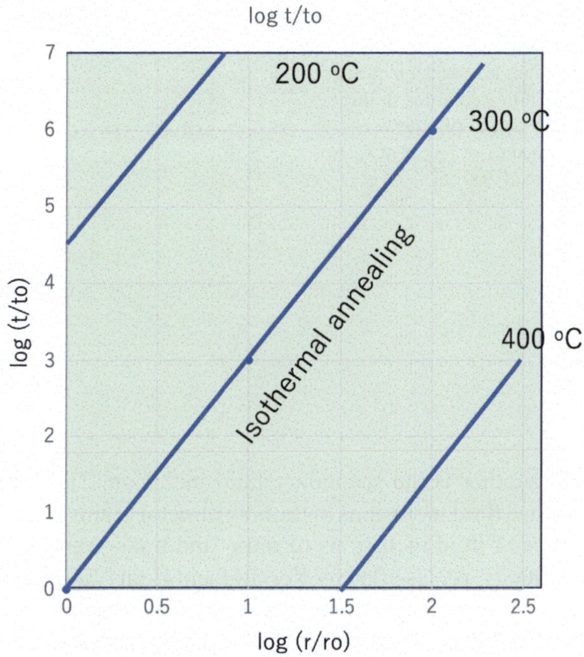

Fig. 3.35 Inferred relation of annealing time scale and size of fluid inclusion normalized by starting condition (r_o, t_o)

On the other hand, the activation energy of the K is approximated to be that of the grain boundary diffusion of Si^{4+} through the pass of healed open crack as mentioned above, and the path of diffusion seems to be microtube formed along the healed interface. The activation energy of Si^{4+} ion along the interface has been estimated as 80 kcal/mol by means of natural occurrence of shape transformation process of quartz inclusion in albite single crystal (Toriumi 1979). Therefore, it seems that the terms of K at 300 °C and 400 °C are estimated as,

$$K_{300}/K_{400} = \exp(-40,000)(1/573-1/673)$$
$$\log(K_{300}/K_{400}) = 4.4. \tag{3.36}$$

And then, we have,

$$\log r_1/r_2 = 1.5 \text{ at the same time.} \tag{3.37}$$

Therefore, the relation of the $\log t$ versus $\log r$ for the 300 °C is shown in Fig. 3.36, and it is applicable for the annealing time of the fluid inclusion arrays formed by healing the microcracks of type I mode in quartz grains. Therefore, it is available that the timing of the shear crack jog and open type crack sealing by quartz and albite during the metamorphism.

There are abundant fluid inclusions in metamorphic rocks and granitic rocks. The fluid inclusion is classified into two types: one is the primary fluid inclusion,

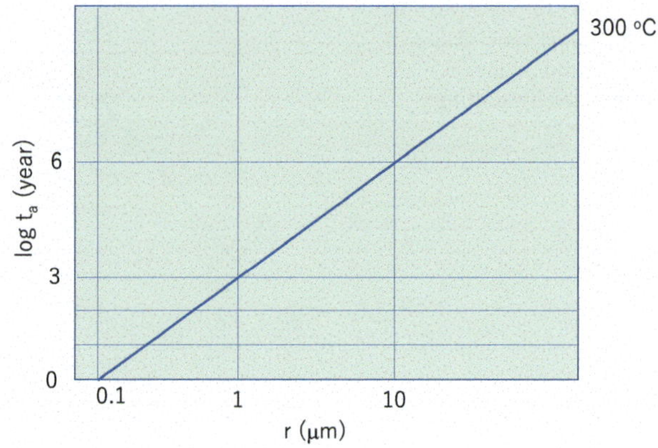

Fig. 3.36 Inferred relationship between fluid inclusion size and annealing time of healed crack in the single crystal of quartz, assumed $t_o = 1$ year at $r_o = 0.1$ μm at 300 °C

and another is the secondary fluid inclusion. The former commonly occurs as the isolated fluid inclusions in the host mineral grains interior, but the latter does as planer arrays of fluid inclusions of intra- and trans-grains (or inter-grains) pattern of their continuity as classified by Kerkhof and Hein (2001). It seems that the secondary fluid inclusion array is formed by the healing process of microcrack filled by microfilm of fluid that is composed of H_2O or CO_2 with various content of hydrogen, methane, CO, and many kinds of ionic and molecular species (Kerkhof and Hein 2001). These fluid inclusions experienced the later metamorphism, metasomatism, ore formation process, and then cooling, and subsequently in these inclusions, various minerals and gaseous bubble are nucleated and formed, passing under their oversaturation conditions.

In this study, it is convenient to take data acquisition that the annealing kinetics is only governed by decreasing the interfacial free energy of the fluid inclusion—single host mineral boundary surface, and furthermore, that it is represented by the average size and spacing of the micro fluid inclusions in their array. It is possible that these data are measured under the optical microscope by high magnitude using object lens of × 100. In the case of the inclined healing microcrack, various focus depth images can help us to determine the above data.

The observed data were obtained from micro fluid inclusions of the quartz grains precipitated in the microcrack and shear crack jog, in the low grade greenschists of the Sanbagwa metamorphic rocks of central Shikoku and Kanto Mountains, to investigate the annealing time after microcrack formation related to the various shear crack slippage in the subduction plate boundary zone. The studied greenschists are composed of albite, quartz, epidote, chlorite, actinolite, sphene, and magnetite, and the metamorphic temperature and pressure conditions of these schists are inferred to be about 300–400 °C and 1–0.5 GPa during the retrograde metamorphism (Okamoto and Toriumi 2004). Meanwhile, it has been shown that the sealing of quartz and albite of the shear crack jog and micro-open crack took place in this retrograde stage of the metamorphism, and thus the micro-open crack healing should occur during this retrograde metamorphism in the subduction boundary zone.

The studied samples of quartz grains precipitated in the open crack and micro fluid inclusion arrays along the healed open microcracks are shown in Fig. 3.37. It shows that the spherical to rounded-irregular shaped micro fluid inclusions are arrayed along the healed crack in the various portions of the single quartz grain. Considering that the length of the healed cracks is in the range from 0.1 to 0.5 mm, micro fluid inclusions over a few microns size should increase in their size during annealing after healing process as mentioned previously.

The sizes of fluid inclusions and distances between two inclusions are plotted in the diagram as shown in Fig. 3.38. The data points appear several clusters from small size to large size with increasing spacing of them. In Fig. 3.39, the cluster of small sized fluid inclusions appears in the outer part of the jog quartz grain, and that of the large ones occurs in the central part of the jog quartz. The medium cluster occurs in the middle part of the jog quartz, thereby suggesting that the time scale of annealing decreases rapidly from central part to outer part of the jog quartz.

The size distribution pattern and cumulative size distribution curve are shown in Fig. 3.40, thereby showing that they manifest obviously the annealing by surface energy regression with ripening of fluid inclusions after healing of microcracks. However, it seems that the frequency patterns are likely to the lognormal distribution one, but not that of Ostwald ripening at the point of long-term tail, suggesting

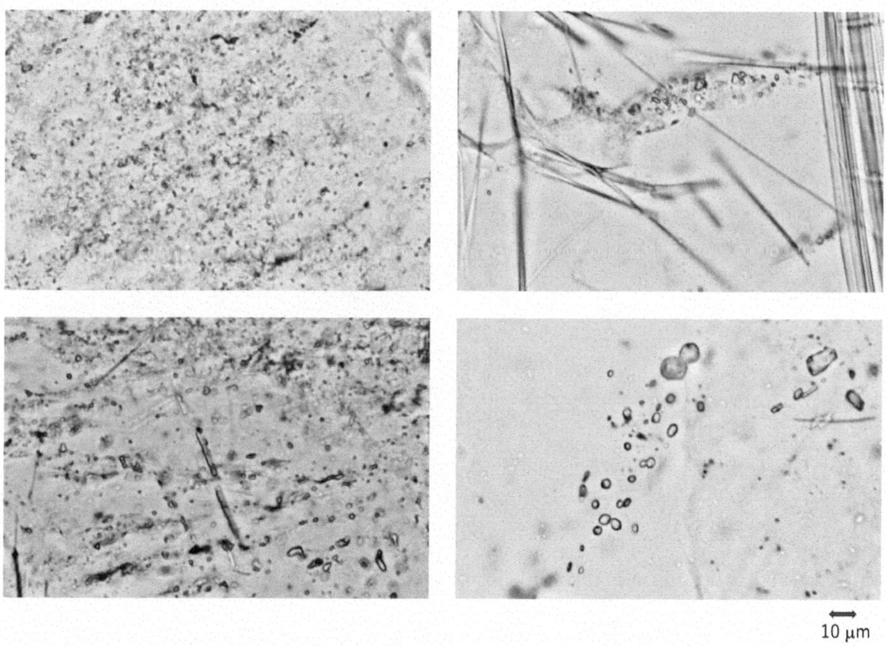

Fig. 3.37 Photographs of variously sized fluid inclusion arrays in quartz grains showing the different annealing time after healing of open crack filled with aqueous fluid

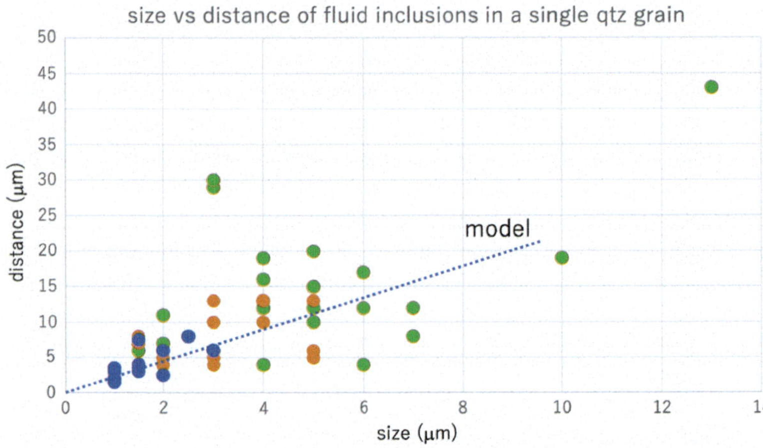

Fig. 3.38 Diagram of fluid inclusion size and their distance of various fluid inclusion arrays of single rock sample from low grade greenschist in Nagatoro, Kanto Mts. Dotted line; model relation

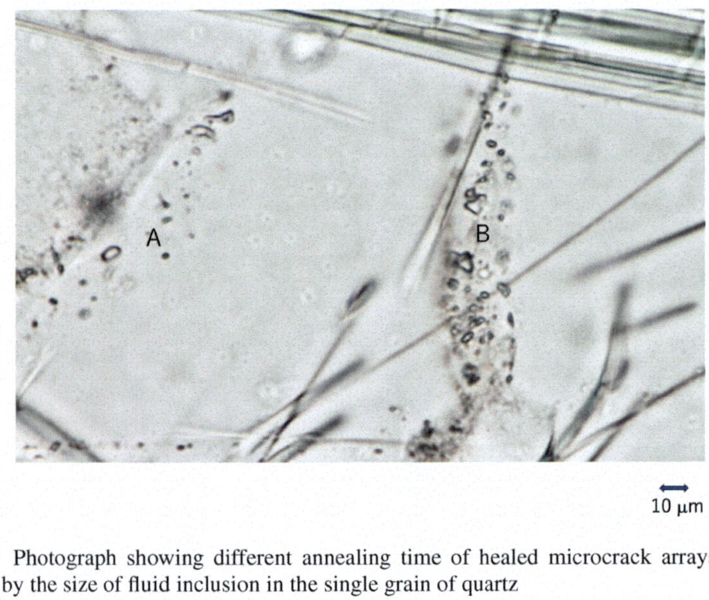

Fig. 3.39 Photograph showing different annealing time of healed microcrack arrays A and B provided by the size of fluid inclusion in the single grain of quartz

the effect of diffusion path through the microtube along the healed crack interface, described previously.

The time scale of the annealing of the middle part fluid inclusion array after healing is possibly inferred from the size difference between the middle to outer parts of sealed jog as shown in Fig. 3.40. From the ratio of the inclusion size of them to be about 3–5, it seems that the duration of the annealing of the sealed cracks after

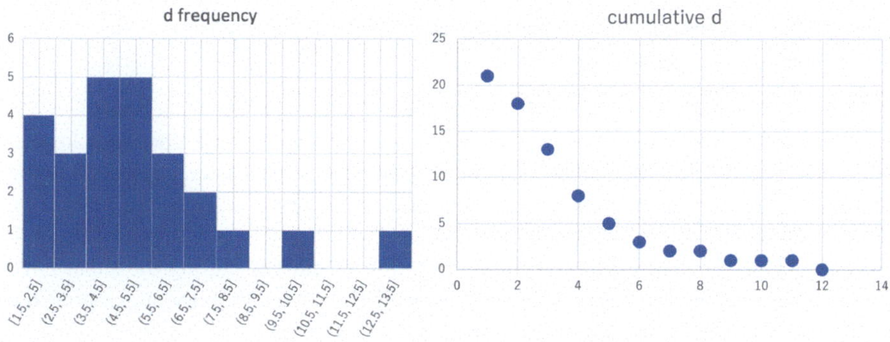

Fig. 3.40 Frequency diagram of size of fluid inclusion in a single grain of quartz (left) and the cumulative frequency (right) of the low grade greenschist of Nagatoro, Kanto Mts

formation of fluid inclusion arrays is about 30 to 300 years, assuming the primary healing time $t_o = 1$ year at 300 °C. Therefore, the annealing should have occurred before the active stage of the open cracks in the process of albite and quartz sealing of them. If this is the case, the studied sealing open cracks filled with albite, quartz, chlorite, and actinolite, these are stable in the greenschist facies conditions to be about 300–400 °C, were formed during the estimated years in the metamorphism.

On the other hand, there is abundant healed cracks in the cracks sealed with albite and quartz containing actinolite, which have abundant arrays of rather large fluid inclusions. Considering the annealing time after healing of open crack in sealing albite and quartz of open crack and shear crack jog, it seems that the sealed cracks were formed within the several to several hundred years during the metamorphism.

There are abundant secondary fluid inclusions in quartz and albite in the sealed open and shear type cracks of the plate boundary metamorphic rocks. It concludes that these secondary fluid inclusions were formed by healing of the open microcracks in the quartz and albite after sealing the open and shear cracks during the metamorphism. The secondary fluid inclusions are arranged on the healed crack and show systematics of size and spacing of the rounded fluid inclusions, thereby suggesting that the annealing by ripening of fluid inclusions took place after healing of the microcracks.

The model of the annealing growth of fluid inclusions after healing of the microcracks which is primarily filled with fluid was proposed in this section, and the results manifest that the stable shape becomes sphere at the initial healing of the microcrack, and the growth law of the inclusions is formulated by the cubic law of the size, being proportional to the time with thermally activation term. The annealing time after the healing of the microcracks shows in the rather wide range from several to several ten micrometers size with ripening frequency distribution of size having long-term tail. The range of the annealing times after open crack healing is from 1 year to several thousand years at the condition of 300 °C.

3.5 Decompression Brecciation of Rocks in the Plate Boundary Zone

As described in the previous chapters, there are common sealed open cracks and their clusters along the shear slip zones of the plate boundary rocks. It seems that the open type cracks sealed by metamorphic minerals were surely made by the stretching of rocks along the shear slip plane as modeled by Yin (2010). It suggests that the decompression fracturing and coeval precipitation of minerals along the shear zone should take place in the process of stretching of the rocks, thereby suggesting the mechanism of fault breccia and cata-clastic rock formation.

Decompression fracturing of deep-seated rocks appears commonly in the wedge mantle rocks near the plate subduction boundary, and the conical seamount composed of serpentinite-carbonate rocks is responsible for this mechanism at the frontal portion of the wedge mantle (Fryer et al. 2002). Furthermore, it is also common that the alpine-type ultramafic rocks partly replaced by serpentine minerals occur as the brecciated serpentinite. In these brecciated rocks, the rather large rock fragments contain abundant open cracks sealed by serpentine minerals and carbonates, and matrix of the interstice of blocks are occupied by the same minerals, thereby showing that the decompression fracturing should be coeval with the precipitation of these minerals, associated with reaction by H_2O–CO_2 fluid.

On the other hand, it is obvious that the shear slip deformation in the deep-seated rocks in the plate boundary zone take place commonly in the compressive stress environment of the plate boundary, and thus the decompression brecciation of rocks seems to be a short-term and micro-tectonic process accompanied with the compressional deformation as like as the open crack formation filled with aqueous fluid and mineral precipitation. However, it is rather strange that the orientation of the sealed open cracks in the brecciated blocks and breccia themselves appears random unlikely to the microcrack clusters along the shear slip zone. It seems that such random orientation of sealed open cracks in the blocks and rocks manifests the isotropic expansion situation of the decompression process, thereby suggesting that the elastic wavy of the decompression should expands outward from the center of the decompression cores formed by the local extensional stress something like large jog formed by cross slip mentioned in the previous chapter.

In this section, the author intends to study the very short-term decompression process resulting in the formation of random orientated decompression breccia.

The decompression breccia is commonly found in the schistose rocks as the massive and disorientated-brecciated rock that composed of large amount of very fine-grained matrix and variously sized angular fragments of the same rock type as surrounding rock mass. It shows in general that the various fragments are randomly distributed and are fractured partly in most case. The small fragments often appear surrounding the large fragment, and orientation of the long axis of them shows random. It is noticeable that the matrix, which are often mixture of the very fine grains of surrounding rocks and newly precipitated carbonates, quartz, albite, and chlorite,

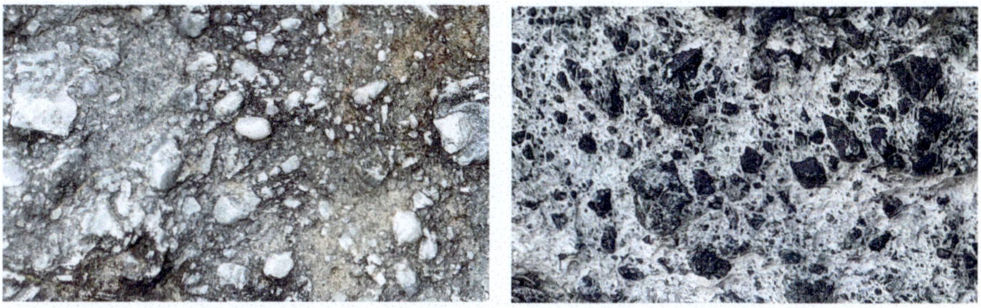

Fig. 3.41 Photographs of decompression breccia of low-grade basic and pelitic schists (left) and serpentinite (right) in the Sanbagawa metamorphic rocks of Kanto Mts. Matrix of breccia is composed of very fine-grained schist and serpentinite, and carbonate

Fig. 3.42 Photographs of decompression breccia of serpentinite in Sarutagawa, Shikoku, showing matrix-dominant breccia (left) and breccia with matrix dominant part and breccia-dominant part

appears no fabric, thereby suggesting the isotropic compaction due to dewatering as shown in Figs. 3.41 and 3.42.

The wall rock surrounding the decompression breccia are commonly fractured by open crack sealed by minerals mentioned above, and associated shear cracks having jog and void sealed by the same minerals, thereby showing the inflow of the aqueous fluid into both the open cracks and the decompression volume, hereafter, called as big jog in this book. Therefore, the big jog is primarily formed by cross slippage of the complex shear cracks, and it is filled with aqueous fluid and precipitated minerals together with various sized breccia derived from surrounding rocks.

In many cases, it shows that the decompression becomes unidirectional by moving the fault wall, and so the volume increment is to be one dimensional. On the other hand, if the decompression wall is cylindrical to spherical, the secondary fragmentation of between breccia should be two and three dimensional, respectively, thereby showing the volume increment of them is higher than the case of unidirectional fracturing. It seems that the cumulative frequency distribution pattern of the breccia manifests probably the geometry of the decompression style in the deep-seated plate boundary rocks as shown in the later sections.

On the other hand, there are abundant big jogs formed by differential slippage of several blocks separated by cross slip plane as shown in Fig. 3.43, and it suggests that the big jog formation is derived from accommodation excess volume filled with aqueous fluid supplied from surroundings. Furthermore, it should be obvious that the aqueous fluid entrained into big jog is probably oversaturated for some kinds of minerals, such as carbonates, quartz, albite, and chlorite, judging from the occurrence of the mineral assemblages of the sealed open cracks just around the jog. Therefore, it seems that the decompression breccia in big jog is composed mainly of fine-grained particulates derived from surrounding rocks and newly precipitated grains of minerals above in the matrix, and of brecciated angular fragments. They appear random and non-foliated distribution in the jog except the boundary zone to the surrounding rocks.

The mode of the grain size distribution of the decompression breccia is important for reconstruction of the geometry of the big jog formation in the deep portion of the plate boundary. Hayakawa (1996) investigated the universality of breccia size distribution formed by the physical model simulation of decompression fragmentation in which the rock units are connected by spring and decompression is carried out by block motion outward from the surface of the system. The simplified model likes that if the elastic stress of the block of the surface is over the strength of spring, the block should be separated from the surface, and elastic wave invades into the system

Fig. 3.43 Photograph of decompression brecciation of low grade greenschist of Yawatahama, Shikoku by multiple faults and related large jogs. DB; decompression breccia, orange arrows show the displacement of block

from the newly formed surface. The successive fracturing should propagate to the system interior, and decompression fragmentation continues making various-sized breccia. In his simulation, no secondary collision and fragmentation is assumed, but the successive brecciation has been clarified with no assumption of the material constant and system size, thereby suggesting that the resultant size distribution and its change with time are to be universal. His simulation is the case of unidirectional, that is the one-dimensional decompression case, but it applies to the two- and three-dimensional cases by the geometrical extension.

The cumulative density distribution of the simulated breccia appears in the case of one-dimensional case (Hayakawa 1996),

$$\begin{aligned} M(m) &= m^{-1.7} \text{ at primary phase} \\ M(m) &= m^{-0.64} \text{ at later phase,} \end{aligned} \tag{3.38}$$

and,

$$\begin{aligned} M(d) &= d^{-5.1} \\ M(d) &= d^{-1.98}, \end{aligned} \tag{3.39}$$

in which M is the cumulative frequency distribution function, index of m and d is mass and size of the block, respectively.

On the other hand, in the case of the two- dimensional cylindrical decompression, the index of the power law mentioned above changes to $-2 \sim -3$ in the primary phase, and to -0.33 in the later phase (Hayakawa 1996). These results are possibly applicable for the geological fragmentation by decompression as shown later.

Therefore, the natural cases of the plate boundary rocks and uppermost mantle serpentinite of the Sanbagawa metamorphic rocks were examined: The samples are carefully selected from the fine outcrops of the brecciated serpentinites in the Sarutagawa route, Shikoku, and in the Kanazaki near Nagatoro, Kanto Mts., and from the brecciated pelitic schists of the Nagatoro area, Kanto Mts. Sample of the Sarutagawa is the large block dropped from the slope above the river, and of the Kanazaki is the outcrop of the large mass embedded in the lower grade schists of the Sanbagawa metamorphic belt. The breccias of both serpentinite masses show angular and surrounded by fine-grained matrix composed of serpentine and carbonates (calcite). The rather large breccia has commonly open cracks with random orientation which are sealed by serpentine and carbonates.

On the other hand, the brecciaed rocks derived from pelitic and quartzose schists occur the outcrops, and the boundary between brecciated mass and surrounding non-brecciated schists can be observed clearly as shown in Fig. 3.44. The one locality of the outcrop appears the occurrence of chimney like mass with cylindrical boundary (Fig. 3.45), and the other does the complicated but platy boundary. In these cases, there are abundant sealed open cracks filled with fine-grained matrix in the rather coarse fragments as shown in Fig. 3.45, thereby showing the origin of fragmentation of the surrounding rocks.

Fig. 3.44 Photographs of decompression breccia of pelitic and quartzose schists in Nagatoro, Kanto Mts., showing the side view (left) and top view (right). The former displays the boundary with surrounding non-brecciated schists

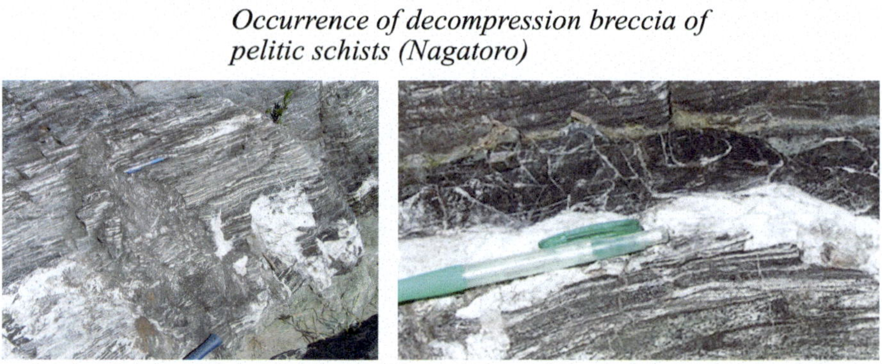

Fig. 3.45 Photographs of chimney like breccia mass (left) and related open crack array sealed with calcite (right) in the top of chimney like breccia. Nagatoro, Kanto Mts

The cumulative frequency patterns of fragment size for these samples were measured on photographs in the range from 0.5 to 30 mm in length. The cumulative distribution of the fragments is represented primarily by the frequency density of the 1 mm window in length, and then the cumulative mass frequency distribution is obtained by transformation to the cubic value of the length. The brecciated mass often displays a little bit of inhomogeneity of the distribution pattern of breccia size and content, but in this study, the central part of the mass showing the average features of them has been measured for clarifying the characteristic cumulative frequency pattern of the decompression breccia.

In Figs. 3.46 and 3.47, it shows that the logarithmic diagram of the cumulative mass frequency distribution of fragment size and fragment mass of the studied decompression breccia of low-grade schists and serpentinites in the Sanbagawa metamorphic belt from Shikoku and Kanto Mts. It is obvious that there are classified into two patterns in the cumulative mass frequency distribution of them in terms of the index of the power law: one is concentrated in the value about − 1.67, and the other about − 2.5 in the mass-cumulative frequency power law as shown in Fig. 3.47. It seems important that the power-law index of the schist-derived breccia is just same as that of the serpentinite-derived breccia, thereby suggesting that the decompression fragmentation process appears universal as denoted by Hayakawa (1996). Furthermore, it should be noticed that the difference in the power-law index seems to be responsible for the geometrical dimensionality of the fragmentation, that is the value of − 1.67 from unidirectional fragmentation along with jog surface, but that of − 2.5 from center-concentration one along with cylindrical surface.

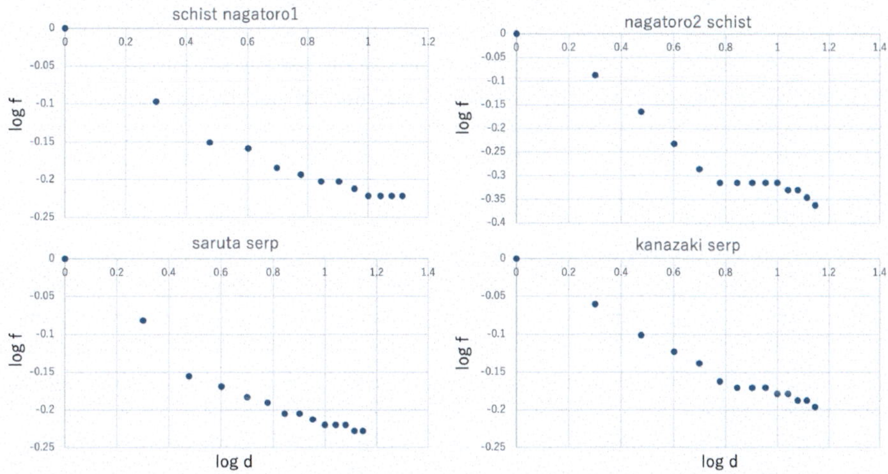

Fig. 3.46 Cumulative frequency diagrams of fragment size (d) of decompression breccia (top two) in Nagatoro, Kanto Mts., and of decompression serpentinite in Sarutagawa. Shikoku (bottom left) and Kanzaki, Kanto Mts. (bottom right)

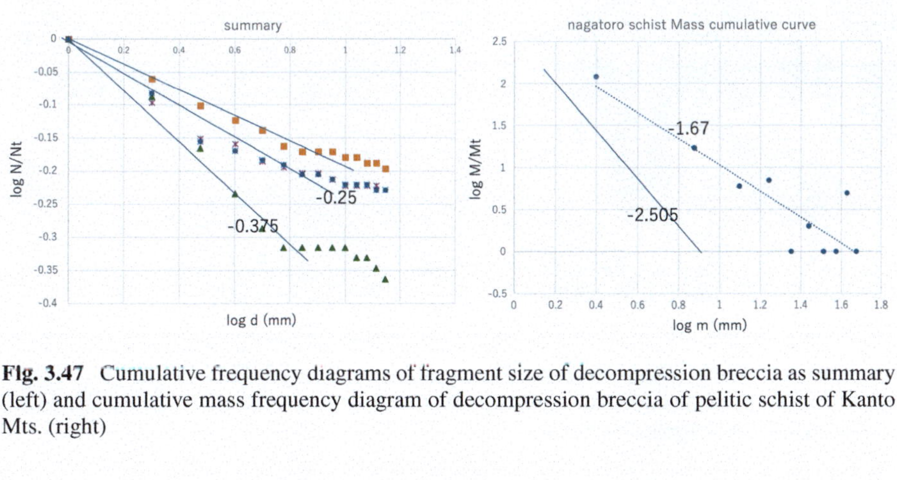

Fig. 3.47 Cumulative frequency diagrams of fragment size of decompression breccia as summary (left) and cumulative mass frequency diagram of decompression breccia of pelitic schist of Kanto Mts. (right)

It is still strange that many sealed open cracks do not show the fragmentation, judging from no fine-grained breccia within them. As mentioned above, in the deep-seated metamorphic rocks, if the open crack once is formed, the volume of crack space should become decompressed state. Instead, the volume of the open crack is now cemented by precipitated minerals such as albite, quartz, chlorite, and sometimes calcite and actinolite in the low-grade schists (e.g., Toriumi 2022), although there is common occurrence of very fine grains of titanite and epidote in the core and sometimes boundaries of albite and quartz.

In this section, the description and measurements of grain size distribution will be shown. The small-scale open crack studied here is filled with fine grains of breccia and matrix composed of albite, chlorite, and fibrous actinolite, and breccia is mainly composed of epidote and titanite (sphene) which is derived from surrounding host schist. Considering that the host schist contains abundant actinolite, chlorite, epidote, sphene, albite, and quartz, the sealing minerals in the open crack studied here are precipitated from the aqueous solution, which is equilibrium with host rock minerals, that is opening and sealing of this open crack is just coeval with plate boundary metamorphism. The grain size of the epidote and titanite is about 10–20 microns both in the host schist and breccia in the crack, thereby suggesting no overgrowth of them after brecciation (Fig. 3.48).

The results are shown in Fig. 3.49, thereby showing that the cumulative mass frequency pattern displays the index of the power law to be about − 1.6. It is noteworthy that this power index is very likely to that of the cases of serpentinite and schist decompression breccias as clarified in the previous section. It shows that the decompression brecciation should occur in the process of crack opening during the main metamorphism.

By the way, it was noted that the sealed open crack is composed of chlorite, actinolite, and albite, and that the breccia in the sealed crack is mainly composed of

Fig. 3.48 Photographs of micro breccias of decompression fragmentation with open crack sealed with fibrous actinolite, chlorite, albite, and epidote in low grade greenschist. Left; crossed polar, right; open polar. Asemigawa, Shikoku

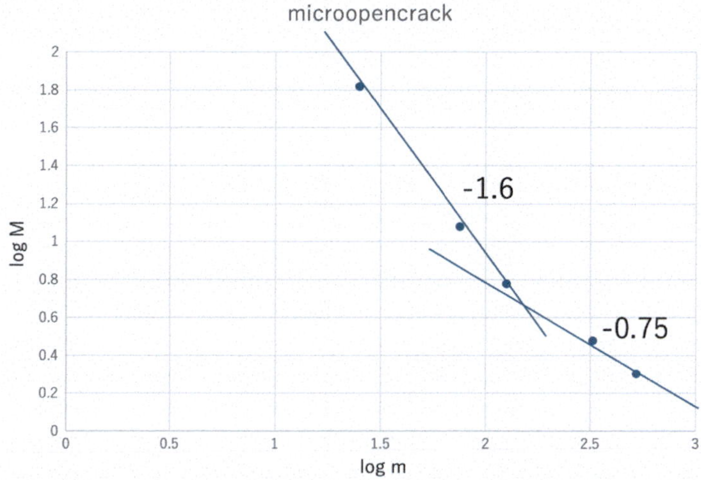

Fig. 3.49 Cumulative mass frequency diagram of the micro breccia size of decompression breccia in the sample of Fig. 3.48

epidote and titanite but not actinolite, chlorite, and albite. It, therefore, seems that the actinolite, chlorite and albite may be dissolved into aqueous solution, and these are newly precipitated from the solution in the open crack volume. That fibrous growth shape of actinolite seems much different from tabular actinolite in host schist is consistent with this model. In addition, it is something interesting that many coarse grains of albite in the albite—chlorite band contain abundant very fine grains of epidote and titanite, being likely to the occurrence of fragmented and fine-grained epidote and titanite in the core of albite of the sealed open crack mentioned above, suggesting the same origin of the fine-grained epidote and titanite included in sealing albite of opening jog.

It has been discussed in the previous chapter that the plate boundary above the subducting oceanic plate displays commonly the waveform interface and jog-like step structure (horst and graben) as seen in the Nankai trough and the Japan trench, respectively, and the former is possibly derived from Kelvin–Helmholtz instability of the rather ductile shear zone and the latter is primarily formed by the normal fault before subduction in the outer rise. Furthermore, the subducted oceanic seamount has been found near the Nankai trough by Kodaira et al. (2004), and then it shows that stress configuration of the accretion prism around the subducted seamount changes partly from compression to extension regime.

It seems that the stress configuration of the plate boundary zone should be controlled by the geometry of the boundary interface, and that the rear area of the step and/or mound of the slab surface is to be extensional and frontal zone to be compressional at the time of the rapid (seismic) and slow slip (SSE) at the subduction plane. So far as the inhomogeneous stress configuration takes place at the time of the seismic slip, the rear zone of the big jog-like step and waveform mound should be suffered from the short-term decompression situation, and then the rocks of the subduction boundary zone must be heavily brecciated in the manner of decompression fracturing as described previously. It seems that these decompression breccias having abundant hydrous materials in the horst and graben may be extruded into the plate boundary by the unbending force of the subducting slab to form the low viscosity and low frictional coefficient layer. Accordingly, it suggests that the plate boundary layer contains abundant aqueous fluid and very fine-grained matrix with angular various sized breccias derived from oceanic crust and uppermost mantle.

Otherwise, it is possibly thought that during the subduction of the big-jogged slab, the decompression tectonics successively occurs between the slab and overriding arc crust and wedge mantle, and it may give rise to brecciation of wedge mantle and the crustal materials above the slab. At the time of decompression, the extensional faults likely to the composite normal faults attaining the bottom of the ocean would be filled by the brecciated fragments and their fine particulate matrix involving large amounts of aqueous fluid (H_2O–CO_2). The particulate matrix materials containing breccias are probably fluidized and explosively erupted onto the bottom of the sea, suggesting the formation of the conical seamount composed of serpentinite-carbonate mud flow (Fryer et al. 1990). On the other hand, if the decompression occurs in the plate boundary in the wedge mantle region, the fluidized serpentine-crustal rock particulate matrix should be settled in the plate boundary metamorphic rocks, being likely to the metamorphosed serpentinite mass described previously in the Sanbagawa metamorphic belts.

The regional metamorphic rocks are commonly formed in the subduction boundary zone of the plate, and therefore, they are suffered from rapid and slow shear slip deformation together with recrystallization of metamorphic minerals under the increasing (prograde) and decreasing (retrograde) pressure and temperature processes. In the shear slip deformation, the recrystallization should be associated with hydration and dehydration of aqueous fluid, and thus the actual recrystallization of metamorphic minerals is represented by crystal growth from aqueous fluid in the grain boundary and finite size of void filled with fluid, and by the dissolution into them with fracturing of preexisting mineral grains. In this section, it is provided that the decompression by jog formation and microcracks controlled by shear crack growth and propagation should be generated locally in the open crack and jog and make the surrounding rocks to heavy brecciation.

The brecciated rock fragments and finely particulate matrix occupy the jog and open cracks with precipitated matrix, and the cumulative size distribution of them displays characteristic pattern, being like the simulated pattern. Inversely, it seems that the natural occurrence of brecciated rocks with large amount of fine matrix may be derived from the rapid decompression products of surrounding various rocks, involving mantle serpentinite and crustal rocks.

References

Beck C, Schlogl F (1993) Thermodynamics of chaotic systems. Cambridge University Press

Cashman KV, Marsh BD (1988) Crystal size distribution (CSD) in rocks and the kinetics and dynamics of crystallization: 2 Makaopuhi Lava Lake. Contrib Mineral Petrol 99:292–305

Fryer P, Saboda KL, Johnson LE, Mackay ME, Moore GF, Dtoffers P (1990) Conical seamount: SeaMARC II, Alvin submersible, and seismic-reflection studies. In: Fryer P, Pearce JA, Stokking LB et al (eds) Proceedings ODP, Init. Repts., 125: College Station, TX (Ocean Drilling Program), pp 69–80

Hayakawa Y (1996) Impact fragmentation of an ideal brittle crystal. Phys Rev 1353:14828–14839. https://doi.org/10.1103/physrevb.53.14828

Kerkhof AMV, Hein UF (2001) Fluid inclusion petrography. Lithos 55:27–47

Kodaira S, Hidaka T, Kato A, Park J-D, Iwasaki T, Kaneda Y (2004) High pore fluid pressure may cause silent slip in the Nankai trough. Science 28(May):1295–1298

Okamoto A, Toriumi M (2004) Optimal mixing properties of calcic and subcalcic amphiboles: application of Gibbs' method to the Sanbagawa schists, SW Japan. Contrib Mineral Petrol 146:529–545

Toramaru M (2019) Vesiculation and crystallization of Magma: fundamentals of the volcanic eruption process. University of Tokyo Press

Toriumi M (1975) Petrological study of the Sanbagawa metamorphic rocks—the Kanto Mountains, Central Japan. Univ Museum Univ Tokyo Bull 9:99p

Toriumi M (1975b) Mechanical differentiation in the metamorphism. J Geol Soc Japan 81(8):505–511

Toriumi M (1979) A mechanism of shape-transformation of quartz inclusions in albite of regional metamorphic rocks. Lithos 12:325–333

Toriumi M (1981) Rounded mineral inclusions in upper mantle peridotites. Phys Earth Planet Interior 27:39–46

Toriumi M (1986) Mechanical segregation of garnet in synmetamorphic flow of pelitic schist. J Petrol 27(6):1395–1408

Toriumi M (1994) Cluster formation of Cr-spinel during magmatic differentiation in the upper mantle. In: Takaki R (ed) Research of pattern formation. KTK Scientific Publication, pp 239–257

Toriumi M (2022) Geochemical mechanics of the earth and deep neural network modelling. Springer

Uno M, Iwamori H, Toriumi M (2015) Transition from dehydration to hydration during exhumation of the Sanbagawa metamorphic belt, Japan, revealed by the continuous P-T path recorded in garnet and amphibole zoning. Contrib Min Petrol 170:33. https://doi.org/10.1007/s00410-015-1185-9

Yin HM (2010) Fracture saturation and critical thickness in layered materials. Int J Solids Struct 47:1007–1015. https://doi.org/10.1016/j.ijsolstr.12.016

Chapter 4
Recent Variation of Global and Regional Correlated Seismicity Rates

Abstract The global correlated seismicity rates from 1990 to 2023 show the temporal variations likely to those from 1990 to 2021, but it is noteworthy that there is rapid high-level period before Tonga-Hunga-Tonga giant eruption. On the other hand, the global correlated seismicity dynamics is characteristic in the structure of the global plate boundaries, which are composed of three arms terminating the global ridge boundary. The simulation of the three-arm slider-block network under external periodic force displays the periodic behavior of correlated displacement and the multi-states dynamics against the periodic external force, thereby suggesting the similarity of the pitchfork bifurcation of global seismicity against the rate of earth rotation. Non-volcanic tremor, low frequency earthquake, and repeating earthquake are represented by rupture process of periodic-arrayed jog (vamp) on the shear slip surface associated with slow to very slow slip events. On the other hand, it is needed that the rapid extrusion process of the subducted huge metamorphic mass in the plate boundary zone is governed by the velocity-independent frictional drag along the large-scale shear zones in the episodic extensional stress environment in the wedge mantle.

Keywords Global correlated seismicity from 1990 to 2023 · Three-arm slider-block network and multi-states nonlinear dynamics · Slow slip event and non-volcanic tremor on shear slip jogged surface

4.1 Global Correlated Seismicity Changes from 1990 to 2023/3

The global seismicity of M4-5 earthquakes is taken from USGS data archives. The source data of earthquakes in the global earth were sorted by the time window (10 days) and space (latitude 2.5°, longitude 5°) window. The localities of the plate boundary zones are from NW Pacific to SW Pacific, SE Pacific, Central America, N Pacific around the Pacific Ocean, and from SW Philippine Sea to E Indian Ocean, Himalaya, W Indian Ocean, and Mediterranean Sea to central Atlantic Ocean as

M. Toriumi, *Physics of Geochemical Mechanics and Deep Neural Network Modeling with Diffusion Augmentation*, Advances in Geological Science,
https://doi.org/10.1007/978-981-97-9376-1_4

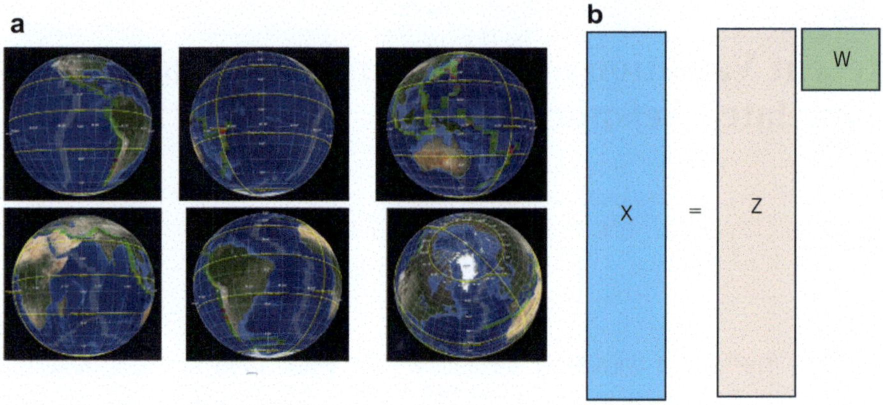

Fig. 4.1 Sampling grids of the seismicity rate along the subduction zone and spreading ridge in the world (green and gray box in A) and data matrix of multivariate timeseries of seismicity rate in the world X decomposed into correlated seismicity matrix Z and orthogonal matrix W by singular value decomposition method (B)

shown in Fig. 4.1. The segments of the spreading boundary are from North Pole, MidAtlantic Ridge, Central Indian Ridge, East Pacific Rise to North American Ridge (Fig. 4.1). The depth range is from 0 to 700 km. The number of M4-5 earthquakes, N_i, are transformed to logarithmic seismicity rate $x_i(t)$ as,

$$x_i(t) = \ln(1 + N_i(t)) \tag{4.1}$$

in which i means the number of the gridded cell studied. The timeseries $x_i(t)$ is represented by data matrix of $\{x_{ij}\}$ in which j is the time step with ten days interval. As proposed in the previous books, the present data matrix is decomposed by Singular Value Decomposition method (principal component analysis) as,

$$X = ZW \tag{4.2}$$

in which X is data matrix, Z is the characteristic feature matrix, and W is the orthogonal matrix from eigen vectors of X as shown in Fig. 4.1. The matrix Z represents the correlated seismicity rate vectors of z_1, z_2, and z_3 to z_n (n is the total number of the cells). The global correlated seismicity timeseries such as z_1, z_2, and z_3 to z_n displays the characteristic spatial and temporal variations of the global background seismicity, showing both the secular trends and large spikes derived from giant earthquakes (Toriumi 2021, 2022). Furthermore, in this study, the clear signals derived from the large volcanic activity were recognized.

The timeseries of the global and regional correlated seismicity contains some amounts of noises likely to Gaussian noisy function as noticed by the previous works, and the temporal variation of the local variance can be found. It follows that the correlated seismicity timeseries manifests approximately the temporal change of the Gaussian probability density having variational means and variances. The phase maps of correlated seismicity rates appear several clusters showing temporal change, thereby suggesting the Langevin type stochastic dynamics concerning the correlated seismicity rates as proposed in the previous books.

In the previous books, the author clarified that the timeseries of the z_1 component in the global correlated seismicity display the secular variation with several stepwise increase from 1990 to 2021, and phase map of z_1 and z_3 appears 4 clusters in which one has large variance accompanied with giant earthquakes of 2004 Sumatra and 2011 Tohoku-Oki. In this volume, the timeseries from 2021 to 2023 were added in the previously sorted data matrix, and the same data processing has been applied for the extended data matrix from 1990/1/1 to 2023/3. The timeseries of the global correlated seismicity rates of M4-5 are shown in Figs. 4.2, 4.3 and 4.4. The recent data of z_1 and z_3 are plotted in the cluster of the period from 2004 to 2021, thereby showing the global correlated seismicity dynamics keeps its high levels involving large variation about the z_3 timeseries. In the previous works, the author identified three clusters of data in z_1–z_3 phase map; A, B, and C with the advancement of time steps, and thus the recent timeseries data of them clearly belongs to cluster C, in which the giant earthquakes took place twice as shown in Fig. 4.2.

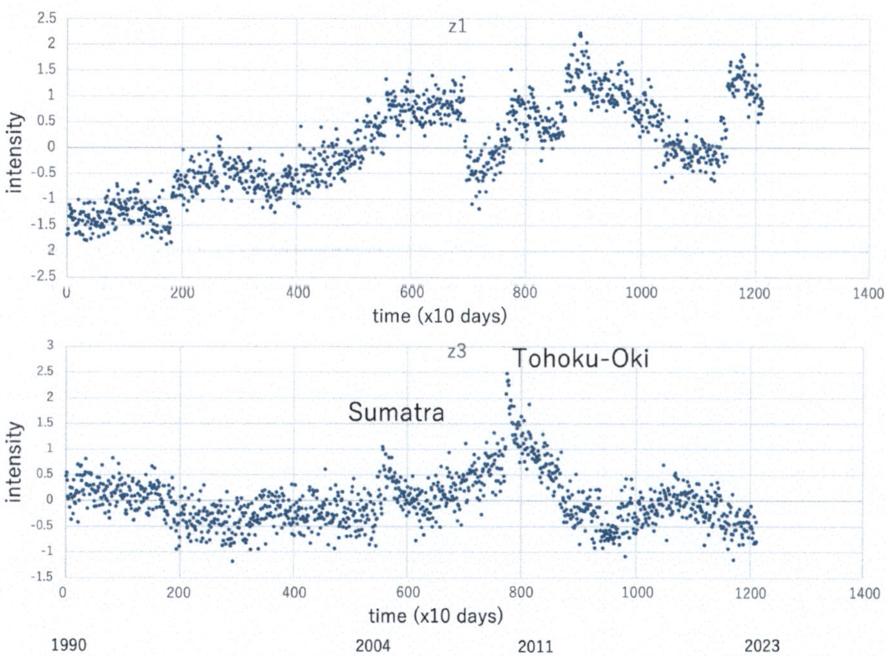

Fig. 4.2 Global correlated seismicity rate from 1990/1/1 to 2023/3/29 of z_1 (top) and z_3 (bottom) against time step (× 10 days). The peaks at 2004 and 2011 appear at Sumatra and Tohoku-Oki giant earthquake

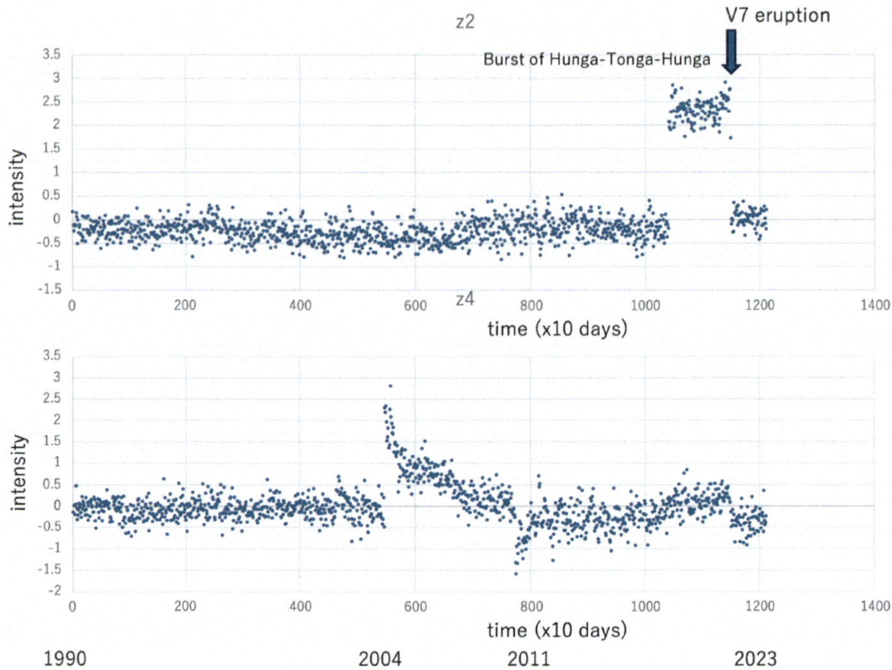

Fig. 4.3 Global correlated seismicity rate from 1990/1/1 to 2023/3/29 of z_2 (top) and z_4 (bottom) against time step ($\times$ 10 days). Signal of giant volcanic eruption at Tonga-Hunga-Tonga appears the sudden high intensity in the z_2 timeseries

On the other hand, it has been pointed out that the relation between the z_1 timeseries and the secular trend of earth rotation (length of day) displays the pitchfork type bifurcation with spin-down of the earth rotation (Toriumi 2022). For LOD rate decreases continuously from 2018 to 2023, that means spin-up of earth rotation, it seems that the z_1 of the recent timeseries after 2020 decreases with the time advancement. However, the z_1 component keeps the high level as shown in Fig. 4.5, thereby suggesting that the mode of bifurcation above has more complicated features involving other controlling parameters.

During the last decade, it is important that at the time of three years before the Hunga-Tonga-Hunga volcano (HTH) eruption (2021 Dec.), the z_2 component rapidly jumped at high level as seen in the z_2 timeseries (Fig. 4.2). The high level z_2 kept during about three years and rapidly dropped at the low level before. In addition, it seems that the eruption stopped at 2022/1/15, being just at the z_2 drop time. The locality cell showing the major components of the z_2 timeseries is just near the HTH as seen in the intensity map of the z_2 component (Fig. 4.3). Therefore, it should be obvious that the before giant eruption of HTH, the seismic burst appeared in the region near the HTH volcano, and this phenomenon continued for three years just before the giant eruption of January 2022, and after eruption it ceased suddenly.

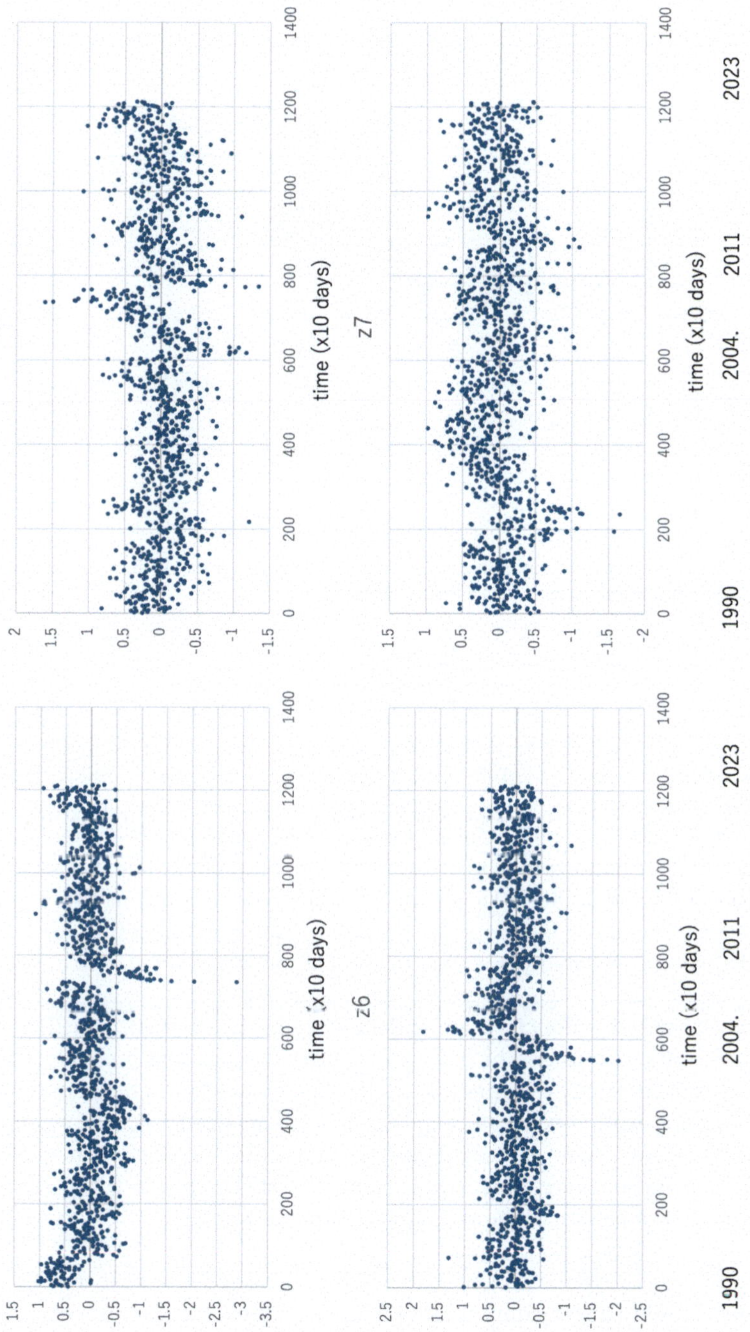

Fig. 4.4 Global correlated seismicity rate from 1990/1/1 to 2023/3/29 of z_5 (top left), z_6 (bottom left), z_7 (top right), and z_8 (bottom right) against time step (× 10 days)

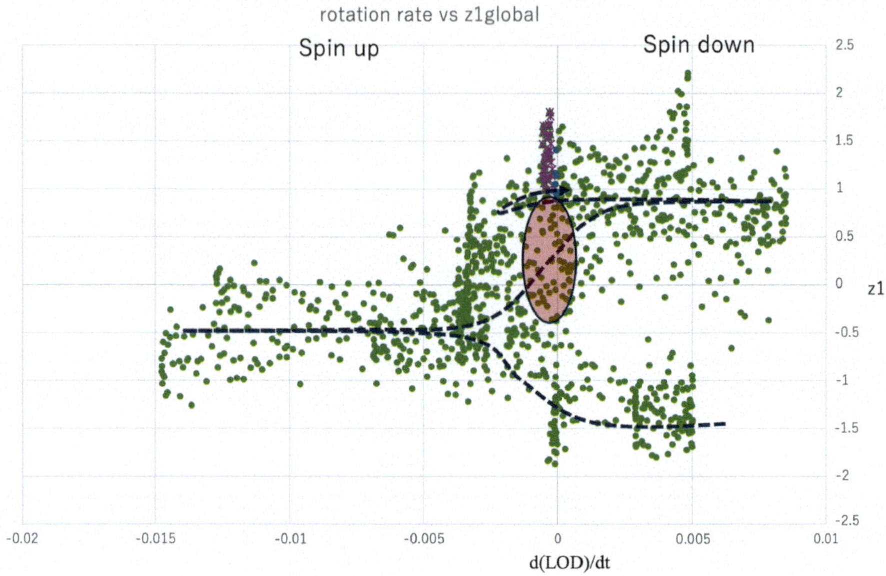

Fig. 4.5 Phase diagram of global correlated seismicity z_1 and rotation rate of the earth, showing the pitchfork type bifurcation pattern from 1990 to 2023

In the phase diagram of z_1–z_3 as described earlier, the clusters A, B, and C are identical, and they show a temporal order from A to C through B (Fig. 4.6). Besides, as noticed in the previous books, the cluster D that displays the large variance of density distribution of data occupies the domain between B and C clusters, and the data of the cluster D is temporally successive after B, and giant earthquakes of Sumatra and Tohoku-Oki occurred during D phase. The data of the D cluster having large variance of distribution probability begin enough before the giant earthquakes, Sumatra 2004 and Tohoku-Oki 2011, and thus the large variance seems to represent the characteristic nonlinear dynamics as proposed by the author (Toriumi 2021, 2022) that display the unstable state of nodal point as tangential relation of z_1 and z_3 nullclines.

On the other hand, it is noticed that the data of z_1 and z_3 from 2021 to 2023 are clearly plotted in the C cluster which displays small variance and probably stable node of the global seismicity dynamics. In addition, it shows that in the period from 2015 to 2021, the transition between B and C clusters happened repeatedly without experience of D cluster and no giant earthquakes in the world. In the phase map of the z_1 and z_3, there is no signature of the giant eruption of HTH (Hunga-Tonga-Hunga) volcano, suggesting a pattern of the volcano-seismic phenomena.

It seems that the characteristic signals near 2023 Turkey Earthquake (M7.8, 2023/2/6) do not appear in every component of the global correlated seismicity rates, z_1 to z_8, though the z_1 shows a sudden increase at 570 days before the Turkey earthquake (Fig. 4.4). It, however, may show that the slight changes in the intensity diagram of

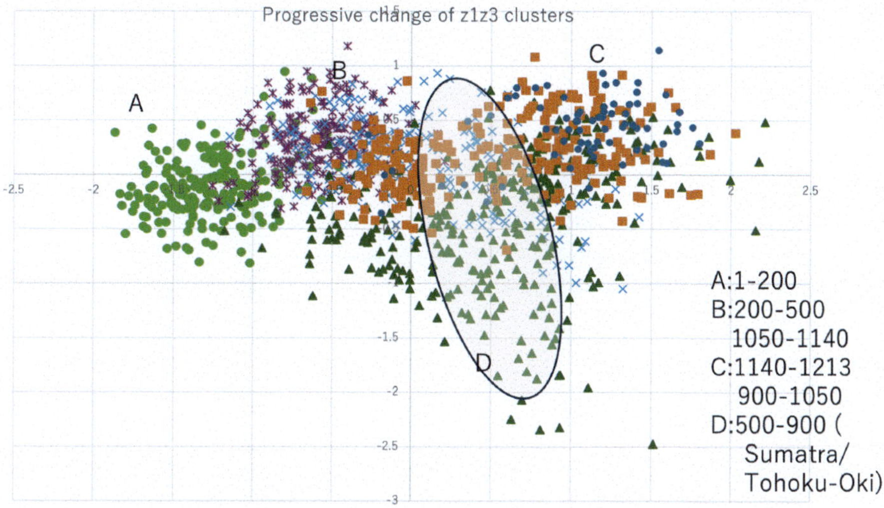

Fig. 4.6 Phase diagram of global correlated seismicity of z_1 and z_3 from 1990 to 2023, showing the sequential change of clusters A to C, and D having large variance between B and C

gridded locality for correlated seismicity components can be recognized in Fig. 4.4, thereby suggesting that the weak increase of the global correlated seismicity z_2 to z_6 manifests the rapid seismicity increase in the region of Turkey. It may appear that the signals in the local intensity of correlated seismicity appear in the early phases of the global correlated seismicity timeseries from 1990 to 2021 as shown in Fig. 4.7.

On the other hand, in the previous book (Toriumi 2022), it was pointed out that the seismicity of the gridded locality is represented by the characteristic global network structure of the seismicity, and the network structure changes temporarily, being synchronized with the change of the clusters in the z_1 versus z_3 phase map from 1990 to 2021. Thus, it will be investigated whether the network structure of the global correlated seismicity z_1 and z_3 from 1990 to 2023 changes like the previous pattern or not. Therefore, in this book, the author applied the singular value decomposition for the separated matrixes from the whole logarithmic seismicity matrix as illustrated in Fig. 4.8. The separation of the whole matrix was carried out in the three phases: the first (A) is from 1 to 500 step, the second (B) from 501 to 780, and the third (C) from 781 to 1212 step. The correlated seismicity and intensity of the gridded localities from each seismicity matrix are obtained, and those are shown in Figs. 4.9, 4.10 and 4.11.

It is important that the timeseries of the global correlated seismicity in A, B, and C are clearly different from those of the whole matrix mentioned above, thereby suggesting that timeseries of their correlated seismicity are very sensitive for the giant earthquake and continuous activity of abundant small earthquakes, and thus, they depend on different localities contributing the intensity of the global correlated seismicity in every partial matrix, A, B, and C. It seems that the timeseries of the global correlated seismicity in the period of A phase displays gradual change with

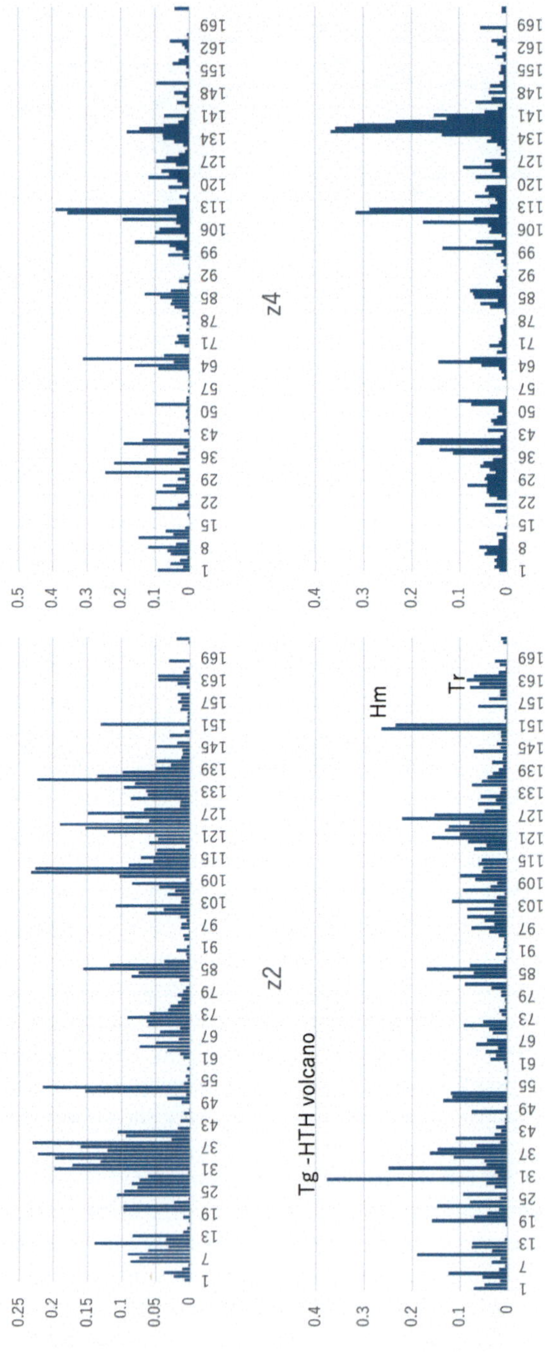

Fig. 4.7 Intensities of gridded locality of the plate boundary zone for the global correlated seismicity, z_1 to z_4 from 1990 to 2023

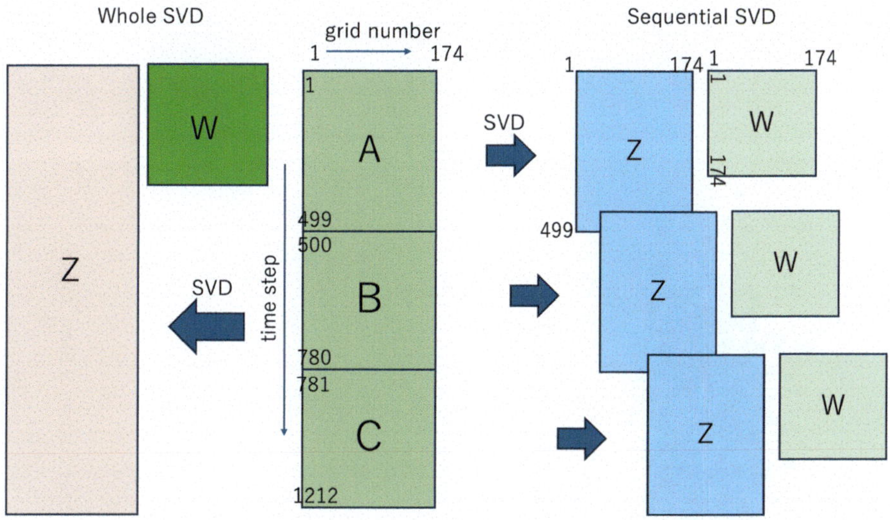

Fig. 4.8 Illustration of sequential singular value decomposition by splitting whole global seismicity matrix to three partial seismicity matrix A, B, and C

somewhat periodic waveform, being different from the timeseries in the B and C. During B and C phases, the giant earthquakes of 2004 Sumatra and 2011 Tohoku-Oki took place, and the latter shows the high seismic activity. To illustrate the difference of the network structure of the global correlated seismicity of z_1 from A to C, the maps showing the basic components of the gridded localities along the global subduction plate boundary are indicated in Figs. 4.12 and 4.13, thereby suggesting that the network structure of the A phase is composed of Solomon (sl)—Tonga-Kermadic (tk)—Chile (ch)—Mexico (mx), the second B of Java-Sumatra (sm)—Tonga-Kermadic (tk)—Chile (ch), and the third C of Himalaya (hm)—Luzon (lz)—South Sandwiches (ss)—Tonga-Kermadic (tk). On the other hand, the z_3 component shows the different patterns such that the first stage A represents the network of Chile (ch)—Alaska (ak)—Kamchatka (kt)—Taiwan (tw), and the second one B does the network of Sumatra (sm)—Northern Japan (nj)—Izu Bonin (ib)—Chile (ch), and the third C the network of nj—tw—ch. These networks of z_1 and z_3 are slightly different from those of the networks inferred by the whole matrix case, as suggested previously. In addition, it seems that the z_3 of the second phase B is represented mainly by the correlated seismicity of Sumatra-Java, Northern Japan, and Chile, thereby showing the aftershocks of the giant earthquakes over M9, though the z_1 does not contain the contribution of the Northern Japan in the B phase. Therefore, it is probable that the network transition of the global correlated seismicity from the whole data matrix should be thought of the background cross correlation in the framework of the z_1 and z_3 components, as like as cluster change in the phase map between z_1 and z_3 (Fig. 4.6).

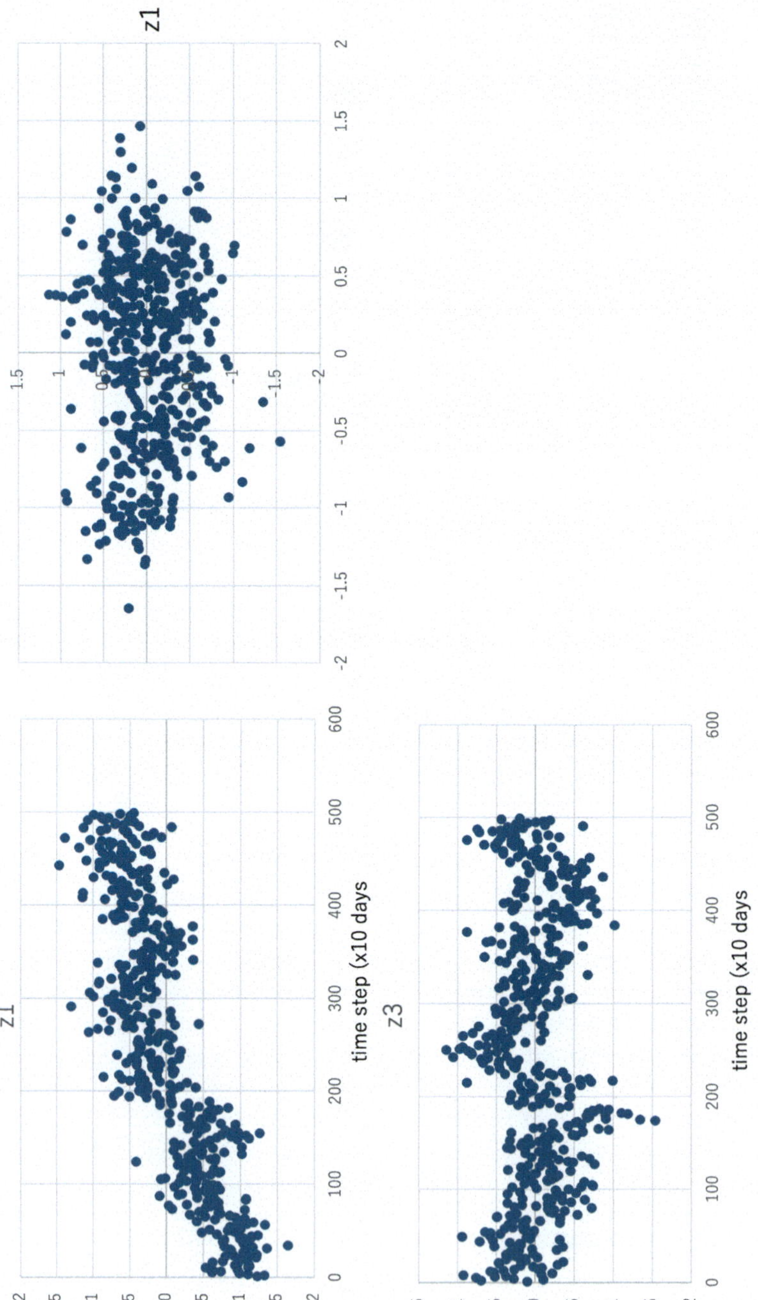

Fig. 4.9 Temporal variation of partial global correlated seismicity rates z_1 and z_3 of segment A, and phase diagram of them (right)

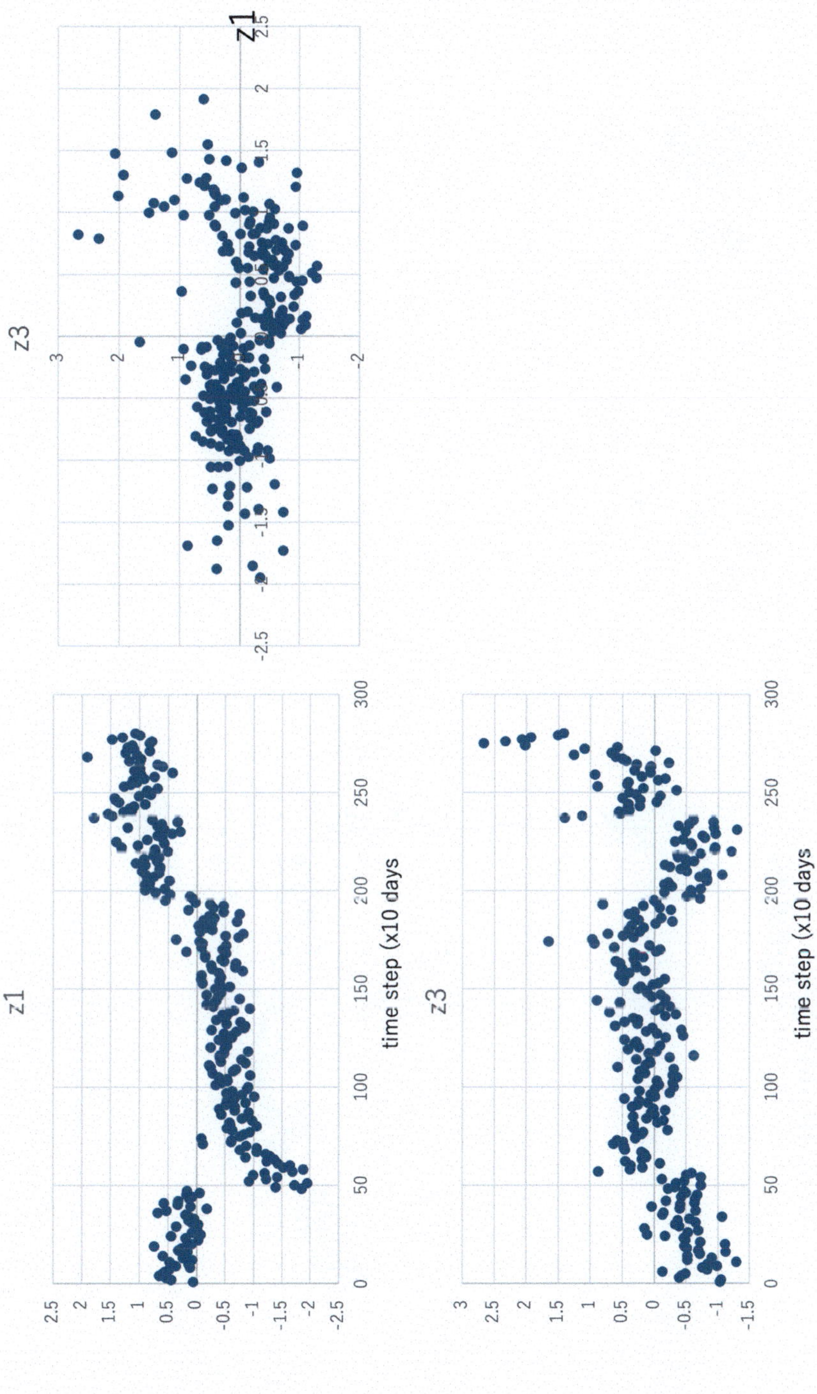

Fig. 4.10 Temporal variation of partial global correlated seismicity rates z_1 and z_3 of segment B, and phase diagram of them (right)

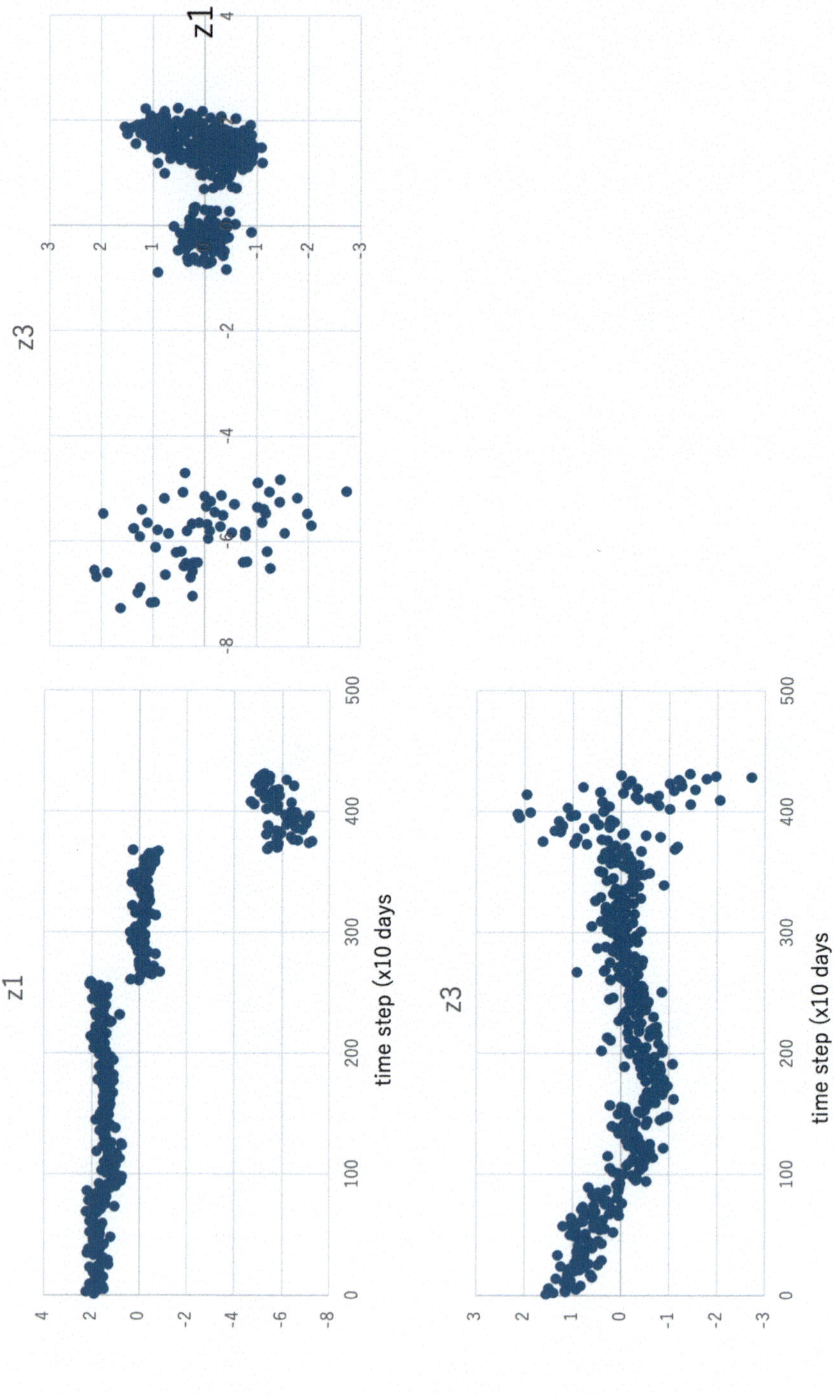

Fig. 4.11 Temporal variation of partial global correlated seismicity rates z_1 and z_3 of segment C, and phase diagram of them (right)

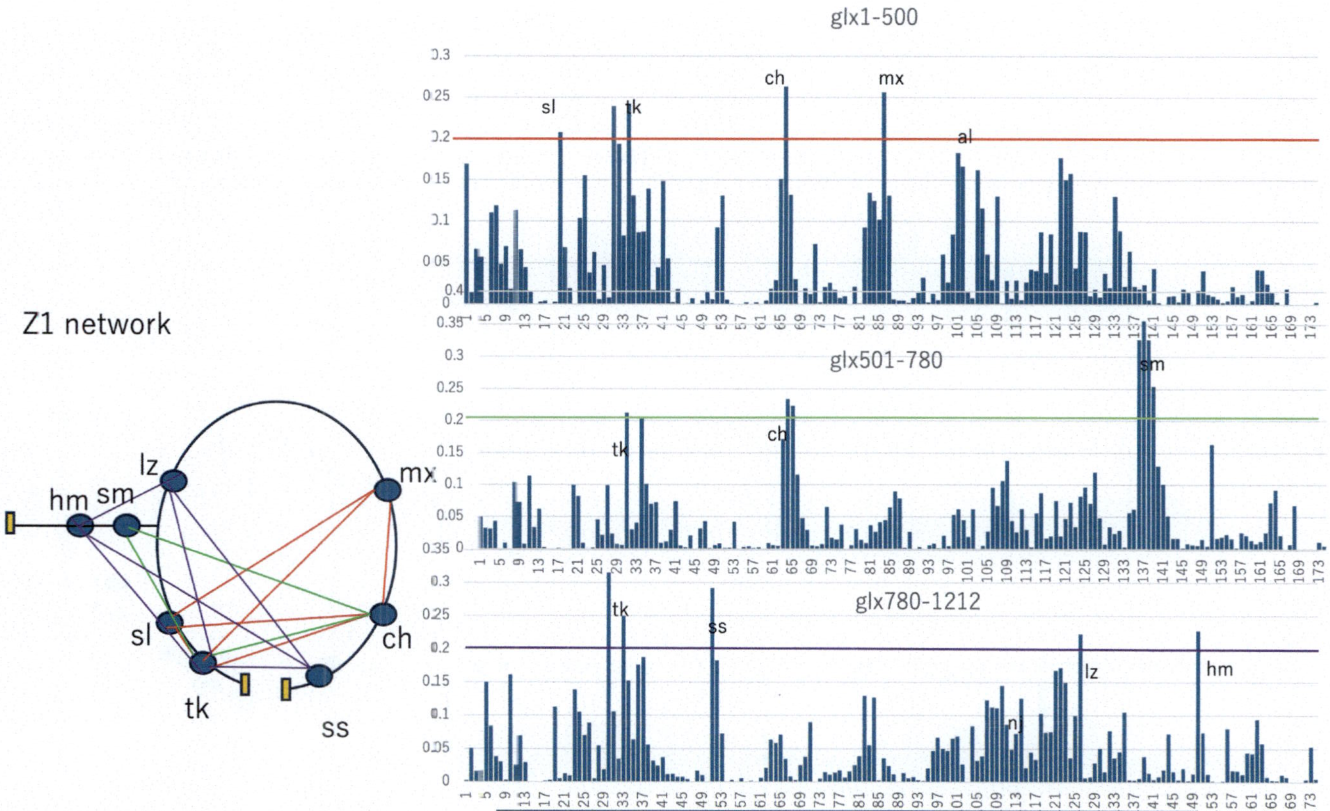

Fig. 4.12 Intensities of gridded locality of the plate boundary zone for the partial global correlated seismicity z_1 from A, B, and C (right figures), and temporal change of the correlated seismicity networks (left) corresponding to A (orange), B (green), and C (violet), which are connected the localities having the intensity of z_1 component over 0.2

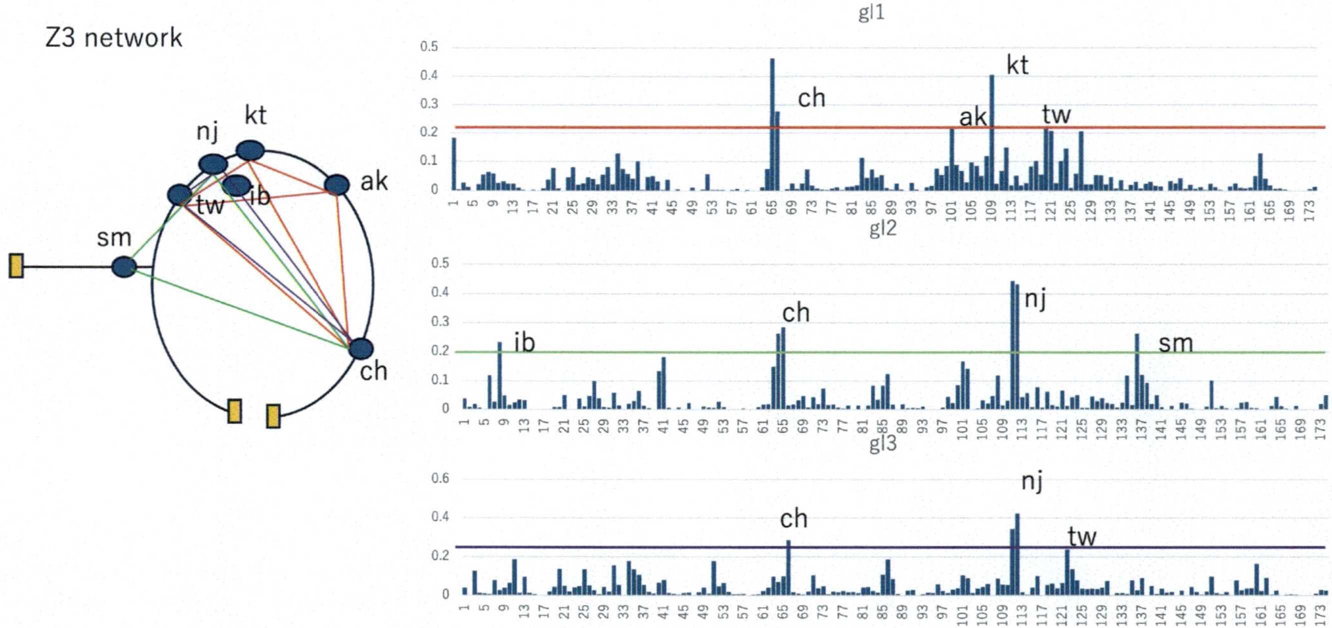

Fig. 4.13 Intensities of gridded locality of the plate boundary zone for the partial global correlated seismicity z_3 from A, B, and C (right figures), and temporal change of the correlated seismicity networks (left) corresponding to A (orange), B (green), and C (violet), which are connected the localities having the intensity of z_1 component over 0.2

4.2 Regional Correlated Seismicity of Japanese Island from 1998/1/1 to 2023/3/30

As discussed in the previous book, the timeseries of the micro-seismicity rates of the gridded localities around Japanese islands from M1 to M2 and M2 to M3 were investigated from 1998 to 2023/3, and data matrix of the logarithmic seismicity rates are decomposed into orthogonal matrix W, and correlated seismicity matrix Z by the singular value decomposition (SVD). The characteristic timeseries of the regional correlated seismicity matrix Z are approximately represented by the features component z_1 to z_3, but other components z_2 and z_4 too as shown in Figs. 4.14 and 4.15, respectively. The timeseries of regional correlated seismicity rate of the Japanese islands region has been discussed in detail by the previous works (Toriumi 2021, 2022), and it is concluded that the secular trend of the regional correlated seismicity displays the stepwise and monotonous increase from 1998 to 2011, but after 2011 Tohoku-Oki giant earthquake, it shows up and down with rather large amplitude until 2021. In this period, it should be noticed that the up and down patterns appear repeatedly a sharp increase and gradual decrease of z_1 timeseries. On the other hand, it is noteworthy that the timeseries of z_3 shows cyclic change with 12 to 15 years and 3 to 5 years periodicity that changes longer after 2011. The intensity of the regional correlated seismicity rate of the Japanese islands is shown in Fig. 4.16, thereby suggesting the strong correlation between Okinawa-Kyushu and northeastern Japan region.

In the phase map as displayed in Fig. 4.17 between z_1 and z_3 in the previous works, it is distinctive that there are four clusters that change from one to another with advancing the time, and the mode of change should be governed by the simple nonlinear dynamics of z_1 and z_3. In addition, it should be important that the giant Tohoku-Oki earthquake took place at the time of C clusters of micro-seismicity with large variance of data probability involving the post-seismic activity.

Considering that the phase map of the correlated seismicity of Japanese islands region manifests the appearance of distinctive clusters of A, B, C, D, and E in the previous works, it should be available that these five clusters represent the stable nodes of the nonlinear dynamical system of z_1 and z_3, with changing its parameters with time (Toriumi 2022). Then, it may be available that the phase map of them shows the same mode of occurrence of the stable nodes in the case extending to the data of 2021–2023.

The timeseries of z_1 from 2021 to 2023 processed in this work does not follow the expected trend discussed in the previous book (Toriumi 2021), so that as shown in Fig. 4.13, during recent several decades, it is obvious that the z_1 component changed suddenly at the high levels in the time step of 830 and maintains about that level still present. It shows that the z_2 (Fig. 4.15) also decrease rapidly at the same time, being likely to the z_1 trend. Nevertheless, the other lower components such as z_3, and components display the continuous change with advancing time (Fig. 4.15). Therefore, it seems that the secular trends of z_1 and z_3 continue after 2021, which appear the high level of z_1 and gradual decrease in z_3, thereby suggesting

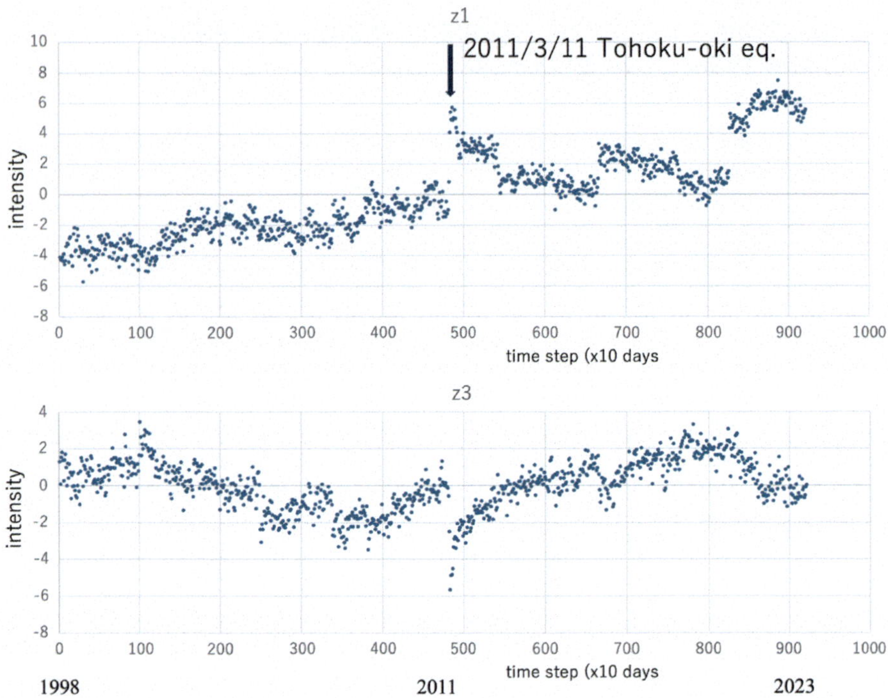

Fig. 4.14 Temporal variation of regional correlated seismicity rates z_1 and z_3 of Japanese islands region from 1998/1/1 to 2023/3/29, showing the rapid increase in 480 (Tohoku-Oki earthquake) and 835 steps

the rather different pattern of the clusters in their phase map shown in the previous book (Toriumi 2022).

In the phase map of them mentioned above, it is a little bit of surprising that the timeseries of z_1 and z_3 in the phase map from 1990 to 2023 are plotted in the somewhat different clusters as shown in Fig. 4.17, suggesting more complicated nonlinear system than the previous model. Besides, it needs to be addressed that by adding the new data in the singular value decomposition processing of seismicity data matrix, the correlated seismicity rates of the decomposed characteristic matrix sometimes change their values. In the case of this correlated seismicity rates, it is probable that the change of the z_1 and z_3 timeseries adding the data from 2021 to 2023 should be due to a little high level of the recent seismicity rates in the several gridded localities such as Fukushima-Oki. It seems that in the phase map of z_1 and z_3 in Fig. 4.17, the recent correlated seismicity rates are plotted in the cluster a few shifted to higher portion of the z_1 compared with previous works. It seems that recent high level of z_1 together with low level of z_2 are derived from high seismic activity along the subduction zone from Fukushima-Oki to Nemuro-Oki through Tokachi-Oki, and Kii Peninsula to Shizuoka-Oki regions as shown in Fig. 4.18. Thus, it suggests that

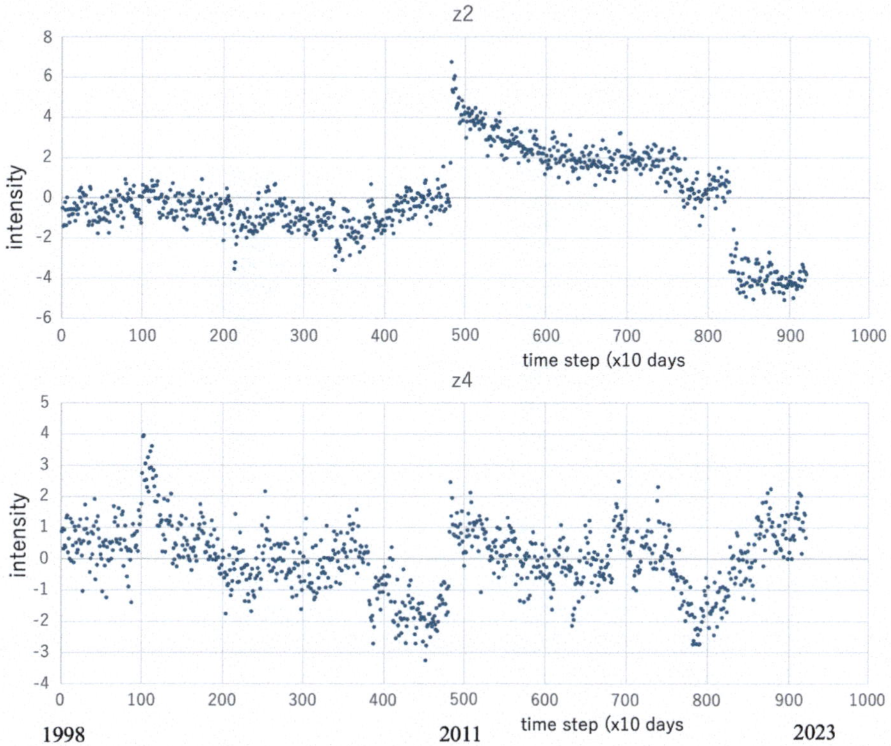

Fig. 4.15 Temporal variation of regional correlated seismicity rates z_2 and z_4 of Japanese islands region from 1998/1/1 to 2023/3/29, showing the rapid increase in 480 step and decrease in 835 steps

the seismic activity of the NE Japan from Fukushima to NE Hokkaido becomes rather high level, being correlated with that of the eastern domain of the Nankai trough.

It should be noticed that the correlated seismicity rates in Japanese islands depend upon the mechanical interaction among the gridded localities, being represented by the network system dynamics as mentioned previously (Toriumi 2022). Especially, it is provided that the most active in seismicity is the NE Japan, and it is correlated with seismicity of the Ryukyu region, but intermittently it changes to correlated seismicity with Nankai trough region. The transition of the seismicity network from NE-NA type to NE-RY-NA occurred in 2016 with the key regions of Yakushima-Oki, Chiba-Oki, and central Kyushu as shown by previous works of network dynamics analyses of seismicity in Japanese islands (Toriumi 2022). On the other hand, it should be important that the recent rapid increase of z_1 appeared in Fig. 4.16 is probably derived the network change from NE-RY-NA to NE-NA type, suggesting the backward transition likely to 2016 transition mentioned above by the change of Chiba-Oki and central Kyushu region.

By the way, it seems that as seen in the change of seismicity of gridded localities, there is a general increase of the local seismicity rates from Fukushima-Oki

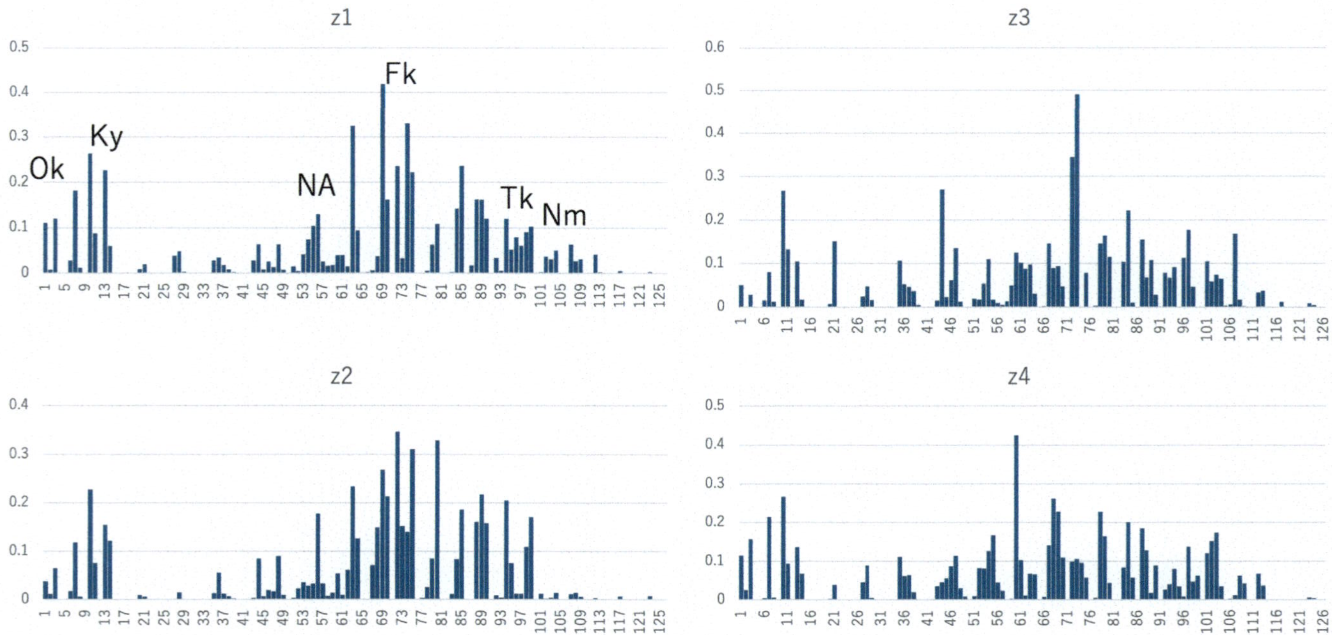

Fig. 4.16 Intensities of gridded locality in regional correlated seismicity rates z_1 to z_4, showing that major components of the regional correlated seismicity are of Ryukyu region and northeaster Japan region

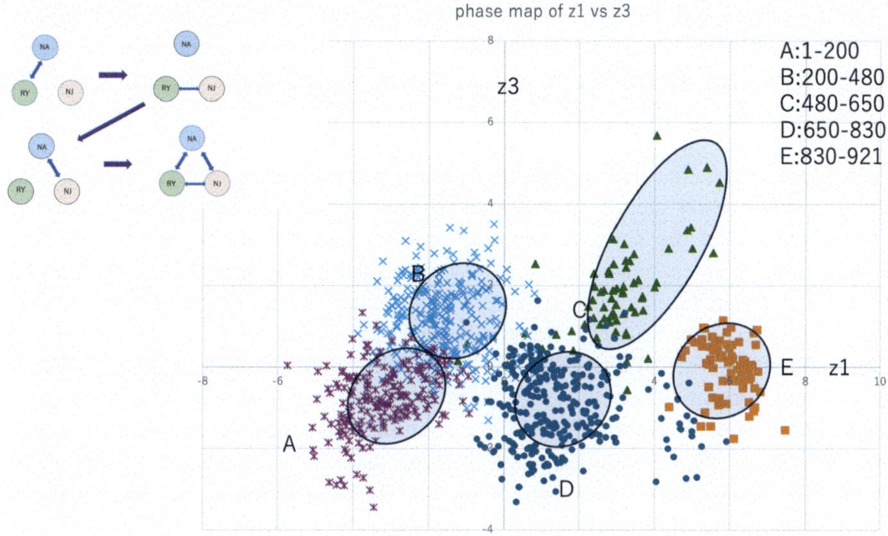

Fig. 4.17 Phase map of regional correlated seismicity z_1 and z_3 in Japanese islands region, showing the five clusters sequentially changing from A to E, and C with large variance

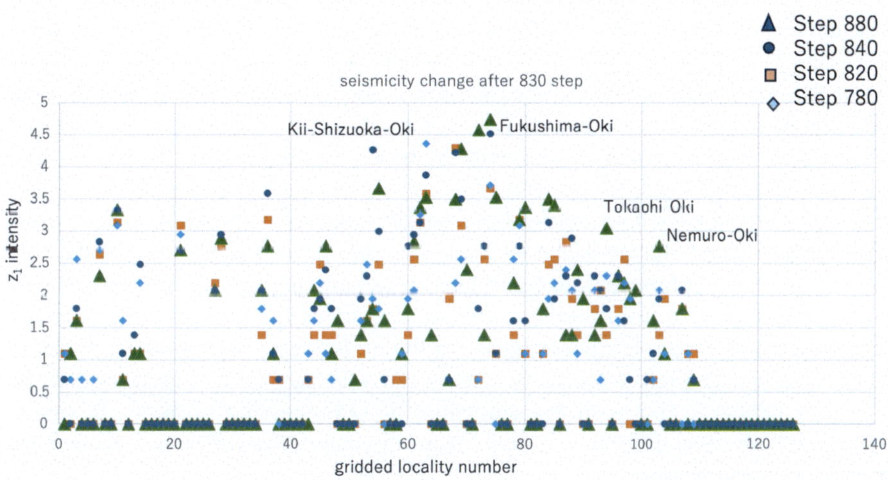

Fig. 4.18 Changes of intensity of gridded locality in the regional correlated seismicity z_1 from 780 to 880 steps, showing large increase during these time steps in noted gridded localities in this figure

to Nemuro-Oki in the NE Japan displayed in Fig. 4.18. The transition of above-mentioned network makes new phase of NE-NA system which is strengthened by high seismicity rates in Hokkaido to Kuril regions. This change of the northward shift of the local seismic activity is probably one of post-effect from the 2011 Tohoku-Oki earthquake.

On the other hand, as discussed in the previous section on the global seismicity network, it will be investigated that how the network structure of the regional correlated seismicity of Japanese islands region changes during the period from 1998 to 2023, as shown in the temporal change of regional seismicity rate of Japanese islands by sequential SVD method (Figs. 4.21 and 4.22). Furthermore, in Figs. 4.22 and 4.23, it shows that the high intensity localities of the regional correlated seismicity z_1 and z_3 vary clearly in the three periods by the locality sets over intensity 0.2. To understand the network geometry of these locality by regional correlated seismicity of Japanese islands, the network of seismically active localities studied is shown in these figures above, thereby displaying the localities of Tottori (60), Noto-Oki (73), Japan Sea (72), Nankai Trough (55), Kinki (63), Chubu (69), Kanto (69), Off-Kanto (70), Off-Fukushima (74), Off-Sendai (84), Off-Tokachi (94). From the regional correlated seismicity z_1, it shows that the network system of the first stage (1–450 step) is represented by the mutual connection among Tottori, Japan Sea, Off-Akita (79), Off-Sendai, and Off-Noto, and that of second stage (451–700) by Noto, Off-Fukushima, Kanto, and Kinki. On the other hand, in the third stage, the network changes to the connection of Nankai, Japan Sea, Off-Kanto, Off-Fukushima, Off-Sendai, Off-Tokachi, and Off-Akita.

In the previous book (Toriumi 2022), the seismicity network of Japanese islands was studied by regional correlated seismicity of the whole seismicity matrix from 1998 to 2021, but not by the results of the sequential SVD for the regional correlated seismicity, as like as the case of global seismicity network. The difference between the results of sequential SVD for seismicity network and the correlated seismicity one of the whole seismicity matrices is presumably derived from the fact that the latter represents the background seismicity, though the former the in-situ correlated seismicity of each segmented seismicity matrices in the phase A, B, and C. The network system of regional correlated seismicity in A to C phase is lack in the connection between Ryukyu and northeastern Honshu seen in the whole SVD results. From the sequential change of the network structure from A to C, it means that the progressive change of the regional correlated seismicity network of Japanese islands displays the sequence from northern seismic activity of the background seismicity and later central Japan–northeast Japan, to the wide range background activity in Japan as shown in Figs. 4.18, 4.19, 4.20, 4.21, 4.22 and 4.23. It appears that the timeseries of the z_1 and z_3 of the sequential SVD processing in the A, B, and C suggests the obvious difference derived from the distinctive difference of network structure of these phases.

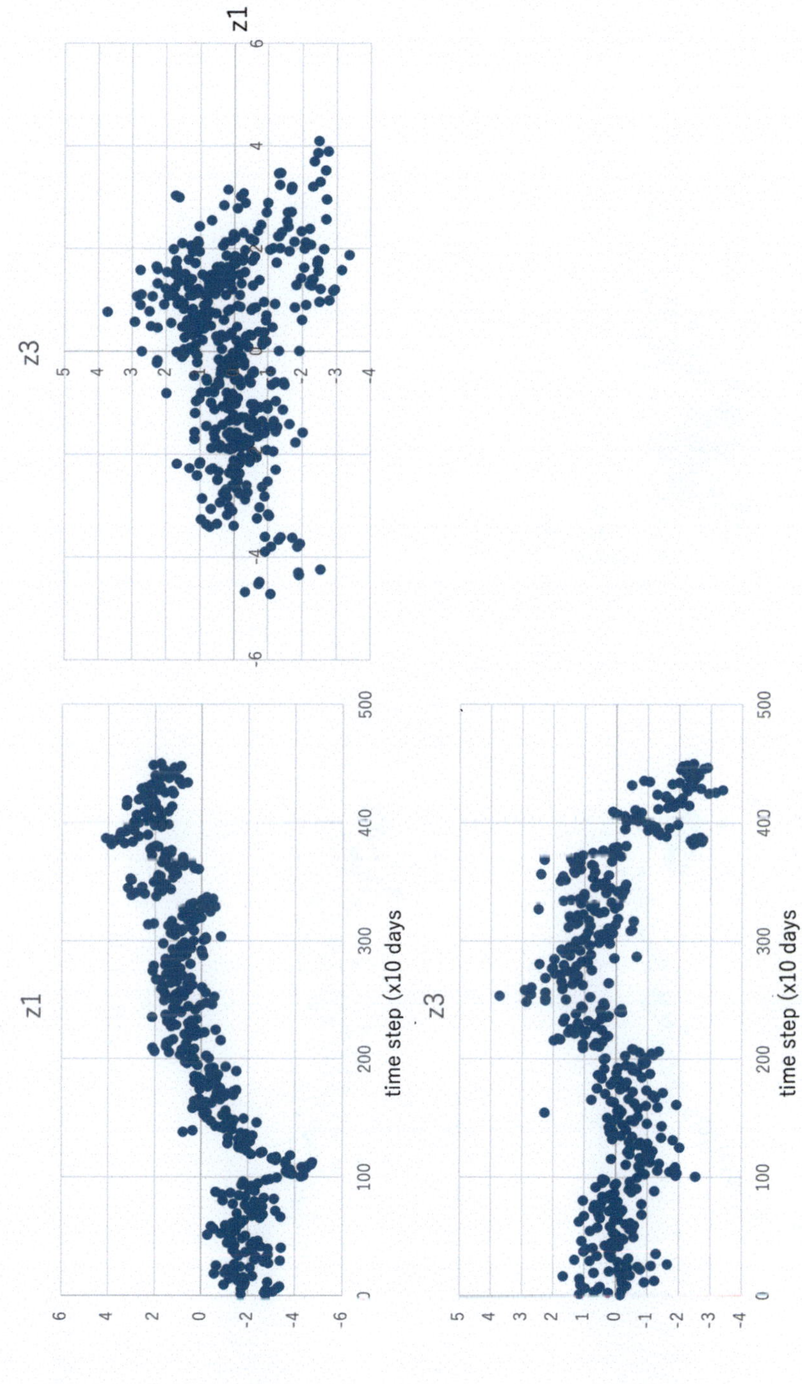

Fig. 4.19 Temporal change of regional correlated seismicity in segmented timeseries from 1 to 450 steps of z_1 and z_3, and phase map of them

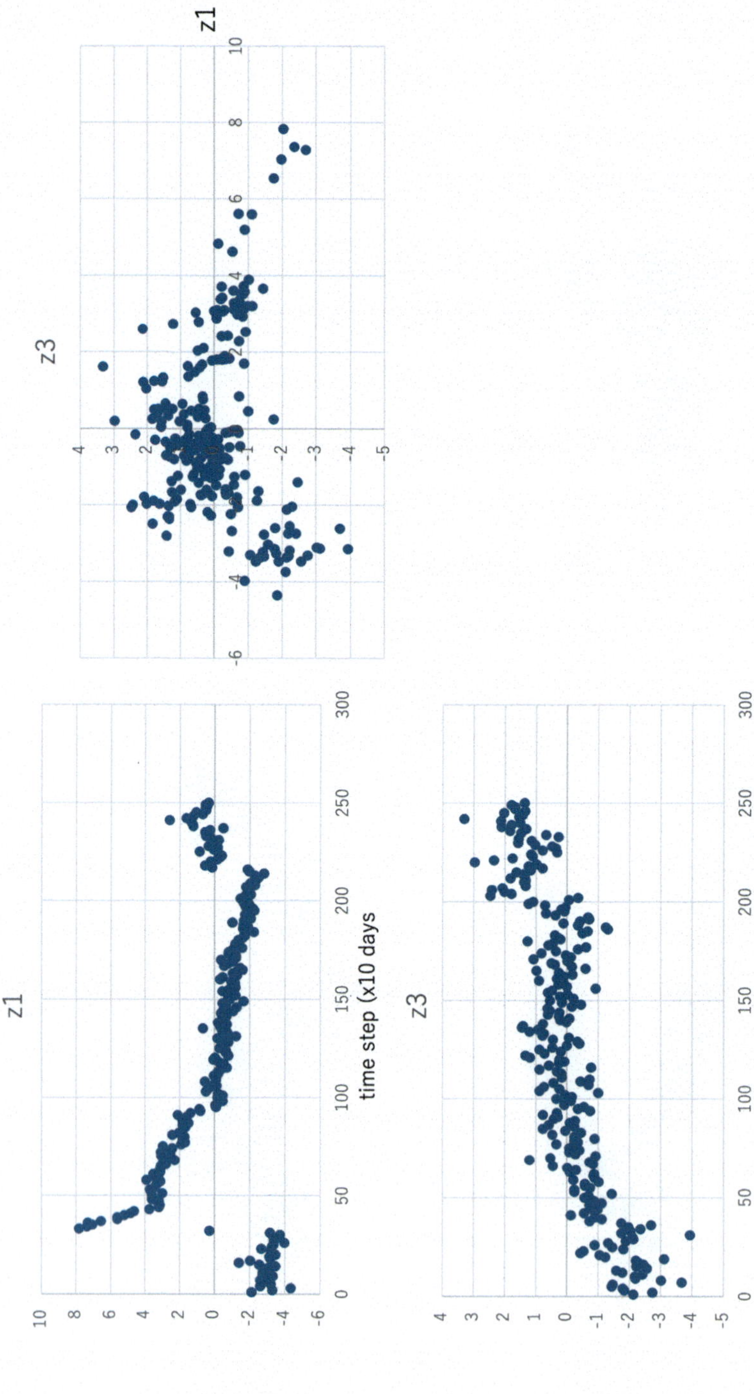

Fig. 4.20 Temporal change of regional correlated seismicity in segmented timeseries from 451 to 700 steps of z_1 and z_3, and phase map of them

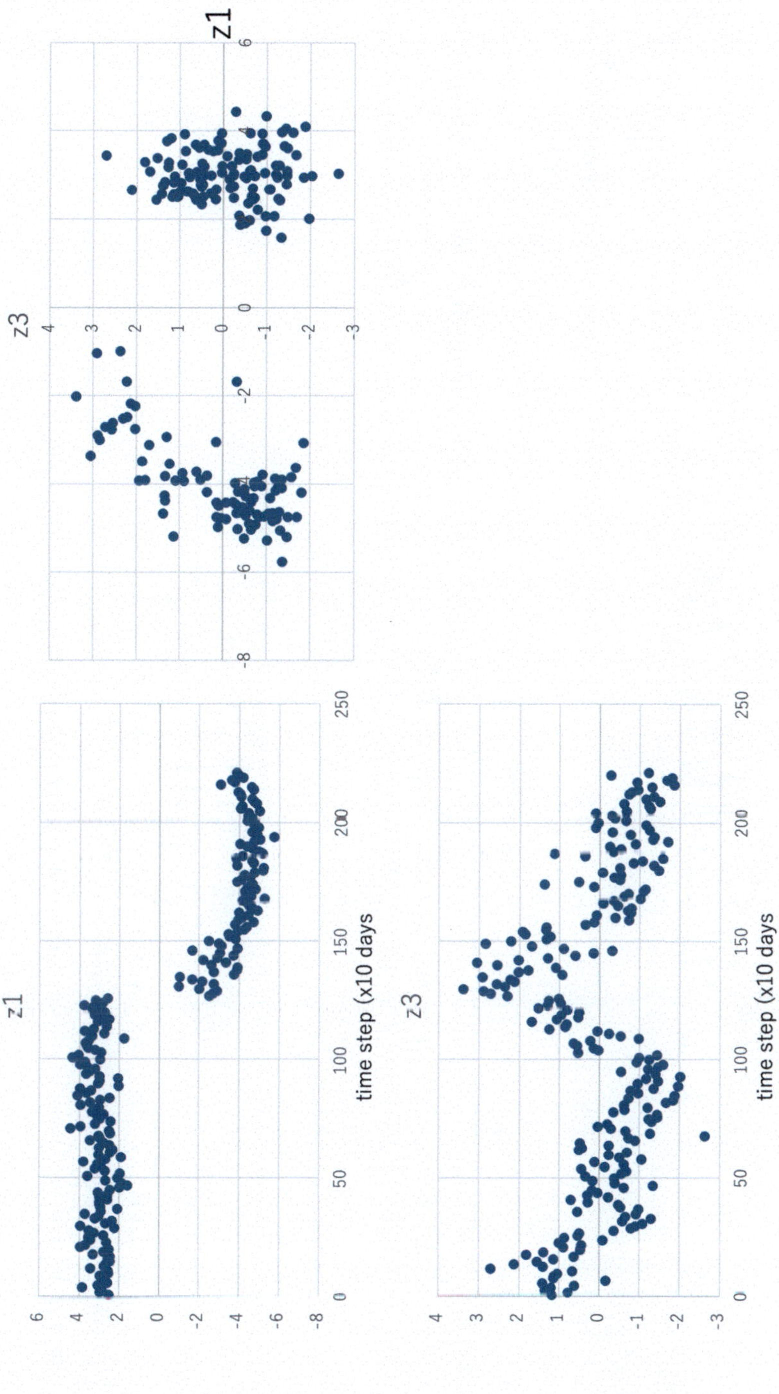

Fig. 4.21 Temporal change of regional correlated seismicity in segmented timeseries from 701 to 920 steps of z_1 and z_3, and phase map of them

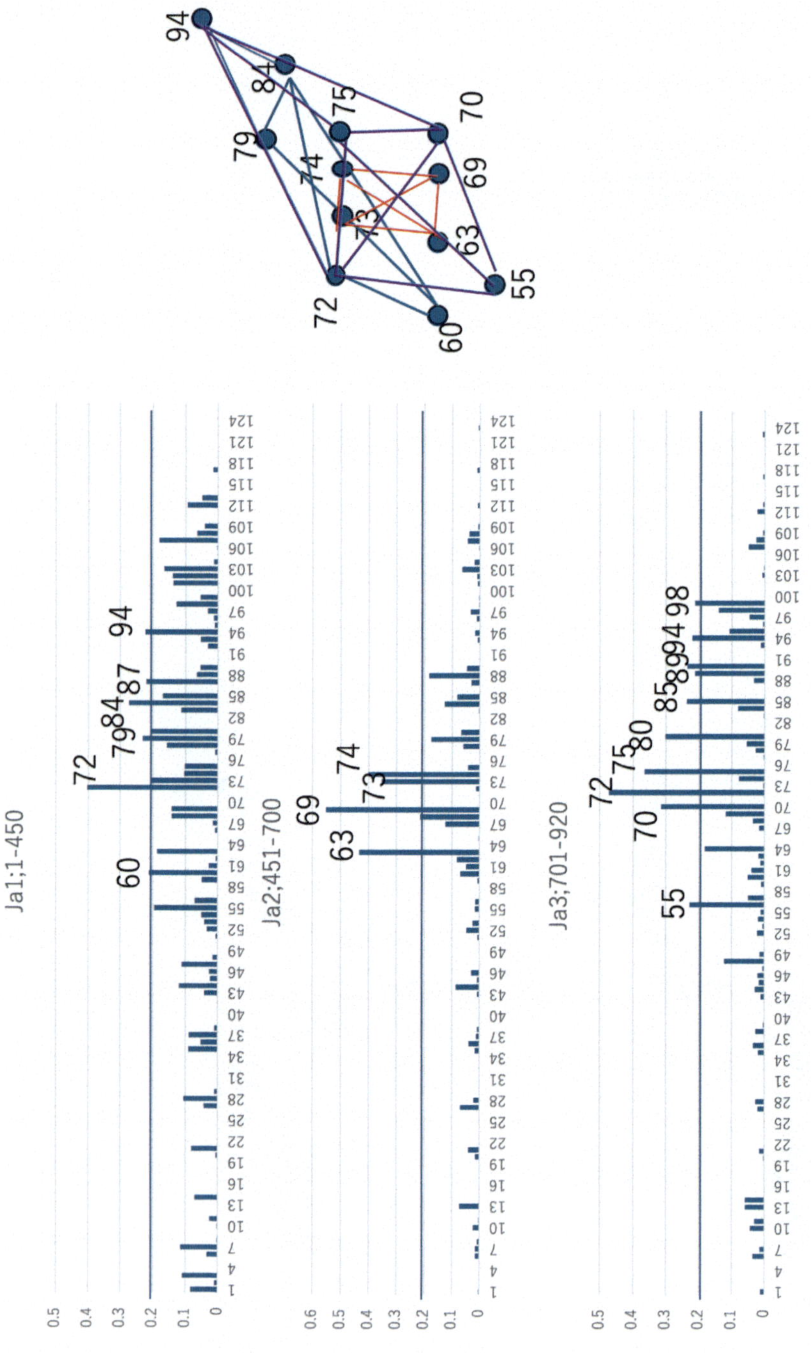

Fig. 4.22 Changes of intensity of gridded locality in the regional correlated seismicity z_1 in three stages of Figs. 4.19, 4.20 and 4.21, showing the regional correlated seismicity network change illustrated in right figure

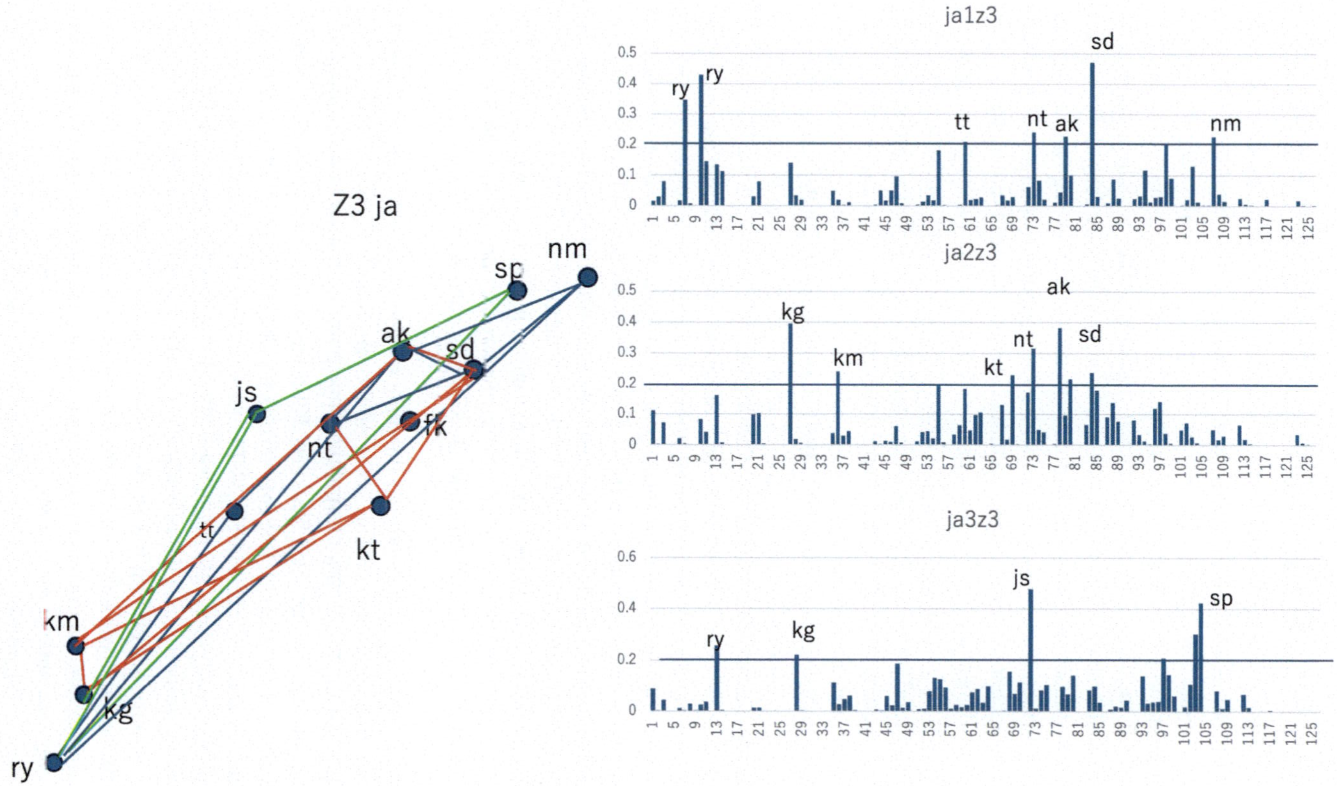

Fig. 4.23 Changes of intensity of gridded locality in the regional correlated seismicity z_3 in three stages of Figs. 4.19, 4.20 and 4.21, showing the regional correlated seismicity network change illustrated in le:t figure

4.3 Physics of Global Seismicity, Earth Rotation, and Fluid Earth

The global correlated seismicity and regional correlated seismicity display the long-term secular variations with annual and tri-annual change in their intensity as clarified in the previous books and previous chapters in the book. The secular variations of the global and regional correlated seismicity appear the similar pattern of an increasing trend from 1990 to 2023, and the correlated partial b-value, which is defined by the gradient of Gutenberg-Richter law in the small earthquake range, display the cyclic change with sub-annual to tri-annual periodicity, suggesting that the global and regional correlated seismicity rates manifest an important interaction between the solid earth mechanics, fluid earth dynamics, and the celestial mechanics.

Johnson and Burgmann (2017) have proposed a new model of regional seismic activity related to the rainfall and snowfall in the Japanese islands and northern California regions, based on the timeseries analysis of these precipitation and seismicity big data. They noticed that there are many coherencies between the season of the precipitation by rain and snow and the small seismicity rates in the various localities of regional areas, thereby suggesting that the water content in the deep underground and even in the uppermost mantle increases in the period of the rainy and snowy seasons in Japanese islands and northern California.

On the other hand, the author proposed a different model that the annual to tri-annual change of the correlated seismicity, not local seismicity, is responsible for the periodic variation of the subduction velocity of the oceanic plate which is governed by the various slow slip events occurred along the plate subduction boundary. In this model, the aqueous fluid filling small open cracks is emitted in the coupled state and is absorbed in the decoupled state of the slab with hanging wall, corresponding to the slow slip and inter slow slip periods. Furthermore, it is noticed that the annual to tri-annual periodicity of the plate subduction seems to be consistent with the cyclic occurrence of non-volcanic tremors and slow slip events along on the subduction zones.

Until now, it may be considered that the plate motion does not keep the constant velocity in the time scale within 1 to 100 years, even though the long-term average velocity is estimated about 10 to 100 mm/y. Then, the author tried to estimate the plate convergent velocity between Tsukuba, Kanto area, and Tasmania, S. Australia using the data of very large base line interferometry (VLBI) from 2012 to 2022 in the previous book (Toriumi 2022). In Fig. 4.24, it shows that the relative plate velocity of them displays clearly periodic fluctuation, thereby indicating the cyclic change of plate subduction velocity, and the periodicity is about 1 year. It should be surprised that this periodicity is very similar to that of the annual variation of the length of day, that is the angular velocity of the earth as shown in the previous book (Toriumi 2022), suggesting that the correlated seismicity change should be strongly affected by the celestial mechanics of the earth.

In this chapter, the author intends to propose the celestial mechanics model of the global seismicity dynamics which is controlled by the fluid earth dynamics involving

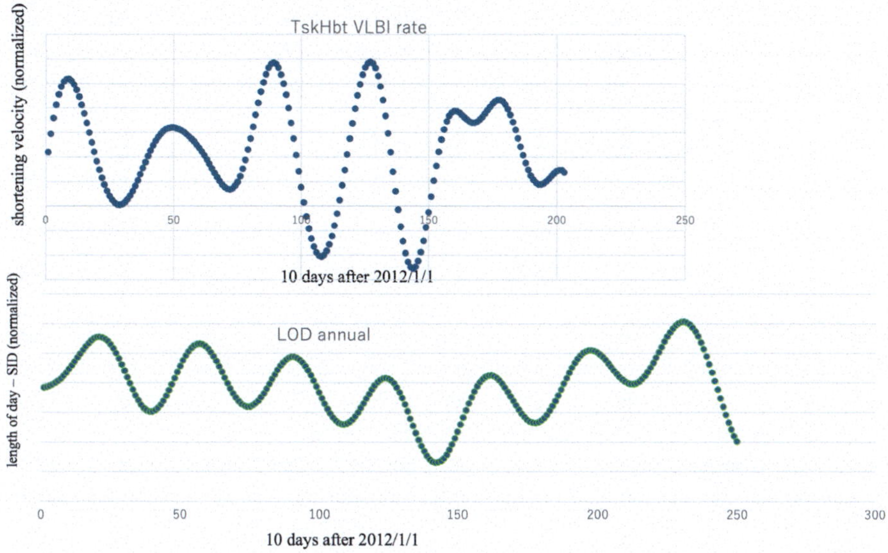

Fig. 4.24 Temporal variation of changing rate in the long distance between Tsukuba and Hobart stations of VLBI (**a**) and length of days (**b**; LOD, migration average) of the Earth from 2012/1/1. The variation in a shows about 400 days periodicity, but b does annual variation

spin-up and -down of earth rotation, elliptical change of earth shape by satellite gravity, plate motion, and sea-level change.

Let us start the summary of physics of the earth rotation and its moment of inertia. The earth is composed of atmosphere, ocean, crust, upper mantle, transition layer, lower mantle, core mantle boundary layer, outer core, and inner core. Among these layers, the fluid earth is represented mainly by atmosphere, ocean, and outer core, and they are possibly contributable to the short-term change of the inertia of moment of the earth. For the outer core is surrounded by solid boundary of the inner core and core-mantle boundary, the change of inner core rotation and the elliptic ratio of the whole mantle may change the total inertia moment of the earth. On the other hand, the total volume of oceanic water may vary in time by the climate change through melting and solidification of land and polar ice. In addition, the Chandler Wobble, that is the drift of rotation axis of the earth, is also possible contributor to the earth rotation change.

It is obvious that there are several modes of periodic component of the LOD timeseries (Fig. 4.24). Except for the secular variation over 10 years, it may be sure that many modes of contribution of the LOD variation should be responsible for atmospheric flow through change of angular momentum of the earth as,

$$M_e = M_{\mathrm{atom}} + M_{\mathrm{ocean}} + M_{\mathrm{core}} + M_{\mathrm{mantle}} + M_{\mathrm{crust}} \tag{4.3}$$

Now, we must look at the change of LOD, and then we ignore possibly the change of M_{ocean} and M_{core} as discussed by Jadin and Yamazaki (1995). Furthermore, it seems that the change in M_{mantle} may be the time scale over several ten years in the present situation of long-term plate tectonics framework, but in the later, we should look at the effect of short-term variable plate motion to the M_{core} and M_{mantle}.

According to the Jadin and Yamazaki, the LOD change (msec in day) represents as,

$$\Delta\text{LOD} = 1.68\Delta\text{AAM} \tag{4.4}$$

in which AAM (unit; 10^{26} kg m^2 s^{-1}) is the atmospheric angular moment of the earth, and constant term of 1.68 comes from the total angular moment coming from zonal mean wind of the atmosphere as illustrated in Fig. 4.25. The coherency between change in LOD and atmospheric angular moment change is very evident in the range of timescale from several days to one year as shown by Jadin and Yamazaki (1995), and later Courtillot et al. (2022). The results by Courtillot et al. manifested clearly that the annual and semiannual periodic change of LOD is just responsible for the change in distance between sun and the earth through the spatial–temporal variation of the ocean surface pressure, and then induced change of intensity of the mean zonal wind. Therefore, it seems that the secular variation of the LOD from 1990 to 2022 described previously is responsible for the long-range variation of the perturbation of earth orbit by celestial mechanics.

By the way, the change of LOD, that is change of angular velocity of the earth, should result in the change in ellipticity of the rotating earth, and the rate of its change is possibly estimated Levin et al. (2017) as,

$$\varepsilon = \omega^2 R_o / g_o \tag{4.5}$$

in which ε and ω are ellipticity and angular velocity of the earth, respectively. R_o and g_o are the radius and gravity of the equatorial. Then, the change of ellipticity is related to the angular velocity in the non-dimensional form as,

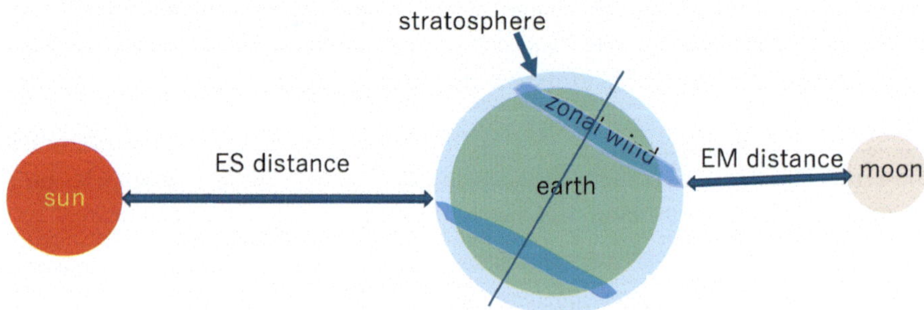

Fig. 4.25 Illustration of atmospheric zonal wind of the earth affecting the annual variation of earth rotation (LOD) together with the tidal effect from the moon

$$d\varepsilon/\varepsilon = 2d\omega/\omega. \tag{4.6}$$

For the change of LOD is related to the change of angular velocity as,

$$d\omega/\omega = -dLOD/LOD \sim 10^{-8} \tag{4.7}$$

Furthermore, kinetic energy of rotating earth holds,

$$E = I\omega^2/2 \tag{4.8}$$

in which I is the momentum inertia of the earth.

Then, the change of angular velocity should result in the change in the kinetic energy of the earth, indicating that the change of AAM due to the velocity change of the zonal wind of the celestial mechanics of Sun and Earth arises the kinetic energy of the earth itself.

The rate of kinetic energy change becomes as,

$$dE/E = 2d\omega/\omega + dI/I \tag{4.9}$$

Considering the constant density and its distribution of the earth, the change of kinetic energy is approximated to be (Levin et al. 2017),

$$dE = 2Ed\omega/\omega \sim 2 \times 10^{29} \times 10^{-8} = 2 \times 10^{21}\ \mathrm{J/year} \tag{4.10}$$

Considering the total moment magnitudes per year in the earth may be estimated around 10^{21-22} J, above value seems to be near the release rate of the seismic moment in a year (Levin et al. 2017). Therefore, it is possible that the annual and near annual variations of the global and regional correlated seismicity rates are derived from the kinetic energy change derived from the change in sun–earth orbital distance through AAM.

On the other hand, the interaction of the uppermost of the outer core and the lower mantle is potential to the contribution to the earth rotation by dipole change and toroidal convection of the outer core layer proposed by Yukutake (1973). He estimated the contribution of the outer core contribution is less than 1–2 microseconds/day, thereby suggesting much less than observed LOD annual change. However, the annual change in the earth rotation derived from atmospheric zonal winds and sun–earth distance change may affect much the toroidal convection system of the outer core, and it suggests that the rotation of the inner core may vary with annual change of the LOD by the change in the eddy viscosity of the outer core interior. It may be noticed that recent seismic study of the inner core suggests the change in relative rotation of the inner core against the lower mantle. Recently, by Gillet et al. (2010), it has been suggested that the periodic variation of the LOD is synchronized with the geostrophic zonal flow of the outer core which gives rise to distinctive variation of core magnetic field with 6 years periodicity.

On the other hand, Jadin and Yamazaki, and Courtillot et al. noticed that the long-term secular variation of the LOD is not resulted from the AAM, and it may be due to the momentum change of the solid earth interior. Toriumi (2022) in the previous book proposed that the secular change of LOD is tightly connected with the change of the global correlated seismicity rates of the plate boundary zones of subduction and spreading of the oceanic plate. Thus, it is necessary to be checked that the change of the angular velocity change, that is the change of LOD, can be compensated with the change of the moment of inertia of the whole earth in the condition of constant kinetic energy denoted previously.

Then, we have,

$$\mathrm{d}E/E = 2\mathrm{d}\omega/\omega + \mathrm{d}I/I = -2\mathrm{dLOD}/\mathrm{LOD} + \mathrm{d}I/I \tag{4.11}$$

The change of the moment of inertia due to average density change in the uppermost layer of the mantle holds,

$$\begin{aligned} I + \mathrm{d}I &= (8\pi/15)\rho(R - \mathrm{d}R)^5 + (2/3)R^2 \times 4\pi R^2 \mathrm{d}R(\rho + \mathrm{d}\rho) \\ &= (8\pi/15)\rho R^5(1 + 5(\mathrm{d}\rho/\rho)(\mathrm{d}R/R)) \\ &= I(1 + 5(\mathrm{d}\rho/\rho)(\mathrm{d}R/R)) \end{aligned} \tag{4.12}$$

in which $\mathrm{d}\rho$ and dR *are* density difference due to plate subduction and obduction, and the thickness of the density change by plate motion per unit duration in the uppermost layer of the mantle, respectively.

Thus, we obtain,

$$\mathrm{d}I/I = 5(\mathrm{d}\rho/\rho)(\mathrm{d}R/R) \tag{4.13}$$

From above equation, angular velocity change becomes,

$$\begin{aligned} \mathrm{d}\omega/\omega &= -(1/2)\mathrm{d}I/I \\ &= -(5/2)(\mathrm{d}\rho/\rho)(\mathrm{d}R/R) \end{aligned} \tag{4.14}$$

Therefore, it may conclude that small amount of density decrease near the shallow solid earth should increase the angular velocity, that is decrease of LOD, and increase of density should decrease the angular velocity in the constant depth of density change by plate motion.

Let us consider the cold subduction of the oceanic plate, and the density of the subduction region becomes large slightly, but in the ridge region it should become small very slightly. It is available that the thickness of the oceanic plate may correspond to the dR in the above equation, and the density change of this depth range of the mantle is probably estimated from the volume ratio of the newly subducted and formed oceanic plate in the surficial mantle, by using the volume expansion rate of olivine. As stated in the previous sections, the short-term (1 min) plate subduction like the Sumatra giant earthquake (M9.2) did affect the earth rotation change within

10^{-11} order of $\delta\omega/\omega$, which means the order of microseconds in LOD change, and this value is much smaller than that of the annual and secular change of the LOD to be about milliseconds.

On the contrary, it may be possible to be considered that the earth spin-up and -down forced externally by the celestial and atmospheric drag should affect the ellipticity of the earth by change in centrifugal force of the earth, and the elastic strain energy by shrinkage and bulging of the equatorial zone. Moreover, it seems that the elastic plates of the surficial earth should be deformed elastically and plastically, and thus the plate subduction zone may store the strain energy to activate the seismicity. It should be surprising that the strain energy estimated from the angular velocity change about 10^{-8} may be much larger than that of the giant earthquake about 10^{-11} denoted previously.

By the way, the elastic strain by externally exited earth rotation is not totally transferred to the plastic strain, but the elastic stress partially contributes to the growth and nucleation of the cracks and shear slip activation on small faults. The ratio of plastic strain energy against the total strain energy, which is equal to the kinetic energy change of angular momentum of the earth, is possibly inferred from the density and velocity of shear slip of the preexisting small shear cracks of type II and III. As discussed in the previous book, the low velocity motion of the shear crack occurs even in the small and repeated stress conditions (Wei 2010). This type of plastic deformation of mineral and rock is responsible for the dislocation multiplication and its hydroxyl reaction at the crack tip region reducing the resistance force against dislocation motion.

Here, assuming Maxwell matter that the energy dissipation by micro plastic deformation due to micro shear cracks and dislocation multiplication by applied stress increment, $\delta\sigma$. The dissipation energy, w_d in the duration of time, T by plastic deformation holds,

$$w_d = 1/2\delta\sigma T \mathrm{d}\gamma/\mathrm{d}t \tag{4.15}$$

in which $\mathrm{d}\gamma/\mathrm{d}t$ is the strain rate. On the other hand, elastic strain energy, w_e hold,

$$w_e = 1/2\delta\sigma^2/E \tag{4.16}$$

in which E is the elastic modulus. Then, the ratio, R of plastic dissipation and elastic energy becomes,

$$R = w_d/w_e \sim (E/\delta\sigma)T\mathrm{d}\gamma/\mathrm{d}t \tag{4.17}$$

Thus, we obtain $R \sim 10^{-2}$, considering $T(\mathrm{d}\gamma/\mathrm{d}t)$, and $E/\sigma \sim 10^4$.

Therefore, the plastic strain energy must be stored in the earth interior together with the plate interior of the surficial earth.

The change of the earth rotation excited by celestial mechanics effect through atmospheric zonal wind seems to effect on the short-term plate motion, especially on the seismicity of the plate boundary zone. The elastic strain energy together with an elastic one distributes to the surficial major plates having oceanic subduction boundary, Pacific Plate, Australian Plate, and Cocos, and Philippine Plate. These giant plates have different position and migration vector on the spherical earth surficial layer: Pacific Plate situates in the wide region spreading from southern hemisphere to northern hemisphere, and Australian Plate and Cocos Plate do the region from southern hemisphere to equatorial region. Even if the homogeneous elastic strain energy in these large plates is possibly derived from the rotation change of the earth, the absolute elongation along the longitudinal orientation of them takes different length according to the NS extension of the plate. On the contrary, the different elastic elongation along the EW orientation of the plate should be generated. These different strain in the orientation may be concentrated in the subduction zones of the plate.

Therefore, it seems that both of annual and semiannual variation of earth rotation and its rates should bring the subduction velocity change and resultant seismicity rate of the subduction zone at the northern and southern ends of the subduction boundary. These conditions are also satisfied in the eastern and western ends of the subduction zone. Furthermore, considering the relative plate subduction velocity is kinematically controlled by the rotation of the plate with its own axis, and thus the different vectors of the plate motion by own plate rotation and by the earth rotation may cause the change in subduction orientation at the trench axis, that result in the rotation of the plate with axis vertical to the plate.

By the way, it is suggested in the previous chapters that the global seismicity network dynamics controls the secular variation of the global correlated seismicity which is strongly related to the earth rotation change. Thus, it seems that the above-mentioned process of the subduction boundary zone of the giant plates should govern the seismicity network transition in the global scale, and the secular variation of the global correlated seismicity rate.

By the way, the mechanical activity of the global seismicity network is possibly represented by the slider-block model of the characteristic geometry of the linearly connected subduction zone blocks terminated with mid oceanic ridges (Mid Atlantic Ridge and East Pacific Rise) as shown in Fig. 4.26, that is the three-arms network (TAN) centered at near the point west of Papua New Guinea. The present TAN model is slightly different from the previous circular network model in the previous book (Toriumi 2022), for the New Zealand and Chile subduction zones do not connected direction by the subduction zone, but they are terminated to the mid oceanic ridges north of the Antarctica Plate. In this geometry, it is a very interest that the temporal variation of the global correlated displacement excited from the temporal periodic change of the ellipticity of the earth by rate of LOD, that is earth spin-up and spin-down.

In this study, the behavior of the simple slider-block model of the above geometry was simulated by R under the RStudio platform in the following assumptions.

1. The unit block is defined to be $100 \times 100 \times 10$ km size.
2. Each unit block is connected by spring side by side.
3. Each unit block relates to overriding arc block by spring or dash pod.

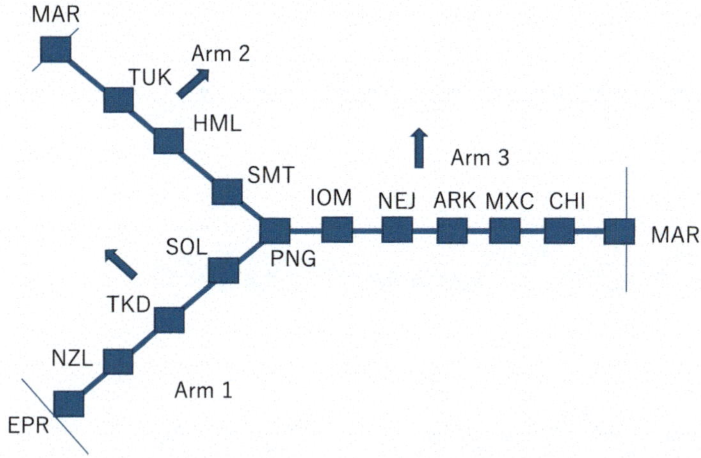

Fig. 4.26 A model of global subduction system composed with three arms: one is from the Papua New Guinea (PNG) to New Zealand (NZL), second is from PNG to Turkey, and third from PNG to Chile (CHI). These arms terminate Mid Atlantic Ridge (MAR) and East Pacific Rise (EPR). Other abbreviations are noted in text

4. Initial displacement and velocity of each unit block is given randomly by Gaussian distribution.
5. Doubly periodic external stress (corresponding to the annual and semiannual variations) gives on unit block.
6. Newtonian mechanics is given for each unit block under various parameters of spring and dash pod.

After simulation, the global correlated displacement of the TAN system was estimated by the singular value decomposition (SVD) of the data matrix, and temporal change of the correlated displacement associated with the periodic external stress is shown in Fig. 4.26. It shows that the temporal change of the block displacement appears scarcely periodic with annual and semiannual variation, and the relative displacement of each block in the three arms does random variation (Fig. 4.27).

It should be noted that the phase map of global correlated displacement of the three arms appears three cycles with three-fold symmetrical pattern of z_1 and z_2 diagram as shown in Fig. 4.28. It is, furthermore, very impressive that the global correlated displacement of the block shows temporal variation of the semiannual and long-term weak periodicity as shown in Fig. 4.29, thereby suggesting that the effect of the periodic external stress operated on the whole network units is obvious but not straightforward. In addition, it seems that the relations between the global correlated displacement and the external periodic stress on the TAN displays the variously repeating maps of the displacement against external stress as shown by Fig. 4.29. It seems, however, to be pointed out that there are three high probability areas in the map of them, thereby suggesting two stable nodes of the low and high

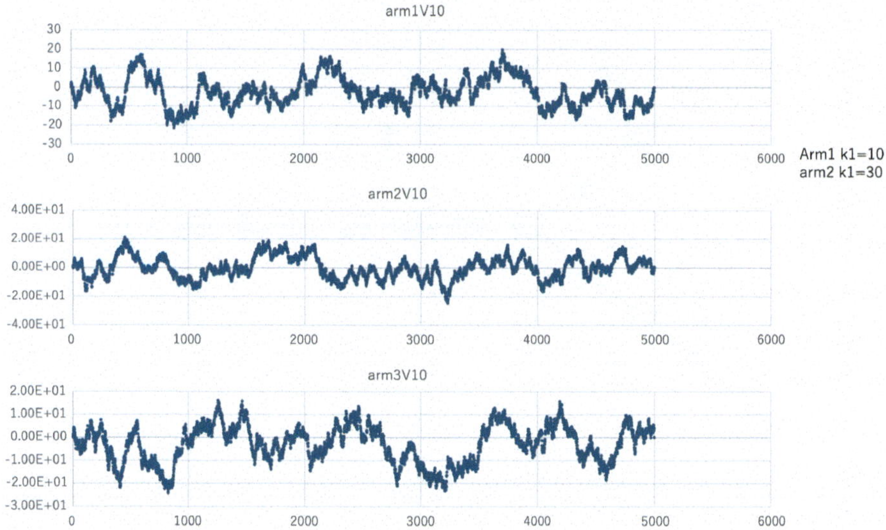

Fig. 4.27 Temporal variation of simulated displacement of 10th nodes in the three arms, showing noisy periodic patterns. Elastic modulus of the arm 2 is larger than that of other arms

correlated displacement in the high external stress regime, and the neutral correlated displacement in the neutral external stress conditions.

4.4 Physical Model of Non-Volcanic Tremor and Very Low Frequency Earthquake

There are many reports and discussions on occurrences and origin of non-volcanic tremor (NVT) in the plate boundary zones after the finding in the SW Japan by Obara (2002). Recent discussions on origin of NVTs are concentrated in the actual process related to the low frequency earthquakes and apparently followed by NVTs (Uchida et al. 2015). Otherwise, it is noteworthy that the excellent denoised waveforms of the NVTs have been clarified to be the clearly modulated waveform continuing in the rather long periods about several days (Hulbert et al. 2020), thereby suggesting that the slow slip or low frequency earthquake event excited the large number of single NVT wavelets, and thus the observed modulation of them are something like small wave sources along on the shear slip plane of low frequency earthquake. These evidence of the NVT, and their occurrences strongly supports the phenomenological model of the NVT derived from LFE proposed by Ide (2008).

On the other hand, it has been shown that the shear slip plane formed by slow slip velocity appears abundant jog by many types of Riedel shear and penetrative shear planes, and wavy interface of the shear plane by Kelvin–Helmholtz instability in the previous chapter. It is very important that the spacing and wavelength of the jog

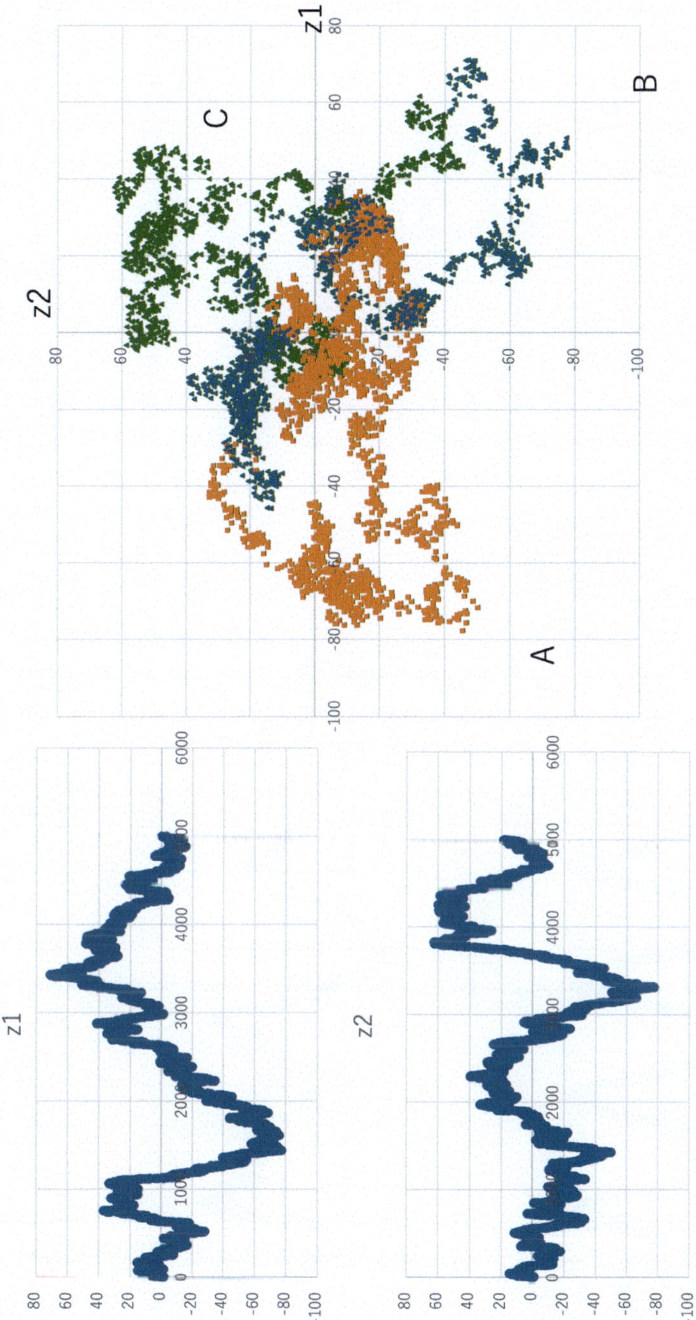

Fig. 4.28 Global correlated displacement of simulated three-arm slider blocks, z_1 and z_2 showing repeated cyclic patterns (left), and phase diagram of them displaying tri-symmetric patterns (right)

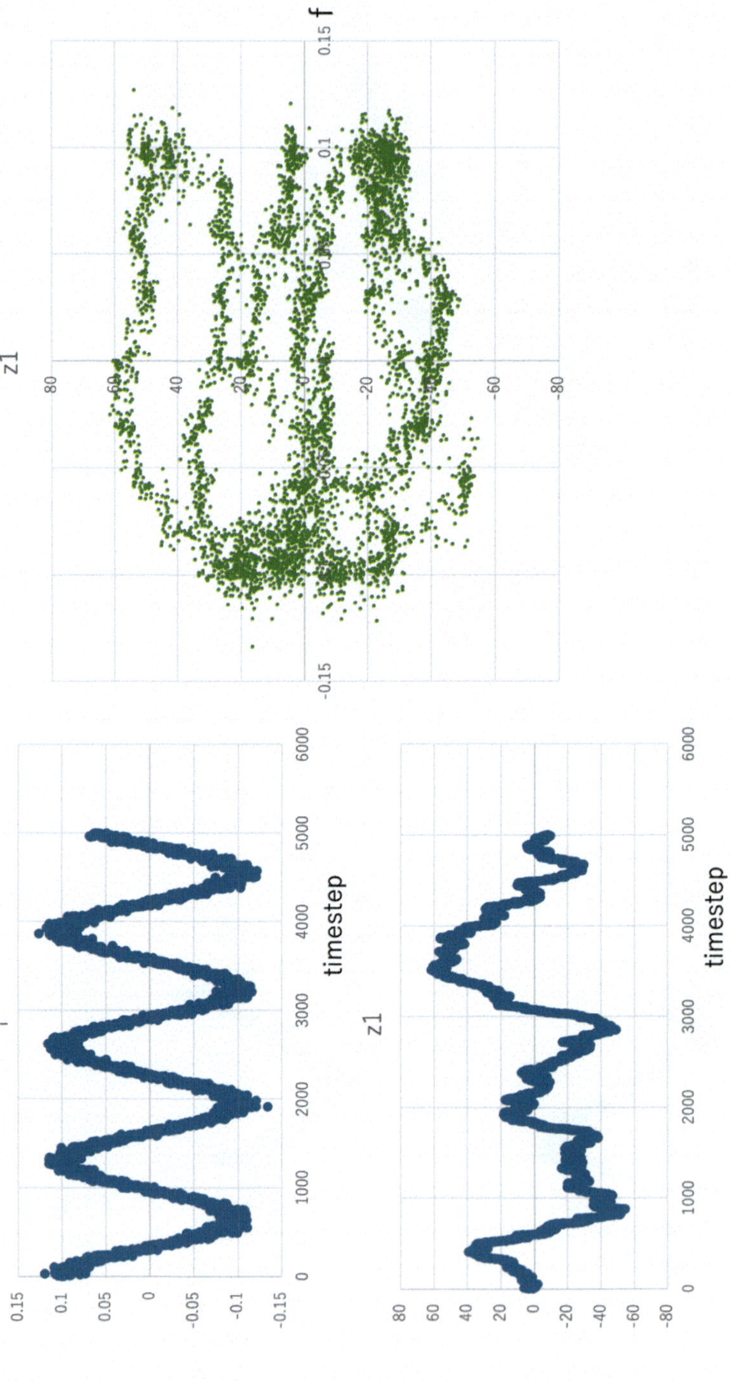

Fig. 4.29 Temporal variations of global correlated displacement (z_1) and given external periodic force with random noise (f). Right figure is phase diagram of f and z_1

are obviously systematic, and they seem to be dependent on the shear slip velocity. The wavy and jog-rich slip interface should physically like the vamp road, and the hanging and foot wall mass should be suffered from cyclic stress during the shear slip motion at the periodic jog like vamp, as shown in Fig. 4.30.

Therefore, in the section, the author intends to propose a physical model of the generation of non-volcanic tremor occurring commonly of the plate subduction boundary. And waveform of the non-volcanic tremor will be compared with the synthesized waveform emitted from the periodic jog along the shear slip interface dependent on velocity.

Let us assume the geometry of the jogged interface of the shear slip surface in the deep-seated rock as shown in Fig. 4.31. The jog wave has wavelength, λ, and amplitude, a, and the spacing of the jog wave is d on the one-dimension model.

Furthermore, it will be assumed that the generated seismic wave emitted from jog rupturing is the simple oscillator as,

$$u = u_o \sin(kx - ft) \tag{4.18}$$

where k is the wave number and f is the frequency.

Consider that the frequency of the simple oscillator should be equal to the ratio of the rupture velocity against the length of the jog,

$$f = v_r / l \tag{4.19}$$

On the other hand, at the observatory of the seismic wave far from the slip plane, the oscillation of the ground is represented by the superposition of seismic waves

Fig. 4.30 Photographs of outcrop showing the periodic jog with kink on the schistosity surface of the low-grade Sanbagawa quartzose schists in Nagatoro, Kanto Mts. Jogs extend sub-normal to stretching lineation (shear orientation) on locally curved (kinked) shear surface, showing periodic distribution with about 1 m distance. The zoom photograph (left bottom) displays the jog space on curved (kinked) schistosity surface filled partly by quartz

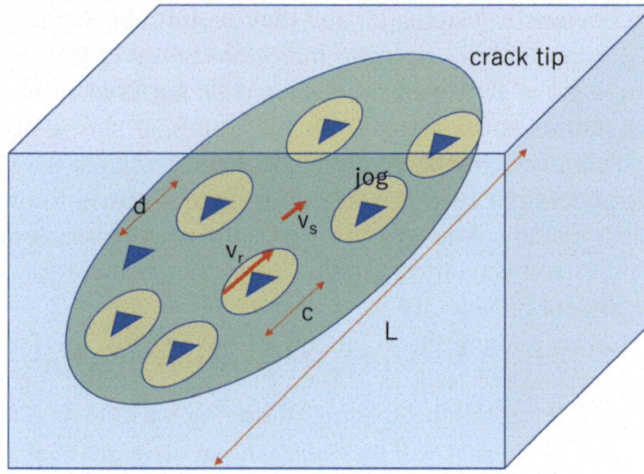

Fig. 4.31 Illustration of a model of periodic jog on shear crack surface. *d*; spacing of jog, *c*; width (wavelength) of jog, v_r; rupture velocity of jog, v_S; slip velocity of shear crack, *L*; length of shear crack

emitted from many jogs along on the single shear slip plane. Therefore, it holds,

$$u_t = \sum u_{o,i} \sin(f(t + id/v_s) + k_i x) \quad \text{for}(i = 1 \sim m) \tag{4.20}$$

in which *m* is the total number of the jog, and in the case that total length of slip plane is *L* and spacing of the jog is *d*, and thus m is equal to the $L/(d + \lambda)$. Then, the phase of the wave from each jog is *id*, and it is,

$$k_i x = id \tag{4.21}$$

Thus, it will be shown that the wave superposition of the simple oscillators from the many jogs at the distant location appears various types of modulation and resonance of their waves. In this work, the author simulated the above superposition of many waves from jogs on the shear slip plane with crack tip velocity to be v_s, and jog size λ and spacing, *d* at given number *m*, as shown in Fig. 4.32. In addition, the length of the jog is assumed to have a frequency distribution of Gaussian type, and thus the frequency of the simple oscillator behaves to be,

$$f = v_r / \left(\lambda + N\left(0, \sigma^2\right)\right) \tag{4.22}$$

In Fig. 4.33, it shows that there are several types of the modulation and resonance of the wave superposition by the difference of *f* and *kx* of the jog geometry: one is the simple modulation clusters type, the second is the simple modulation with intermittent quiet periods, and the third is the random type of modulation. The third type modulation is derived from the Gaussian distribution of jog spacing *d*, but if the constant *d* is the case, the modulation of the elastic wave is apparently of simple type and intermittent type.

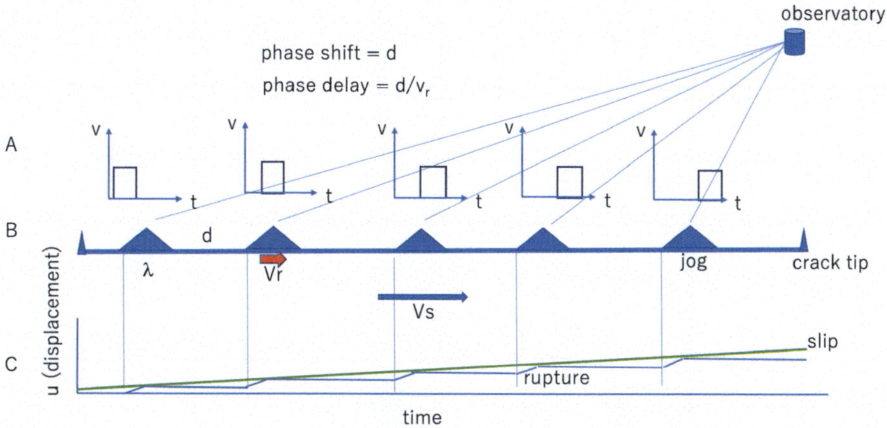

Fig. 4.32 Schematic relations of seismic model derived from bumped (jogged) shear crack slip velocity v_S, and rupture velocity of bump v_r, and wave superposition at observatory. The seismic wave velocity assumes box type against t as shown in (**a**), and displacement with time advance by shear crack motion is displayed in (**c**)

In this study, the wave propagated by the active shear crack sliding across the arrayed jogs on the shear plane is obviously modulated depending on the jog geometry. Thus, it needs that what are the controlling factors of the jog geometry on the shear slip interface. Especially, it should be investigated that how the shear slip velocity governs the wavelength and shear crack jog spacing, and their distribution density function in the highly sheared metamorphic schists.

As described in the previous chapters, it is shown that the monomineralic bands of the highly sheared schists are developed abundantly, and that the sealed jogs occupied by the banding minerals are almost always formed by Riedel and penetrative secondary shear displacement crossing the main shear plane. In addition, it is clarified that the waveform jogs having 1 to 100 mm wavelength are commonly found in the same schists, and those of wavy jogs sealed by albite and chlorite seems to be derived from the Kelvin–Helmholtz instability of the plastic shear deformation across the neutral plane. Otherwise, in the case of the wavy sealed jogs on the simple case, it should be important that the wavelength and wave spacing are obviously equivalent with each other, being likely to the sine function, and thus, the wavy jog pattern is possibly approximated to be due to the Kelvin–Helmholtz instability mentioned above.

By the way, it is universal that the shear jog slip interface should have arrayed jogs with periodic and characteristic length depending on the shear jog slip velocity. In the case of shear jog slip even in the case of slippage within the ductile thin layer, it is clarified that the wavelength and spacing of the wavy interface is determined by the KH instability. So far as the shear slip interface has such arrayed jog (likely to bump in road), the shear jog slip with various velocity should manifest the various peak frequency of the seismic waves.

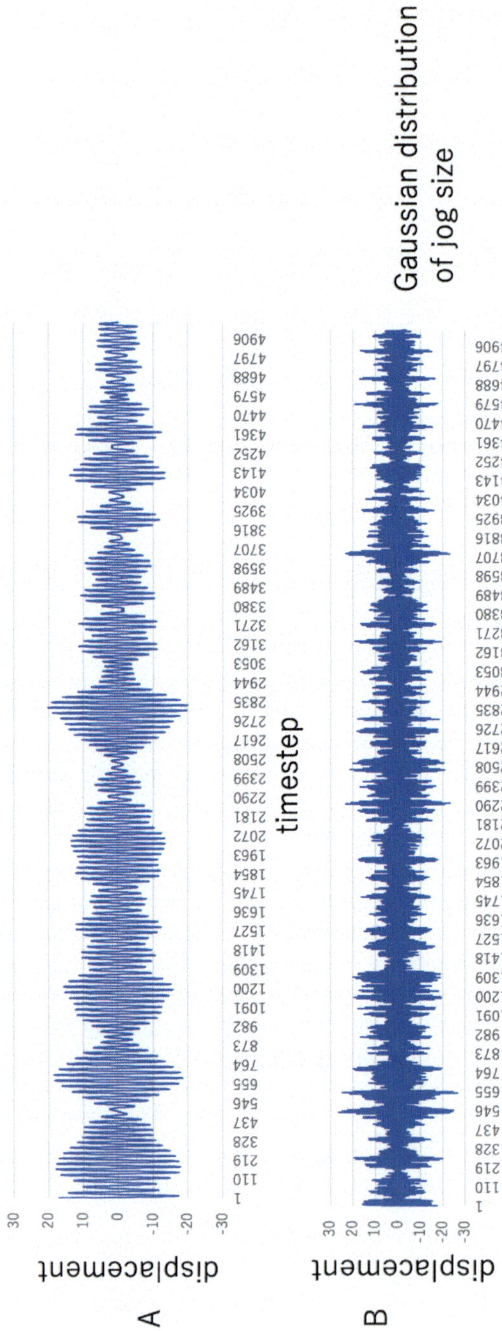

Fig. 4.33 Modulation of elastic wave generated from the arrayed jog on the shear slip surface. **a**; modulation by periodic arrayed jog, **b**; modulation by periodic arrayed jog with random noise

It is very strange that the very thick monomineralic bands and arrayed lenses in the low to high-grade zones of the plate boundary metamorphic belts as described in the previous sections. In the case of basic schists, the bands consist of various shape and grain size of albite and small amounts of quartz and calcite. Considering that there is common occurrence of large bands even in the low-temperature zone, it suggests that the common structures of these large bands and lenses seem to be derived from the long-term and repeating jog-enlargement in the process of syn-metamorphic brittle shear deformation. It may mean that the thick bands and lenses seems to be associated with the long-lived shear slip interface, and with metasomatic bands consisting with chlorite and with epidote outward to the matrix side. The sequence of thick bands may be derived from the sequential precipitation of albite–chlorite–epidote–(actinolite) as discussed in the previous section.

The thick bands seem to be derived from the moving arrayed jogs along the shear slip interface by shear slip motion, and it is still unknown whether thick bands are formed by moving thick arrayed jogs or the thin but multiple stage jog motion. However, looking the fact that there is inhomogeneous pattern of grain size distribution of constituent albite that displays several thin layers of fine grains sandwiched by the coarse grained and medium grained layers as shown in Fig. 4.34, it may be probable that the formation of the thick bands is responsible for the multiple motion of successive arrayed jogs.

Fig. 4.34 Photograph of albite aggregate filling shear crack jog showing multiple layering of fine-grained albite

On the other hand, as there is common occurrence of thin albite bands formed by single motion of arrayed jogs, the thick bands seem surely to be derived from the continuous localization of the jog formation on the single slip plane as simulated in the previous sections about the metasomatic reaction with formation of jog by dissolution mechanism, thereby suggesting the positive feedback mechanism of jog formation by dissolution of matrix minerals, and by cross slip with main shear slip motion. It is sure that from Fig. 4.35, the Riedel shear and penetrative slip shear planes are commonly associated with main shear slip around the thick band of albite in greenschist. Thus, this fact manifests that the flux of the aqueous fluid was concentrated around the shear crack jogs, and then the mineral reactions involving aqueous fluid were active around the jog region during the metamorphism. Furthermore, it should be noteworthy that the mineral reactions and related relaxation of the strain energy and increase of interfacial free energy make the shear crack toughness and critical length for crack propagation (rapid mode of slip velocity) to increase as discussed in the previous section.

Matsuzawa et al. (2004) has clarified that some of earthquakes which occurred near the subducting plate boundary in the northeast Japan trench region display the very similar waveform and source parameters, and thus he concluded that these earthquakes occurred repeatedly at the same position on the plate boundary. They should be called as the repeated earthquake that is the repeater, by Matsuzawa et al. (2004). The repeaters show the regularly periodic rapid seismic slip in the various scales, that

Fig. 4.35 Photograph of outcrop of the shear crack jog filled with albite aggregate showing the Riedel type shear slip (red arrow) associated with main shear crack

is various magnitude, along on the subduction plate boundary, of periodicity from 2 to 5 years in the cases of M3 to M5 (Matsuzawa et al. 2004), thereby suggesting that the repeaters represent the mode transition from growth to propagation of shear crack, which is probably responsible for the accumulation of fluid in shear plane, or of stress accumulation after recovery by shear crack sealing of minerals. However, it may be strange that the repeaters have the same extension and same periodicity of shear crack at the same location on the slab boundary, thereby manifesting the dynamic balancing of sealing recovery and stress accumulation of sealed shear crack region which is controlled surely by the flux of the aqueous fluid through associated open crack and jog.

Next, let us discuss the relation of the timescale of recovery by sealing and fluid flux into the given size of the repeater. The repeater occurs on the plate boundary, and the geometry is assumed to be a square of length, L. The density of small shear crack jog on the repeater area is assumed to be N_j, which is enlarged by shear crack growth, and the jog on shear crack interface are absorber of fluid. As discussed in the previous chapters, the aqueous fluid flowing in the jog makes them to grow over the critical size of Griffith crack, and their jog tip velocity seems to be proportional to rupture velocity v_r, and thus, we obtain,

$$dc/dt = v_s \tag{4.23}$$

where c is the length of shear crack, and K is the stress intensification factor.

On the other hand, the critical condition of the percolation of large number of small shear cracks of growing states by influx of aqueous fluid and differential stress condition is given by critical index to be $\phi = 0.6$, and then we obtain,

$$\phi = N(r_o + c)^2/L^2 \tag{4.24}$$

and,

$$c = v_s t_c \tag{4.25}$$

in which t_c is the critical time for the small crack percolation.

Therefore, it holds,

$$\phi = N r_o^2 (1 + v_s t_c / r_o)^2 / L^2 \tag{4.26}$$

Then, for $v_s\, t_c/r_o >> 1$, and $\left(\phi/Nr_o^2\right)^{1/2} = F$, we obtain,

$$(v_s t_c)/L = F \tag{4.27}$$

Therefore, it holds,

$$\log t_c = \log L - \log v_s + \text{cc}.$$

$$\log t_c = 0.5M - \log v_s + \text{cc}. \tag{4.28}$$

It has been clarified that the magnitude of repeaters is related to the logarithmic time intervals of them summarized by Uchida (2019) as shown in Fig. 4.36, but considering the effect of the differential stress operated on the shear slip plane, the relation should be modified by the stress, as shown in Fig. 4.36, thereby manifesting the availability of the present model of the repeater earthquakes of wide range of magnitude from 1 to 5. Otherwise, it is quite important that the repeaters after 2011 Tohoku-Oki giant earthquake displays the distinctive short intervals of them, but they display the gradual increase with advancing the time, suggesting that the differential stress conditions rapidly increase after the giant earthquake, and it gradually decreases as being formulated in the above equations.

By the way, it has been found in many decades that there are various corner frequencies of the earthquakes ranging from M1 to M7 and they are obviously proportional to the cubic magnitude of the earthquakes (Hiramatsu et al. 2002). Considering that the moment magnitude of the earthquakes is represented by,

$$\begin{aligned} \text{Mo} &= \mu S\, \text{d}u \\ &= \mu L^2\, \text{d}u \end{aligned} \tag{4.29}$$

which S is the coseismic slip area, du is the coseismic slip, and μ is the rigidity of the surrounding rock.

On the other hand, the seismic corner frequency is given by,

$$f_c = v_r/\lambda \sim v_r/d \tag{4.30}$$

in which v_r is the rupture velocity of the earthquake slip, d is the jog spacing, and λ is the length of jog.

By the way, the representative displacement, du by the shear crack is related to the length of the present shear crack at the critical condition against the propagation as,

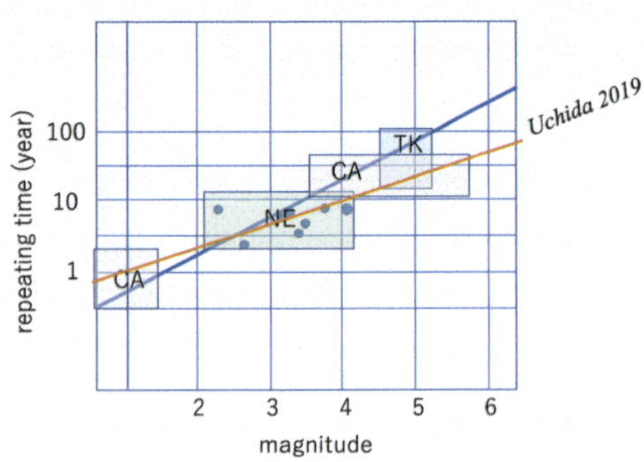

Fig. 4.36 Relation of repeating time and magnitude of the repeating earthquakes (repeater) from Cascadia (CA), Tonga-Kermadic (TK) and northeast Japan (NE) modified from Uchida (2019; red line). Blue line; this model

$$du = \tau L/2\mu \tag{4.31}$$

in which τ is the shear stress acting on the shear crack.

Thus, we obtain,

$$\begin{aligned} \mathrm{Mo} &= \tau L^3/2 \\ L &= nd \end{aligned} \tag{4.32}$$

Therefore, combining the relationship between the slip size of shear crack and the corner frequency f_c,

$$\log \mathrm{Mo} = -3 \log f_c + \log \tau + 3 \log v_r + 3 \log n + \mathrm{cc}. \tag{4.33}$$

This equation is just equivalent with the empirical relation of the seismic moment and the corner frequency proposed by Hiramatsu et al. (2002).

On the other hand, the abundant aqueous fluid lenses in the grain boundaries and shear crack jogs give several mechano-chemical processes in the rocks: the one is the formation of micro-open crack clusters between shear slip plane and jog formation by cross slip, the second is the metasomatic reaction around the crack tip and jog, and the third is aqueous fluid flux into the open crack clusters and jog. Next, the author would like to propose a simple model for self-competing process of the shear crack growth and aqueous fluid flow, reaching to the large shear slip events.

First, let us assume the basic mechanism of the clusters of open microcracks beside the multiple shears slip planes. As discussed in the previous chapter on the microcrack cluster, the spacing of microcracks of type *I*, ds, is represented by the spacing of shear slip plane, d, as (Yin 2010),

$$ds = cd.(c = 1 \sim 0.6) \tag{4.34}$$

Furthermore, the density of the shear crack, ρ is,

$$\begin{aligned} \rho &= nl^2/L^3 \\ L/n &= d. \end{aligned} \tag{4.35}$$

Here, l is the length of shear crack.

And the density of micro-open crack, m becomes,

$$\begin{aligned} m &= (nl/ds)dl/L^3 \\ &= \rho/c \end{aligned} \tag{4.36}$$

in which n is number of the shear slip plane with length of l per given reference volume L^3, and c is the constant (~1).

From above relation of n and l, we have,

$$\begin{aligned} \mathrm{d}\rho &= (l^2/V)\mathrm{d}n + 2(n/V)l\mathrm{d}l \\ V &= L^3 \end{aligned} \tag{4.37}$$

On the other hand, the aqueous fluid flows into the region of the microcrack clusters around the shear slip region, and the fluid flux, q becomes,

$$q = k_o m a^3 \nabla P_\mathrm{f} \tag{4.38}$$

where k is the permeability that is given by $k = k_o m a^3$. Here a is the aperture of the single microcrack. Furthermore, the pressure gradient of the fluid should be represented by the spatial gradient of the low-pressure void volume by microcrack density, and then it should become,

$$\nabla P_\mathrm{f} = k \nabla m (k \text{ is constant term}). \tag{4.39}$$

On the other hand, the increasing rate of the shear crack length is governed by the growth rate of the shear crack under aqueous fluid, and thus, it becomes (Wei 2010),

$$\partial l/\partial t = Kql\sigma^2 \tag{4.40}$$

in which σ is the applied stress, and K is the constant term. On the other hand, the density of the shear crack should be governed by the crack growth from the very small crack nuclei at the grain boundary, and it is, therefore, controlled by the fluid flux in the present region as,

$$\begin{aligned} \mathrm{d}n/\mathrm{d}t &= \beta q (\beta \text{ is a constant}) \\ &= \beta k_o m a^3 \nabla m \end{aligned} \tag{4.41}$$

Therefore, we obtain,

$$\partial\rho/\partial t = (l^2/V) k_o a^3 \beta \rho \nabla \rho + 2(\rho V/l^2) K k_o a^3 \sigma^2 l^2 \rho \nabla \rho \tag{4.42}$$

Finally, the controlling equations are summarized as,

$$\begin{aligned} &\partial\rho/\partial t = A l^2 \rho \nabla \rho + B \sigma^2 \rho^2 \nabla \rho \\ &\partial l/\partial t = (C\sigma^2) l \rho \nabla \rho \\ &A = k_o a^3 \beta / V,\ B = 2KVk_o a^3,\ C = Kk_o a^3 \end{aligned} \tag{4.43}$$

In these equations, A, B, and C are the constant terms.

These are the master equations of the shear slip (crack) density and shear slip (crack) length controlled by fluid flow through microcrack clusters. These differential equations manifest that the shear slip density and shear slip length increases with time. From Fig. 4.37, it is clarified that the high-density band of shear crack appears in the characteristic distance, suggesting that the shear band and related monomineralic bands discussed in the previous chapter should develop in the process of shear crack-fluid flow through microcrack clusters.

Next, we would like to show the evolution of the shear slip plane magnification and elongation during applied stress and aqueous fluid flow conditions. The simulation of those evolutionary tracks is given in the Fig. 4.38, thereby showing the rapid increase and plateau of the shear slip density and length. It is important that the rapid increase of those density and length of the shear crack should be accompanied with shear decompression brecciation of rocks around the shear slip plane because of accompanied rapid increase of open microcrack cluster.

In the previous book, the author has proposed that there are two modes of extrusion mechanism of the metamorphic mass: one is the slow extrusion and exhumation controlled by viscous drag and buoyancy force, and the other is the rapid mode of extrusion governed by the velocity-independent drag by frictional extrusion of hydrated faults from the deep crust and upper mantle. Later, in this book, it has

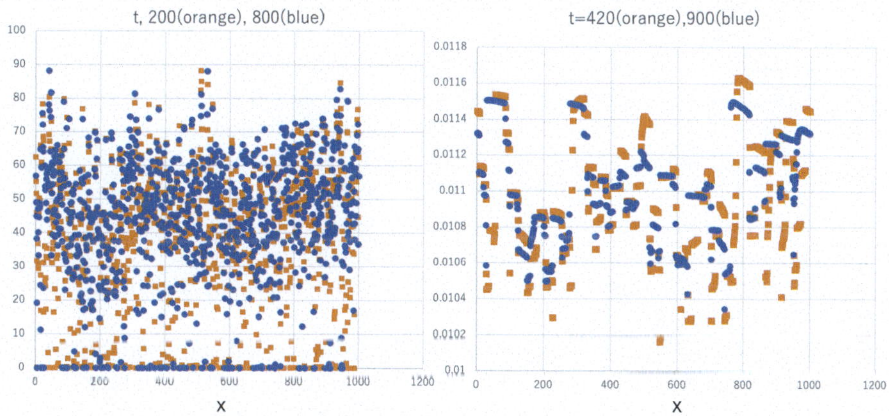

Fig. 4.37 Simulation results of length (left, $t = 200$ and 800 steps), and density ($t = 420$ and 900 steps) of shear crack against position x

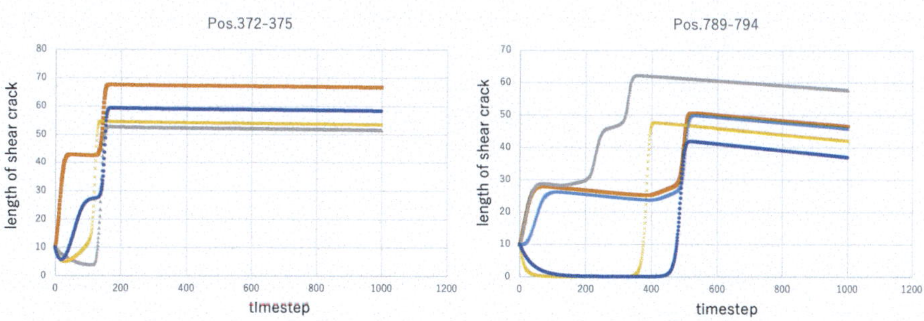

Fig. 4.38 Evolutionary tracks of the length of shear crack for various positions x

shown that the additional condition of the regional extensional stress environment of the upper mantle and the deep crust is needed for actual extrusion of the deep-seated metamorphic mass.

When Obara (2002) has found the non-volcanic tremors in the north of the Nankai Trough of SW Japan, he noticed that the NVT clusters apparently migrates from western Shikoku to the Kii Peninsula within one month. On the other hand, recent studies of NVT and repeating earthquakes during the post-seismic slip after 2011 Tohoku-Oki earthquake clarified that the NVT clusters occurred being associated with generation of very low frequency repeating earthquakes during the after-slip of giant earthquake (Takahashi et al. 2022), and furthermore, it seems that the NVTs took place during rather long period before and after the repeater. Therefore, they concluded that the NVTs are strongly related to the very low frequency earthquakes including the repeaters, and that the spectra of NVTs and related repeaters show very similar pattern with amplitude peak at 2–4 Hz.

Otherwise, it shows that the periodicity of the repeaters is probably derived from the percolation transition at the threshold of the regular distribution of shear crack jogs as discussed in the previous sections, and the common spectra of them are responsible for the up and down motion of the rock masses at those jogs on the shear slip plane. The regional area of the repeater is ranged in the extension of the NVTs distribution. It is probable that the repeater area should migrate by moving its tip front, which is governed by jog kink, and then the migration may take place by moving the kink front of the jog on repeater front as displayed in Fig. 4.39. In the case of the NVTs cluster along the Nankai Trough of SW Japan, the direction of the NVTs cluster migration should occur toward the eastward and westward both. The kink mode of migration mentioned above is checked by the array of NVTs clusters along the direction of NS, thereby suggesting the well-consistency with observations of NVTs by Obara (2002).

On the other hand, it is expected from the model presented above that the large plate boundary earthquakes should be associated with the NVTs clusters due to the oscillation by jog vamping of the hanging wall rocks of the accretionary prism. Although, until now it seems that there is nothing about the distinctive precursors that involve the NVTs along the slip plane of the large earthquakes. However, it shows that there are NVTs clusters around the slow slip event source along on the plate boundary, as exemplified in Nankai Trough (Araki et al. 2017), thereby suggesting that the NVTs and SSE are strongly correlated with each other in the basic process of them. Therefore, it should be needed to investigate that whether the NVTs events just along on the source area of the later large earthquakes having low frequency and regular earthquakes were taking place or not.

In this book, the author intends to propose a prediction model of the large and giant earthquakes by means of deep machine learning using neural network modeling with multiple labeling. To obtain the accurate probability prediction of the large earthquakes, it should be needed that there would like to be enough large amounts of seismic data in the target regions and very long timeseries of seismic activities surrounding the regions. As being well-known, the large earthquakes, which occur mainly surrounding the Pacific Ocean along the subduction plate boundary, are rare

Fig. 4.39 Photograph of kink structure of jog on shear crack surface; kink of jog filled with quartz in low-grade quartzose schist in Nagatoro, Kanto Mts

and short-term event, and bring large-scale disaster against human community and environmental system during long periods. The giant earthquakes over M8 occur in the about several hundred years interval without known precursors until present. However, as far as the large and giant earthquakes emit the huge amount of seismic energy in the very short duration, attaining the 10^{19} J/s, the physical process, that governs the large earthquakes should be undertook before the events, and thus the process should be formulated by several basic but complicated mechanisms.

In the previous sections, it seems to be clear that the repeating large earthquakes are associated with the non-volcanic tremor with low frequency during and before the events. The author has proposed the model that the repeating low frequency earthquakes and non-volcanic tremors are mutually dependent on each other, and thus the non-volcanic tremors occurring before large repeaters are a potential precursor for the use of the actual prediction of the earthquakes. If this is the case, it may be obvious that the characteristics of the non-volcanic tremors as a precursor should be intensely and comprehensively studied, especially the distribution and the change of spectrogram with approaching the initiation of the large earthquakes.

4.5 Intermittent Process of Large-Scale Exhumation of Metamorphic Masses

It has been still unknown that the timing of the exhumation of the metamorphic belts shows intermittent and sometimes periodic along the subduction plate boundary, though plate subduction is still continuous. There seem two basic mechanisms for this: one is the mechanism depending on the tectonic framework of the subduction zone, and second is the mechanism controlled by the changing the parameters of constituent materials such as the accretion of masses and viscosity depending on water content. The framework of the subduction system represents the subduction angle and obliqueness, and geometry of the deeply subducted slab, that is, the stagnant slag or penetrative slab.

On the other hand, the water content of the deeply accreted metamorphic rocks along on the plate boundary may vary with the time due to dehydration of the subducting slab and hydration of the wedge mantle, and then, the mean density and effective viscosity of the metamorphic mass should be lowered by the present process. Considering the critical size of the metamorphic mass, and content of water in the metamorphic mass at the exhumation, the exhumation process may occur intermittently over the period of enough accumulation of the mass and the water content in the deep subduction boundary zone (Fig. 4.40).

By the way, there is geological evidence that large-scale exhumation of the geological masses along the active and passive plate boundaries was very rapid rather than velocity of plate subduction. In the western Pacific margin, the serpentinite masses extrude into the forearc trench break between the forearc basin and trench slope, from the wedge mantle region along the tectonic zone from Kamogawa, Boso Peninsula to Hayama, Miura Peninsula in the Kanto region. In addition, in the forearc region of the

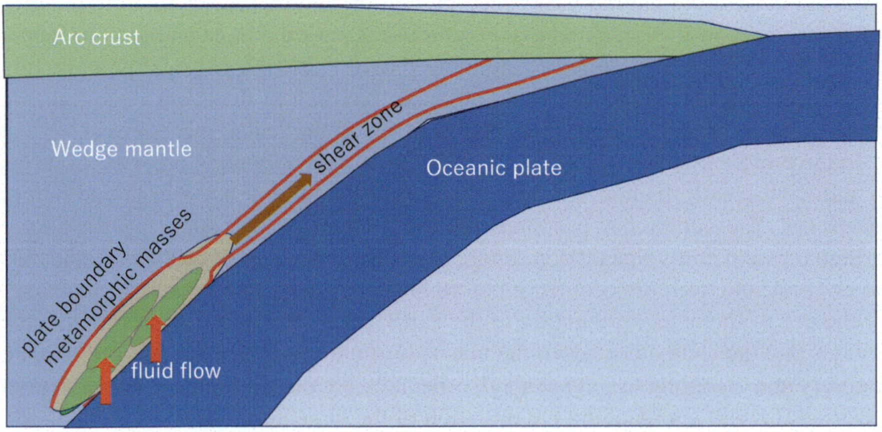

Fig. 4.40 A model of extrusion of the deep-seated metamorphic mass along the major shear zone with entrainment of aqueous fluid coming from dehydration of subduction plate

Izu-Ogasawara-Mariana subduction zone, the long arrays of the conical seamounts composed of serpentine breccia run along the trench slope break parallel to the trench axis (Fryer et al. 1990). Maekawa et al. (1993) found breccia of blueschists containing jadeitic pyroxene and quartz which manifests extrusion of the deep-seated plate boundary rocks along the slab. Furthermore, there are abundant blocks of highly brecciated amphibolite and basaltic rocks in the serpentinite masses along the Kamogawa and Hayama tectonic zone mentioned above. These geological occurrences of the tectonic blocks and breccias of serpentinite masses and metamorphic rocks manifest the rapid extrusion of the deep-seated metamorphic and mantle masses as shown in Fig. 4.41.

Therefore, the large-scale rapid and episodic extrusion process along the subduction plate boundary is surely the basic process in the mechanics of the subduction plate boundary. In this chapter, the author intends to propose the possible model of large-scale and rapid extrusion of the deep-seated geological masses and the intermittency of the rapid extrusion.

There are several models of metamorphic exhumation from the subduction boundary: one is the viscous drag exhumation model, and the other is the friction drag exhumation model in the along the subduction boundary or in the wedge mantle (Toriumi 2022). In both models, the driving force of the exhumation or extrusion should be the gravitational one yielded by density difference and total mass of the

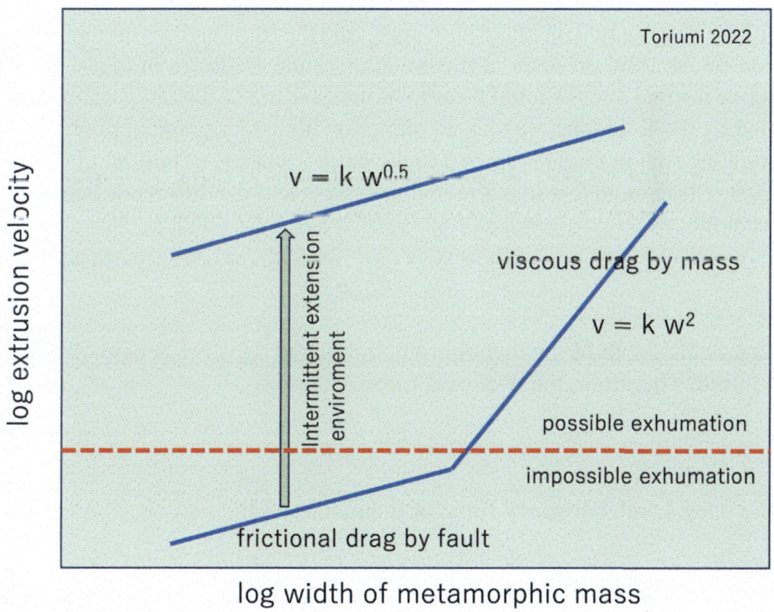

Fig. 4.41 Diagram of extrusion velocity of deep-seated metamorphic mass by viscous drag-controlled extrusion and frictional drag-controlled process in the wedge mantle. Red line shows the boundary of possible and impossible extrusion of metamorphic mass by both mechanisms

extrusion body against the surrounding upper mantle. The subduction channel model of the plate boundary zone is also based on the viscous drag force against the gravitational exhumation force (Cloos and Shreve 1988). However, the author indicated the rapid extrusion of the high-grade metamorphic rocks in the Sanbagawa metamorphic belt, which is controlled by the buoyancy-driven and frictional drag force on the boundary surface between the mass and the surrounding mantle. From the previous book (Toriumi 2022), the viscous drag model is not the case because of rapid extrusion and velocity dependent backward force against the buoyant force. Furthermore, the extremely rapid cooling, which is manifested by the abundant occurrence of dendritic quartz, albite, and feldspars in nano-granite and micro-felsite inclusions in metamorphic garnet of the granulite rocks (Hiroi et al. 2014).

On the other hand, the subduction channel model of the rapid extrusion of the massive metamorphic rocks is still available in the case that the effective viscosity of the massive metamorphic rocks is much low as like as the fluidized clastic sediments, for example, the serpentine mud flow of the conical sea mounts near the Izu-Mariana trench (Fryer et al. 1990). However, the metamorphic mass extruded from the deep plate boundary zone is not such a brecciated mass but almost always massive metamorphic rocks. Therefore, it seems that the rapid extrusion of the plate boundary metamorphic mass should be responsible for the very small frictional force against the gravitational buoyancy force along on the boundary between the metamorphic mass and the surrounding mantle as shown in Fig. 4.40.

By the way, the frictional stress is proportional to the normal stress on the boundary surface of the deep-seated metamorphic mass, and the frictional coefficient is strongly controlled by the fluid pressure in the boundary zone as shown in Fig. 4.42, that is, the effective normal stress should be zero in the case of $P_{\text{fluid}} = P_{\text{rock}}$. This condition is satisfied by the fluid lenses enclosed along the boundary plane of the rocks. If the boundary zone rich in fluid is reached to the earth's surface or bottom of the sea, the fluid pressure is enough less than the rock pressure, and the difference between these pressures holds,

$$dP = \delta\rho g h \tag{4.44}$$

in which h is the depth of the metamorphic mass and $\delta\rho$ is the density difference of rock and fluid. Therefore, the frictional force becomes,

$$2\tau S = 2S\mu\delta\rho g h \tag{4.45}$$

On the other hand, buoyancy force of the metamorphic mass is,

$$Fb = wS\delta\rho_m g \tag{4.46}$$

Therefore, we obtain the net force acting on the metamorphic mass as,

$$\Delta F = Fb - 2\tau S = Sh\delta\rho_{\text{m}} g(w/h - 2\mu\delta\rho/\delta\rho_{\text{m}}) \tag{4.47}$$

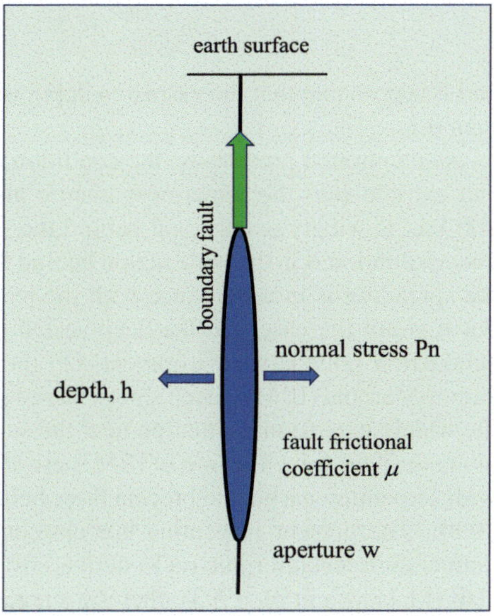

Fig. 4.42 A model for extrusion of deep-seated metamorphic mass by frictional drag in the extensional environment condition

If this force is positive, the metamorphic mass should extrude into earth surface, but it is negative, the mass could not extrude in the mantle. The ratio of $\delta\rho/\delta\rho_{\mathrm{m}}$ is probably about 20, and thus we obtain the critical condition of the extrusion of the mass in the mantle as,

$$w/h > 40\mu \tag{4.48}$$

Considering that w is about 10 km and μ is about 0.01 in the case of fluid saturated boundary zone, then we have,

$$h < 25\,\mathrm{km}.$$

This value is in the range of the crust, and thus, even if the neutral stress condition in the plate boundary zone, the metamorphic rocks situated in the depth over 30 km should not extrude into the crustal portion.

On the other hand, considering that the plate boundary zone is extensional in the case of trench retreat likely to Izu-Mariana, the stress normal to the boundary surface of the metamorphic mass is enough less than $\delta\rho gh$, and it is approximated to be,

$$\begin{aligned} dP &= d\rho gkh \\ k &< 1 \end{aligned} \tag{4.49}$$

where k is parameter of environmental stress condition.

Therefore, it becomes,

$$w/h > 40\mu k \tag{4.50}$$

and it is probable that the extrusion depth is enough larger than 250 km for k is less than 0.1.

As mentioned previously, the condition that the deep-seated metamorphic mass can extrude into the uppermost mantle and lower crust from depth deeper than 100 km, is widely extensional behind the subduction boundary. What tectonics of above situation is in the wide region behind the subduction boundary? First, the back-arc spreading is in accordance with the tectonic situation of extensional tectonics, but it is not the case that the deep-seated metamorphic mass extrudes in the near subduction boundary environment. On the other hand, it is obvious that the Izu-Bonin-Mariana (IBM) trench shows backward migration after Tertiary age, showing the widely extensional situation near the subduction boundary. It is noteworthy that along the near-trench region of IBM there are arrays of conical seamounts consisting with serpentine-carbonate breccia flow, being parallel to the trench axis (Fryer et al. 1990). The present serpentine breccia contains abundant high-pressure and low-temperature metamorphic rocks derived from several ten km depth of the subducted slab (Maekawa et al. 1993), thereby suggesting the extensional situation just near the trench axis.

There are possibly two model of the wide extensional condition behind the subduction plate boundary: one is kinematics of the plate subduction process to be stagnant slab situation within the transition zone, and the other is derived from large volume expansion of the wedge mantle by its hydration and carbonization from subducted oceanic plate.

The former reason why the widely extensional situation appears behind the subduction boundary has been discussed by means of the simulation study by Yoshida et al. (2012), based on the results of seismic tomography around the western Pacific regions by Fukao et al. (1992) who have clearly showed the stagnant slab of the Pacific Plate under the Japan Sea. Yoshida et al. (2012) succeeded the effect of stagnant slab arising the widely extensional situation of the back-arc region involving retreat of the trench axis.

By the way, the stagnant slab of the subducting plate takes place when the slab reaches to the bottom of the transitional layer of the mantle by the simulation of Yoshida et al. (2012). Accordingly, the time required for the stagnant slab generation is clearly represented by the plate velocity and the length of the slab from the subduction boundary to the bottom of the transitional layer as,

$$T_{\mathrm{stag}}(\mathrm{My}) = 0.66 \cos^{-1} \beta / Vp(\mathrm{m/y}) \tag{4.51}$$

Assuming plate velocity (V_{p}) about 0.05 m/y and subduction angle (β) around 30°, the time scale starting stagnant slab becomes around 30 My. This time scale is just the initiation of the extensional situation behind the subduction boundary, suggesting the intermittent process of the extensional tectonics in the wide back-arc region. There is still question when the trench retreat associated with wide extension behind subduction boundary takes cessation. It seems that the lower mantle penetration of the slab corresponds probably with the cessation of the stagnant slab, which occurs at the stage of the vertical subduction into the lower mantle (Fukao et al. 1992). Yoshida et al. (2012) studied the vertical change of subduction angle after beginning the stagnant slab to be about 10–20 My. Therefore, the time scale of the extensional stage near the subduction boundary is to be over 50 My. Considering the time scale of the subduction cycle from beginning to cessation to be about 100 My, the intermittent time scale of the extension tectonics behind the plate boundary may be around 50 My, thereby suggesting the extrusion of the deep-seated metamorphic masses with intermittency of about 50 My.

On the other hand, the volume expansion by hydration (carbonatization) of wedge mantle along the subducted slab may be possible to induce the local extensional stress condition along the subduction zone. Let us consider the case that the volume of 200 × 10 × 1 km of the wedge mantle along the plate boundary is hydrated to form the serpentinite mass, and the density of the initially dry peridotite with density 3.3 g/cc to be changed to antigorite mass with density of 2.6. The volume increase reaches about 80 × 5 × 1 km, and thus the 50 km length of the top of the serpentinite mass should lie on the crust, and the 30 km length of that mass should form the great dyke in the forearc crust as shown in Fig. 4.43. If the deep-seated metamorphic mass is embedded in the hydrated serpentinite mass along the plate boundary, the present metamorphic mass must extrude into crustal portion associated with small amounts of serpentinite fragments as shown in Fig. 4.43.

In many subduction boundary zones, there are several metamorphic belts along the coastal region around the Pacific Ocean. In the southwestern Japanese islands, the Shimanto metamorphic belt, Sanbagawa metamorphic belt, Suo metamorphic belt, and Renge metamorphic belt. These metamorphic ages, that means crystallization periods of the metamorphic minerals but not the extrusion time, are 60–70 Ma, 110–120 Ma, 180 Ma, and 220 Ma (Itaya et al. 2011), respectively. However, the extrusion

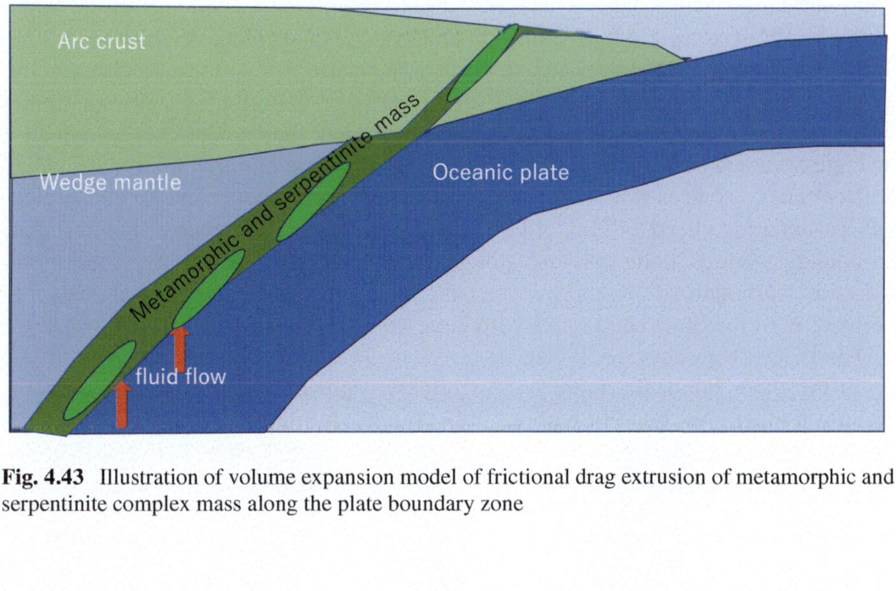

Fig. 4.43 Illustration of volume expansion model of frictional drag extrusion of metamorphic and serpentinite complex mass along the plate boundary zone

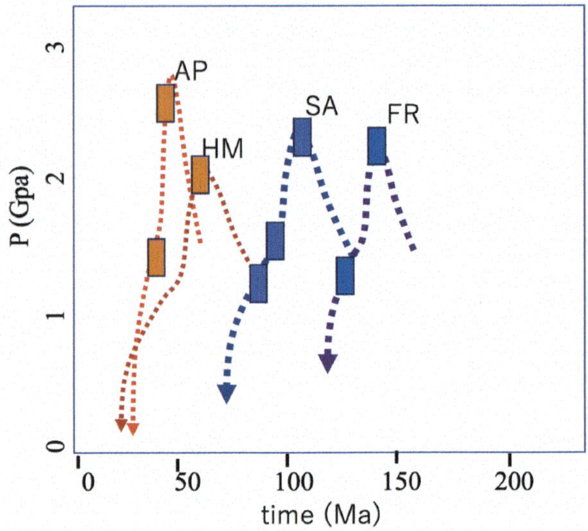

Fig. 4.44 Relation of growth ages of metamorphic minerals and extrusion timing of the large subduction-related metamorphic masses. Red; collision type metamorphic belts, blue; subduction type metamorphic belts. The extrusion ages are the time after main growth ages of metamorphic minerals about 10–30 million years

ages of these metamorphic belts are presumably the ages of 10–30 million years after above ages as shown in Fig. 4.44.

As discussed in the previous book (Toriumi 2022), giant and rapid extrusion of the deep-seated metamorphic masses often took place in the subduction boundary and continental collision boundary zones. The signatures of mineral occurrence that shows short-term annealing and rapid cooling, are the occurrence of rounded mineral inclusion and dendritic morphology of metamorphic minerals, and their distribution appears in the narrow zones in the rather lower metamorphic grade zones (Toriumi 1979, 2022; Hiroi et al. 2014). It shows that the short-term annealing layers of the high-grade zone should extrude into the lower grade metamorphic mass by the fault boundaries. Thus, these boundaries should manifest the evidence of the large displacement over several ten kilometers and large displacement velocity.

By the way, typical high-speed extrusion in the geological context can be seen by the gaseous explosion breccia such as kimberlite pipe and alkali basaltic diatreme forming maar. Furthermore, the large-scale detachment faults manifest geologically the huge amounts of horizontal displacement over several hundred kilometers. This fault boundary zone is now composed with highly brecciated materials cemented with carbonates, quartz, albite, chlorite, and clay minerals, thereby showing that the boundary zones, called as fault gouge, were suffered from strong brecciation, hydration, carbonatization, and metasomatic reactions. The evidence manifests the fluidization of the brecciated rocks with large volume fraction of fluid composed of H_2O and CO_2 (e.g., Dawson 1980).

On the other hand, the rapid and extremely rapid extrusive rocks keep rather medium temperature condition about 300–400 °C during extrusion, and thus, the highly brecciated rocks may appear the composite textures formed in the early high-grade rocks and in the rapid annealing of lower grade matrix of the breccia. It may be clear that the high-grade metamorphic minerals in the core of breccia keep their growth texture, but the low-grade matrix minerals appear the very fine-grained and rapidly grown microstructures in the microcracks and micro-fissures of the high-grade minerals. In addition, it seems that the low-grade minerals should be formed in the fluid-rich environments, thereby suggesting the highly metasomatic alteration textures such as metasomatic bands and replacement texture of hydrous minerals after high-grade minerals.

On the other hand, if the retrograde metamorphism during the rapid or extremely rapid extrusion occurs in the medium grade condition about 300–500 °C and 0.3 to 0.5 GPa, metamorphic rocks should be composed of mineral assemblages of the greenschist to amphibolite facies condition. It shows that the metamorphic peak conditions of present retrograde-metamorphosed rocks look like those of the greenschist to epidote–amphibolite facies, even if the protoliths of these metamorphic rocks were originally eclogite or granulite facies rocks. In many metamorphic belts in the world, there are abundant high -grade metamorphic blocks in the low grade to medium grade metamorphic areas, suggesting that high-grade regional metamorphosed protoliths might have been suffered almost always from large-scale hydration reaction in the low to medium grade metamorphism and metasomatism.

There are many examples that the isolated high-grade blocks are found in the greenschist to epidote–amphibolite or amphibolite facies zones in the plate boundary metamorphic belts, those are UHP metamorphic rocks of Alps (Bousquet et al. 2008), Dabie Shan, Ural, Himalayan, Higo of Kyushu, and HP rocks of Sanbagawa, Franciscan, Klamath, Blue Ridge of E. Australia, and others (Parkinson et al. 2002) as shown in Fig. 4.44. In the Sanbagawa metamorphic belt, it shows that the protoliths of eclogite facies occur apparently as a tectonic block surrounded by epidote–amphibolite to blueschist facies schists, and those are sporadically and strongly hydrated into epidote–amphibolite and greenschist facies mineral assemblages (Toriumi and Kohsaka 1995). Furthermore, it shows that the high-grade rocks of the Sanbagawa belt have suffered from the strong hydration and retrograde metasomatism as shown in Fig. 4.45 under coeval high shear deformation (Toriumi and Noda 1986).

In summary, it seems that these occurrences of strongly and widely metasomatized and hydrated high-grade minerals in the broad metamorphic area manifest the large-scale and high-speed ductile and brittle fault extrusion phenomena in the late-stage metamorphism, that should be called as the extrusion stage of the metamorphic belt. The age of this extrusion stage may be obtained by K–Ar and Ar–Ar chronology on ground mass muscovite and phengite in the matrix (Itaya et al. 2011).

The forearc and back-arc basin develops in the extensional zone along the subduction boundary as described previously, and it should be distinctive that the plate boundary metamorphic mass extrudes into the present basin from the upper mantle depth. There are probably two modes of extrusion of the deep-seated rocks into the crustal portion from the upper mantle: one is the massive extrusion by fault-guided frictional slip control, and the other is the gaseous and fluid-supported fragmentation and extrusion likely to the diatreme explosion.

The latter is characterized by the mechanical mixture of the various sized metamorphic blocks and surrounding uppermost mantle rocks showing angular and sometimes

Fig. 4.45 Photographs of brecciation of the boundary zone of rapid and slow extrusion zones of the metamorphic mass of the Sanbagawa belt, showing the brecciated protoliths of garnetite rocks and surrounding strongly brecciated and sheared metamorphic schists (right), and their shear fault boundary (left)

rounded shape, and surrounding matrix sediments and cemented fine fragments. It seems that the minerals precipitated from gaseous and solution cements in the interstitials of the matrix. On the other hand, the former must be represented by the subsidiary shear faults such as Riedel shear and penetrative shear from the master boundary shear fault, and there should be also abundant open cracks filled with metamorphic minerals formed by large shear slip motion of the boundary faults during rapid extrusion as shown in Fig. 4.46.

Fig. 4.46 Photographs of shear slip controlled decompression breccias of low grade greenschists (right top) and serpentinites (right bottom)

It should be shown that the above mode of emplacement, that is fault-assisted extrusion mechanism, may be likely to the detachment in the case of exhumation of metamorphic core complex formation. In the Canadian Rocky Mountains, the large-scale detachment fault above Paleozoic sedimentary rocks and the high-grade metamorphic core complex with thick brecciated and matrix supported fault gouge. The displacement by the detachment low angle fault reaches over 50 km, thereby suggesting that the boundary gouge must be very low viscosity, so that the fragments and matrix should be supported by the gaseous phases during long-range displacement of the upper hanging wall masses (e.g., Pluijm et al. 2006).

By the way, it may be referred that the rockslide phenomena in the shallow crustal portion are the same mechanical process, as like as the above-mentioned fault-assisted metamorphic extrusion at the points that both are controlled by the mobility of the faults and rock volume, and furthermore, the rapid displacement of the rock mass. Moreover, it seems that the rockslide should be accompanied with less fragmentation of the overriding massive rock, and strong brecciation along the fault boundary. Thus, the geological structure of the rockslide mass and large-scale detachment mass is practically available for reference of rapid extrusion signatures of the metamorphic mass.

It becomes sure that the fragmentation of the overriding mass may be governed by the open crack and related normal fault associated with master shear slip faults. The situation of these displacements by various cracks and shear slips over the main shear slip fault are probably like that of the frontal part of the subduction zone that consists of decollement and accretionary wedge with abundant out-of-sequence thrust and related normal and reverse faults (Kimura and Kinoshita 2009).

These characteristic features of the large-scale basal slip and related faults can be formulated by the rotation derived from the displacement gradient tensor above the basal slip as follows:

$$
\begin{aligned}
D &= \{\mathrm{d}u_i/\mathrm{d}x_j\} \\
&= 1/2\{\mathrm{d}u_i/\mathrm{d}x_j + \mathrm{d}u_j/\mathrm{d}x_i\} + 1/2\{\mathrm{d}u_i/\mathrm{d}x_j - \mathrm{d}u_j/\mathrm{d}x_i\} \\
&= 1/2(\text{strain}) + 1/2(\text{rotation})
\end{aligned}
\tag{4.52}
$$

If the weak frictional coefficient of the master shear slip (Yin 2010), the stress geometry shows the extensional along the shear plane and fracturing of overriding geological masses should be controlled by normal faults (with open cracks), and then consequently the brecciated blocks by these subsidiary faults should rotate by above-mentioned rotation deformation, as shown in Fig. 4.46.

The above brecciation of the geologic masses should be needed for weakening the frictional coefficient of the faults. It seems that in the preceding sections, the extrusion of the deeply seated metamorphic masses is controlled by the concentration by entrainment of aqueous fluid along the boundary faults between the metamorphic masses and surrounding rocks, for drastic decrease of the effective frictional stress depending on the difference between the fluid pressure and rock pressure, thereby indicating sharply that the extrusion and later exhumation of the metamorphic masses is derived from the hydration weakening of the boundary faults.

The hydration of the wedge mantle forming antigorite is required in the conditions under several GPa, but on the other hand, the slab dehydration occurs even over 6 GPa because of lower temperature conditions of the slab (Omori et al. 2002). Thus, it is obvious that the hydrated slab should dehydrate with emission of aqueous fluid toward the overriding wedge mantle, which is in the tensional configuration of the stress derived from downward bending of the slab. At this depth region over the slab, many exotic blocks carried out from the trench by slab subduction should be experienced by eclogite and UHP metamorphism, and then the eclogite blocks start to extrude into upper portion by sudden decrease of frictional coefficient of widely wetted boundaries as discussed in the previous section.

On the other hand, at the depth of about 30–50 km, it seems that the stress configuration changes compression in the wedge mantle over the slab showing the down-dip compression state (Kita et al. 2010). Then, the extruding eclogitic blocks may stop uprising, and accumulate in the wedge mantle of these depth range. Later, the tectonic configuration of the shallow wedge mantle changes to extensional state by intermittent stagnant slab or by slab break-out. And the metamorphic complex consisting of eclogitic blocks and deep accretionary sediments should extrude into the earth surface by the frictional drag extrusion mechanism mentioned above.

Metamorphic rocks are extruded from deep subduction boundary, attaining several hundred kilometers depth, to the solid earth surface. It is still unknown that how speed the metamorphic mass migrates within the wedge mantle. The extrusion mechanism is divided into two processes as proposed in the previous book (Toriumi 2022): the one is the buoyancy-driven and viscous drag process, and the other the buoyancy-driven and frictional drag one. The former process is represented by the velocity-controlled slow exhumation like mantle diapir, but the latter by the velocity-independent rapid extrusion necessary to the low friction fault boundary.

In this section, it has been shown that both processes cannot migrate at most the several kilometers upper level, but not attain the solid earth surface level in the compressional stress environment of the plate boundary zone within the several ten million years. The actual exhumation of the deep-seated metamorphic rocks such as diamond-bearing eclogite is required for the situation of the regionally extensional environment of the wedge mantle. It, however, seems to be contracted with the rapid extrusion of them in the regional extensional environment that the deep-seated metamorphic rocks have suffered from the deep subduction of the sedimentary and volcanic rocks associated with subduction of the plate. The above confusion should be avoided by the temporal difference between the extensional situation and compressional one in the wide region from the fore arc and the back arc the extensional situation in the arc region is probably intermittent within the long-term compression field of the arc as shown in Fig. 4.47.

Therefore, it should be imagined that the various metamorphic rocks which show various equilibration temperature and pressure together with various crystallization ages mean the residence period of them in the deep subduction zone, and that the extrusion and exhumation stage is represented by the final isotopic age of the matrix

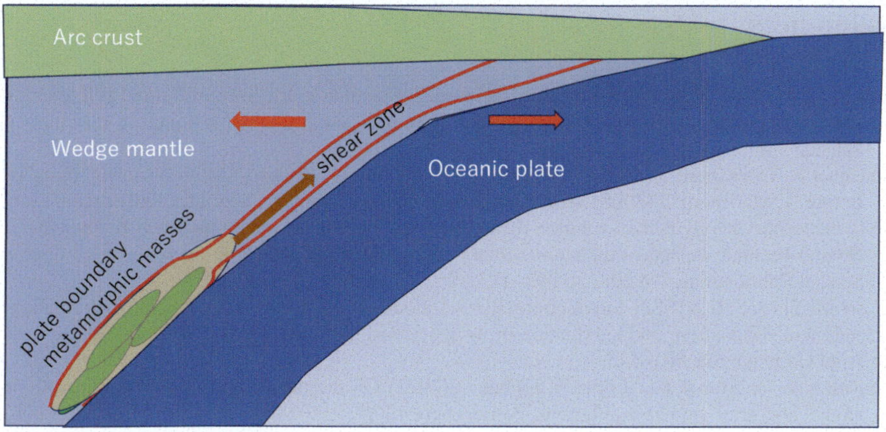

Fig. 4.47 A model of rapid extrusion of deep-seated metamorphic mass due to negative buoyancy along brecciated shear slip zones associated with intermittent extensional environment

in them. After such consideration, it can be found that the extrusion times of the Alps and Himalayan metamorphic belts are about 35 Ma and 30 Ma, respectively. On the other hand, the extrusion times of the Sanbagawa and Franciscan metamorphic rocks are 60 Ma and 120 Ma, respectively.

These ages of extrusion of the highest-grade metamorphic rocks imply apparently the very short period less than 0.1 my (10,000 years by Toriumi 2022) of the massive extrusion and exhumation of various grades of metamorphic belt, and it appears that the wide range of isotopic ages of metamorphism in the single metamorphic belt manifests the long-term stagnant of the various depth in the plate boundary zone. Toriumi and Kohsaka (1995) clarified that the Iratsu and Tohnaru eclogite blocks experienced cyclic pressure and temperature conditions during growth of garnet and omphacite by detailed investigation of chemical compositions of them from the core to the rim and included omphacite grains, thereby showing that these eclogitic blocks were suffered from repeated process of subduction and extrusion within the wedge mantle along on the subduction plate as illustrated in Fig. 4.47.

Furthermore, it seems that the very short-term extrusion event recorded in the highest-grade rocks of the metamorphic belts shows very low frictional resistance and very low density of the extrusion mass at that time. It, therefore, may be suggested that the extrusion process of the highest-grade mass looks like diatreme explosion of brecciated serpentinite at the conical seamount along the Mariana trench.

References

Araki E, Saffer DM, Kopf AJ, Wallace LM, Kimura T, Machida Y, Ide S, Davis E (2017) Recurring and triggered slow-slip events near the trench at the Nankai Trough subduction megathrust. Science 356(6343):1157. https://doi.org/10.1126/science.aan3120

Bousquet R, Oberhänsli R, Goffé B, Wiederkehr M, Koller F, Schmid SM, Schuster R, Engi M, Berger A, Martinotti G (2008) Metamorphism of metasediments in the scale of an orogen: a key to the tertiary geodynamic evolution of the Alps. In: Siegesmund S, Fügenschuh B, Froitzheim N (eds) Tectonic aspects of the alpine-dinaride-carpathian system. Geological Society, London, Special Publications, vol 298, pp 393–412

Cloos M, Shreve RL (1988) Subduction-channel model of prism accretion, melange formation, sediment subduction, and subduction erosion at convergent plate margins, 2. Impl Disc Pure Appl Geophys 128:501–545

Courtillot V, Le Mouel J-L, Lopes F, Gilber D (2022) On the nature and origin of atmospheric annual and semi-annual oscillations. Atmosphere 13(11):1907. https://doi.org/10.3390/atmos13111907

Dawson JB (1980) Kimberlites and their xenoliths. Springer, Berlin. 10.10071978-3-642-67742-7

Fryer P, Saboda KL, Johnson LE, Mackay ME, Moore GF, Dtoffers P (1990) Conical seamount: SeaMARC II, Alvin submersible, and seismic-reflection studies. In: Fryer P, Pearce JA, Stokking LB et al (eds) Proceedings ODP, Init. Repts., 125: College Station, TX (Ocean Drilling Program), pp 69–80

Fukao Y, Obayashi M, Inoue H, Nenbai H (1992) Subducting slabs stagnate in the mantle transition zone. J Geophys Res 97(B4):4809–4822. https://doi.org/10.1029/91JB02749

Gillet N, Jault D, Canet E, Fournier A (2010) Fast torsional waves and strong magnetic field within the Earth's core. Nature 465/6. https://doi.org/10.1038/nature0901

Hiramatsu Y, Yamanaka H, Tadokoro K, Nishigami K, Ohmi S (2002) Scaling law between corner frequency and seismic moment of micro earthquakes: is the breakdown of the cubic law a nature of earthquakes? Geophys R Letts 29:521–524. https://doi.org/10.1029/2001GLO13824

Hiroi Y, Yanagi Y, Kato T, Kobayashi B, Prame T, Hokada M, Satish-Kumar M, Ishikawa T, Adachi Y, Osanai Y, Motoyoshi K, Shiraishi T (2014) Supercooled melt inclusions in lower-crustal granulites as a consequence of rapid exhumation by channel flow. Gondwana Res 25:226–245

Hulbert C, Jolivet R, Gardonio B, Johnson P, Ren CX, Rouet-Leduc B (2020) Tremor waveform denoising and automatic location with neural network interpretation, arXiv:2012.13847v1 [physics.geo-ph]

Ide S (2008) A Brownian wave for slow earthquakes. Geophys Res Letts 35:17. https://doi.org/10.1029/2008GL034821

Itaya T, Tsujimori T, Liou JG (2011) Evolution of the Sanbagwa and Shimanto high-pressure belts in SW Japan: Insight from K-Ar (Ar-Ar) geochronology. J Asian Earth Sci 42:1075–1090. https://doi.org/10.1016/j.jseaes.2011.06.012

Jadin EA, Yamazaki K (1995) Changes in the earth's rotation and the atmosphere angular momentum on intra-annual and decadal time series. Meteol Geophys 45(3):113–120

Johnson CW, Fu YN, Burgmann R (2017) Seasonal water storage, stress modulation, and California seismicity. Science 356(6343):1161–1164. https://doi.org/10.1126/science.aak9547

Kimura G, Kinoshita M (2009) Accretionary prisms and megathrust seismogenic zones. University of Tokyo, Press

Kita S, Okada T, Hasegawa A, Nakajima J, Matsuzawa T (2010) Existence of interplate earthquakes and neutral stress boundary between the upper and lower planes of the double seismic zone beneath Tohoku and Hokkaido, northeastern Japan. Tectonophysics 496(1–4):68–82. https://doi.org/10.1016/j.tecto.2010.10.010

Levin BW, Sasorova EV, Steblov GM, Domanski AV, Prytkov AS (2017) Variations of the Earth's rotation rate and cyclic processes in geodynamics. Geodesy Geodyn 8(3):206–212

Maekawa H, Shozui M, Ishii T, Fryer P, Pearce J (1993) Blueschist metamorphism in an active subduction zone. Nature 364:520–523

Matsuzawa T, Uchida N, Igarashi T, Okada T, Hasegawa A (2004) Repeating earthquakes and quasi-static slip on the plate boundary east off northern Honshu, Japan. Earth Planets Space 56:803–811

Obara K (2002) Nonvolcanic deep tremor associated with subduction in southwest Japan. Science 296:1679–1681

Omori S, Kamiya S, Maruyama S, Dapeng Z (2002) Morphology of the intraslab seismic zone and devolatilization phase equilibria of the subducting slab peridotite. Bull Earthquake Res Inst Univ Tokyo 76:455–478

Parkinson CD, Katayama I, Liou JG, Maruyama S (2002) The diamond-bearing Kokchetav Massif, Kazakhstan, Universal Academy Press Inc., Tokyo

Pluijm BA, Vrolijk PJ, Pevear DR, Hall CH, Solum J (2006) Fault dating in the Canadian rocky mountains: evidence for late Cretaceous and early Eocene orogenic pulses. Geology 34:837–840. https://doi.org/10.1130/G22610.1:1

Takahashi H, Hino R, Uchida N, Matsuzawa T, Ohta Y, Suzuki S (2022) Tectonic tremors immediately after the 2011 Tohoku-Oki earthquake detected by near-trench seafloor seismic observations. Progress in Earth and Planetary Science, 9:66

Toriumi M (1979) A mechanism of shape-transformation of quartz inclusions in albite of regional metamorphic rocks. Lithos 12:325–333

Toriumi M (2021) Global seismicity dynamics and data-driven science. Springer

Toriumi M (2022) Geochemical mechanics of the earth and deep neural network modelling. Springer

Toriumi M, Kohsaka Y (1995) Cyclic P-T path and plastic deformation of eclogite mass in the Sanbagawa metamorphic belt. J Fac Sci Univ Tokyo 22:210–231

Toriumi M, Noda H (1986) The origin of strain patterns resulting from contemporaneous deformation and metamorphism in the Sambagawa metamorphic belt. J Metamorphic Geol 4:409–420

Uchida N, Shimamura K, Matsuzawa T, Okada T (2015) Postseismic response of repeating earthquakes around the 2011 Tohoku-oki earthquake: moment increases due to the fast-loading rate. J Geophys Res Solid Earth 120. https://doi.org/10.1002/2013JBO10933

Uchida N (2019) Detection of repeating earthquakes and their application in characterizing slow fault slip. Progr Earth Planet Sci 6, Article number:40

Wei RP (2010) Fracture mechanics: integration of mechanics, materials science, and chemistry. Cambridge University Press

Yin HM (2010) Fracture saturation and critical thickness in layered materials. Int J Solids Struct 47:1007–1015. https://doi.org/10.1016/j.ijsolstr.12.016

Yoshida M, Tajima F, Honda S, Morishige M (2012) The 3D numerical modeling of subduction dynamics: plate stagnation and segmentation, and crustal advection in the wet mantle transition zone. J Geophys Res 117:B04104. https://doi.org/10.1029/2011JB0089989

Yukutake T (1973) Fluctuations in the earth's rate of rotation related to changes in the geomagnetic dipole field. J Geomag Geoelectr 25:195–212

Chapter 5
Neural Network Modeling of Regression in Nonlinear Dynamics Timeseries

Abstract Convolution neural network and recurrent neural network modeling of nonlinear dynamical systems is applicable for regression of the multivariate timeseries, which displays chaotic temporal variations between two states in the case of logistic and Lorenz dynamics. In the global correlated seismicity dynamics, it is provided that the pitchfork type bifurcation dynamics concerning the relation of rate of earth rotation and the correlated seismicity is possibly understood by the transition between global seismicity network structures under the periodic external force. In the three-arm slider-block network, it shows that the temporal variation of the multivariate timeseries appears multi-states bifurcation and supercritical bifurcation under the periodic external force, thereby suggesting that the global and regional correlated seismicity dynamics is nearly approximated to this three-arm network of the global subduction system.

Keywords Regression of nonlinear dynamical timeseries · Multi-states bifurcation · Global correlated seismicity rate and rate of earth rotation

5.1 Neural Network Regression of Minimal Nonlinear Dynamics

In the previous studies of the global and regional correlated seismicity dynamics, it has been clarified that the present dynamics has minimally two modes of behavior, and thus it should be represented by the simple nonlinear dynamical equation as shown in Fig. 5.1. Besides, it was noted that at the case of the tangential relation between two null clines of the minimal nonlinear equations, the variance of the correlated seismicity suddenly changes very large values, and then the large earthquakes took place in this large variance period as shown in Fig. 5.2 (Toriumi 2022).

The ordinary differential equations of above simple two state dynamics hold,

$$\begin{aligned} \mathrm{d}z_1/\mathrm{d}t &= z_1 - kz_1^3 + z_2, \\ \mathrm{d}z_2/\mathrm{d}t &= h\cos(wt), \end{aligned} \tag{5.1}$$

M. Toriumi, *Physics of Geochemical Mechanics and Deep Neural Network Modeling with Diffusion Augmentation*, Advances in Geological Science,
https://doi.org/10.1007/978-981-97-9376-1_5

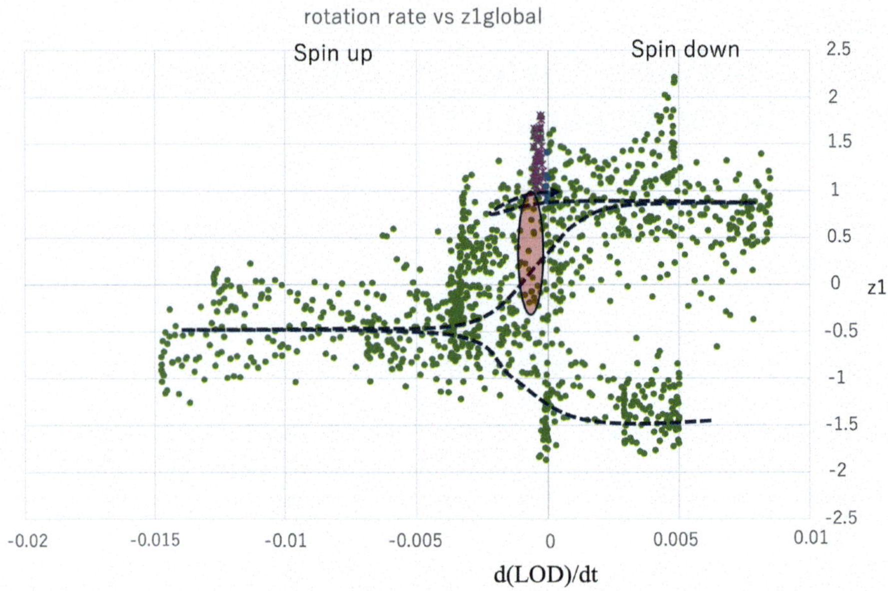

Fig. 5.1 Relation of global correlated seismicity rate z_1 and earth rotation rate d(LOD)/dt of the earth, showing pitchfork type bifurcation dynamics in the period from 1990 to 2023. Violet points; 2022–2023 data, orange; region of the Sumatra and Tohoku-Oki giant earthquakes

in which k and h are constant term, and w is the frequency of the periodicity of z_2.

In this dynamical system, it is important that there are three regimes in the mode of null clines of above differential equations: one is the case that null cline equations have tangential points and one crossing point, and second is the case that null clines have three crossing points, two are stable nodes and one is unstable node. Furthermore, third case is that the null clines have one crossing point which is a stable node of above ODEs. These cases are shown in Fig. 5.3. It seems that in different regimes, solutions of ODEs with noisy drift and events over the critical level have distinctively different timeseries with advancing the time step as discussed on the global correlated seismicity in the previous book (Toriumi 2021).

As shown in the previous book, the deep leaning method to investigate the dynamical behavior of the natural systems has succeeded in approximation of the actual behavior in the case of monotonous dynamical systems, but it is scarcely available for the multi-state nonlinear dynamics controlling the multivariate timeseries. This seems to be due to the reason why deep training of the timeseries is ranged to be a regression within the single process, and testing should be carried out in the range of composite behavior of dynamics, being different from that single process. If it is possible to be carried out in both regimes of two or more modes of dynamical behavior, it seems to be possible that two or more dynamics should mathematically connect them to make the available nonlinear dynamical equations controlling the present system.

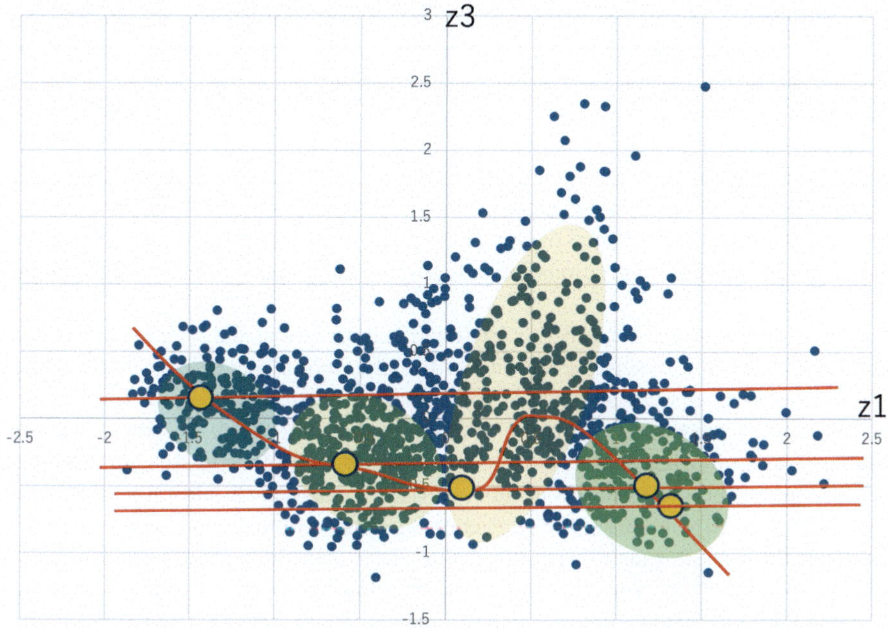

Fig. 5.2 Phase map of global correlated seismicity rates z_1 and z_3, showing the four successive clusters of A, B, C, and D. The centers of them are connected by the inferred null cline of the possible nonlinear dynamics. The node C represents tangential relation of two null clines, and the large variance of C cluster is possibly derived from the present tangential relation

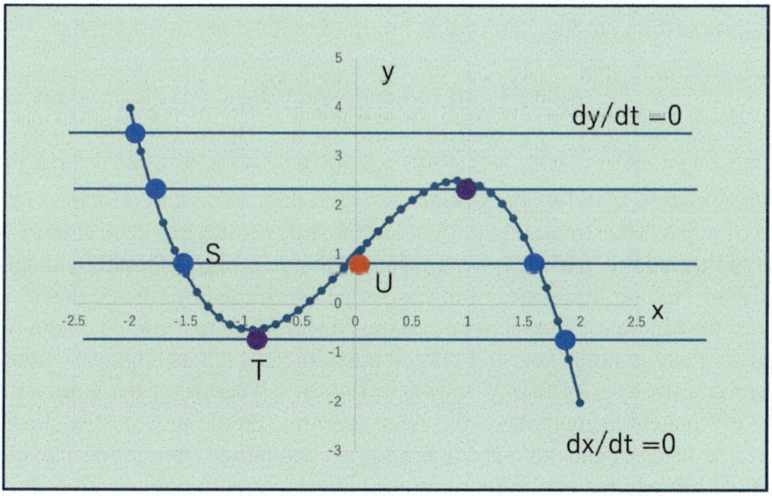

Fig. 5.3 Possible null clines of the nonlinear dynamical equations concerning with global correlated seismicity rates z_1 and z_3, showing the tangential relation. *T*; tangential node, *U*; unstable node, *S*; stable node

By the way, it seems that variances of the solution of nonlinear dynamics with noisy drift term near the tangential point becomes very large, because of very large value of the inverse second derivatives of the potential function driving z_1 of the NLD, thereby showing the drastic change of the dynamical states. In the case of global correlated seismicity dynamics, the author has investigated the possibility of the prediction by the deep learning from the finite range of the time series to the next range of them in the time series having secular variation and drift term. The results show the distinctive difference of the time series from the actual ones because of different type of secular variations in the learning segment and testing one (Toriumi 2021).

In the previous books, it has been concluded that the global and regional correlated seismicity of the micro-seismic activity of the plate boundary regions displays the different series of stochastic stable nodal points in the state space of several correlated seismicity parameters (Toriumi 2022), thereby suggesting that there are two different types of dynamical system in the 1990–2021 timeseries of global correlated seismicity and the transition between their dynamics should take place near 2003–2004 and 2011 to 2012. In the case of Japanese regional correlated seismicity dynamics, the similar transition occurred near 2010–2011 in the times series of 1998 to 2021 (Toriumi 2022). The transition of the correlated seismicity dynamics seems to be coeval with the periods of the giant earthquakes of Sumatra 2004 and Tohoku-Oki earthquakes 2011 (Fig. 5.2).

Thus, in the section, the author intends to investigate the deep learning problems of the minimal nonlinear dynamical system in the case of simulation and natural processes by means of the various data processing from successive data segments across the multivariate time series having the different types of secular trends.

In the proceeding studies, the simple nonlinear ODEs are considered.

$$
\begin{aligned}
\mathrm{d}X/\mathrm{d}t &= f(X) + dN\left(0, v^2\right) \\
f(X) &= aX - bX^3 (a, b > 0)
\end{aligned}
\tag{5.2}
$$

In this equation, a and b are constant, d is the drift term, and N is the Gaussian function of noise. The simulation of this nonlinear dynamics has been carried out on RStudio with available parameter set (a, b), as shown in Fig. 5.4, thereby displaying obviously the chaotic timeseries of the two states with random up and down.

To infer the kinematic model of both states, it should be required to obtain both of two modes of stable nodal sets and related phenomena such as large variation event on background timeseries like X shown in Fig. 5.3. Therefore, the author intends to study the plausible method of the deep learning prediction of episodic events related to the background state-space timeseries depending on nonlinear dynamics, by means of mixed time evolution of multiple state-space dynamics.

In this study, the Gaussian drift noises on $\mathrm{d}X/\mathrm{d}t$ are added in the present NLD, producing the random two state timeseries as shown in Fig. 5.4, and the neural network (NN) deep learning of the finite time step signals with ground truth of one step ahead in the given noisy timeseries segments of X is carried out to obtain the

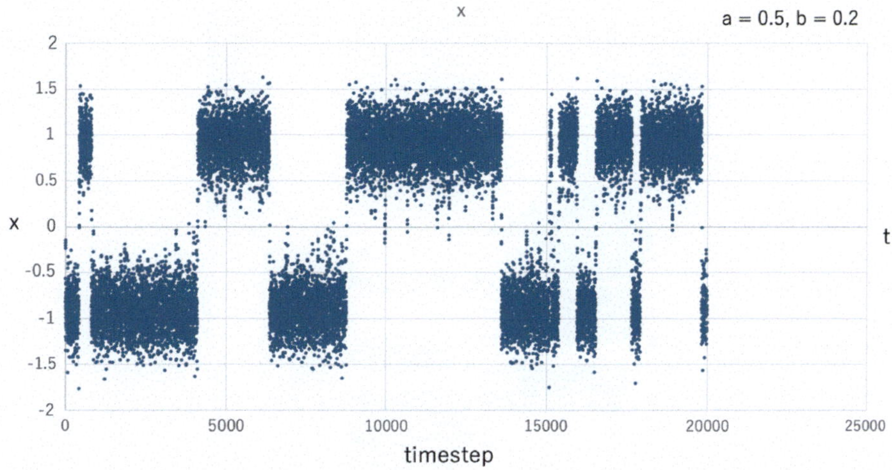

Fig. 5.4 Chaotic temporal variations of two states simple nonlinear dynamics with Gaussian noisy drift. See text

predictable regression by neural network modeling (NNM) for another segment of the timeseries. This validation method of the correlated seismicity timeseries has been conducted, and the important discrepancy between actual timeseries and NN predicted timeseries is not observed in the different stage apart from the training timeseries. To avoid such discrepancy, it will be needed for prediction of global seismicity change that the NN deep learning should involve the various segments of the timeseries in the composite modes like the modes appeared in NLD described previously.

The timeseries of the different modes of NLD solution of X mentioned above is shown in Fig. 5.4, and the deep learning of the single and multiple segments of the timeseries will be performed to obtain the neural network predictors F for inferring the different stage timeseries such as $X_{t+1} = F(X_{t-k}, X_{t-k+1}, X_t)$. In this study, the simple 1dCNN model for the timeseries is applied for evaluation of the complex system controlled by NLD equations. As shown in Fig. 5.5, after the one dimension convolution neural network model (1dCNN) and recurrent neural network model (RNN) predictor F is set up from the timeseries of training segments involving the two states of the whole timeseries simulated by noisy and simple one-dimensional NLD denoted above. Furthermore, it is required for prediction study that the testing timeseries after training timeseries simulated by NLD will be accurately estimated in the enough length. In this study, the author intends to develop the NN deep learning available for this prediction study.

As shown in Figs. 5.5 and 5.6, it seems that the regression timeseries inferred by 1dCNN and RNN predictors after enough training is obviously in good accordance with the simulated timeseries of the testing segment, though the regression timeseries is transformed by regularization to the value of $(1, 0)$ range. Therefore, it seems that if the studied timeseries involving multivariate timeseries from the nonlinear dynamics

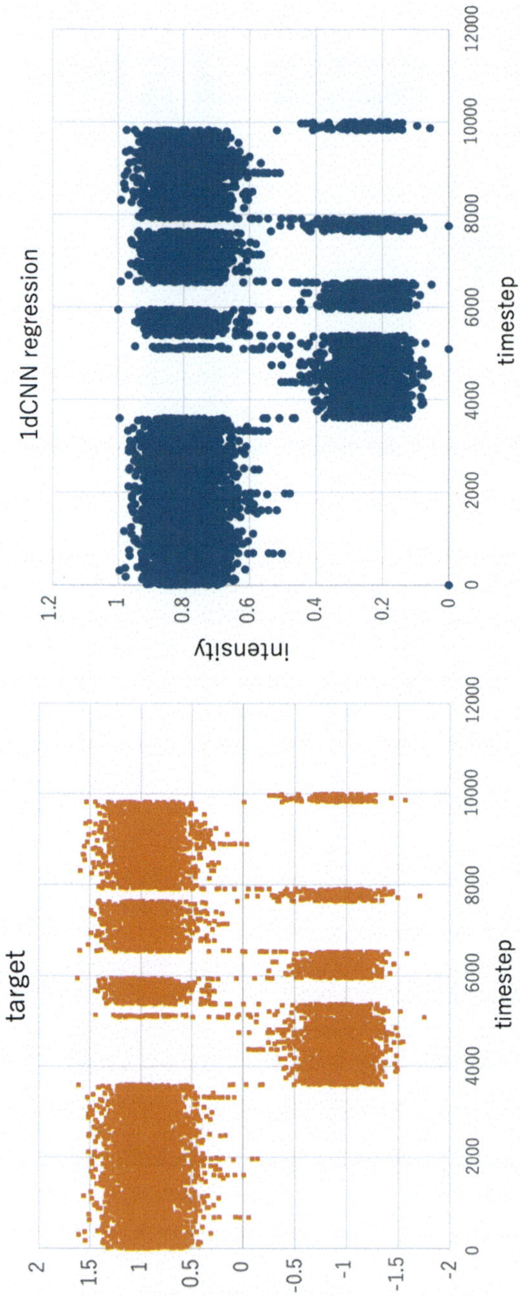

Fig. 5.5 1dCNN regression result of one-dimensional chaotic nonlinear dynamic equation from training segment to testing segment of the timeseries. Orange; testing timeseries, blue; result of 1dCNN regression on testing one using training parameter sets

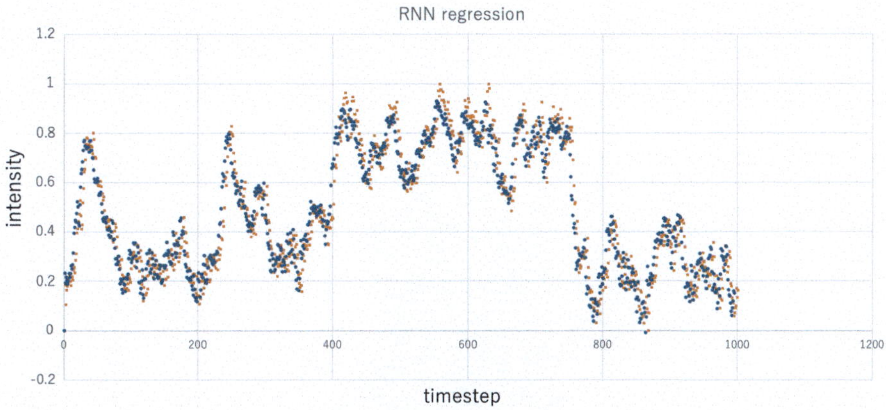

Fig. 5.6 RNN regression result of one-dimensional chaotic nonlinear dynamic equation from training segment to testing segment of the timeseries. Orange; testing timeseries, blue; result of RNN regression on testing one using training parameter sets

contain the variations of different state segments, the NN regression and modeling for prediction study is probably plausible practically.

5.2 Neural Network Regression of Lorenz Dynamics and Composite Nonlinear System

In the three-component dynamical system, the Lorenz NLD is representative in the chaotic systems. The author intends to study the DNN predictability of the simulated events on the Lorenz attractor with Gaussian noises, $N(0, m)$ in order to investigate the natural event precursors such as earthquakes on seismic noisy signals. The parameters of following Lorenz dynamical equations are given as,

$$\begin{aligned} \mathrm{d}x/\mathrm{d}t &= a(y-x) + hN\left(0, v^2\right), \\ \mathrm{d}y/\mathrm{d}t &= x(b-z)-y + hN\left(0, v^2\right), \\ \mathrm{d}z/\mathrm{d}t &= xy-cz + hN\left(0, v^2\right). \end{aligned} \tag{5.3}$$

The noisy attractor of this system appears in the phase maps of (x–y) and (x–z), shown in Fig. 5.7, suggesting that the complicated highly noisy attractors are generated. The timeseries of x, y, and z are extremely complicated and drifted in the high and low levels irregularly as shown in Fig. 5.8.

The convolution neural network (1dCNN) regression of the Lorenz timeseries was carried out by the same method as the previous simple NLD equations as shown in the previous section. The representative results of x, y, and z timeseries are displayed in Fig. 5.9, thereby showing a good agreement with the original behavior of them with

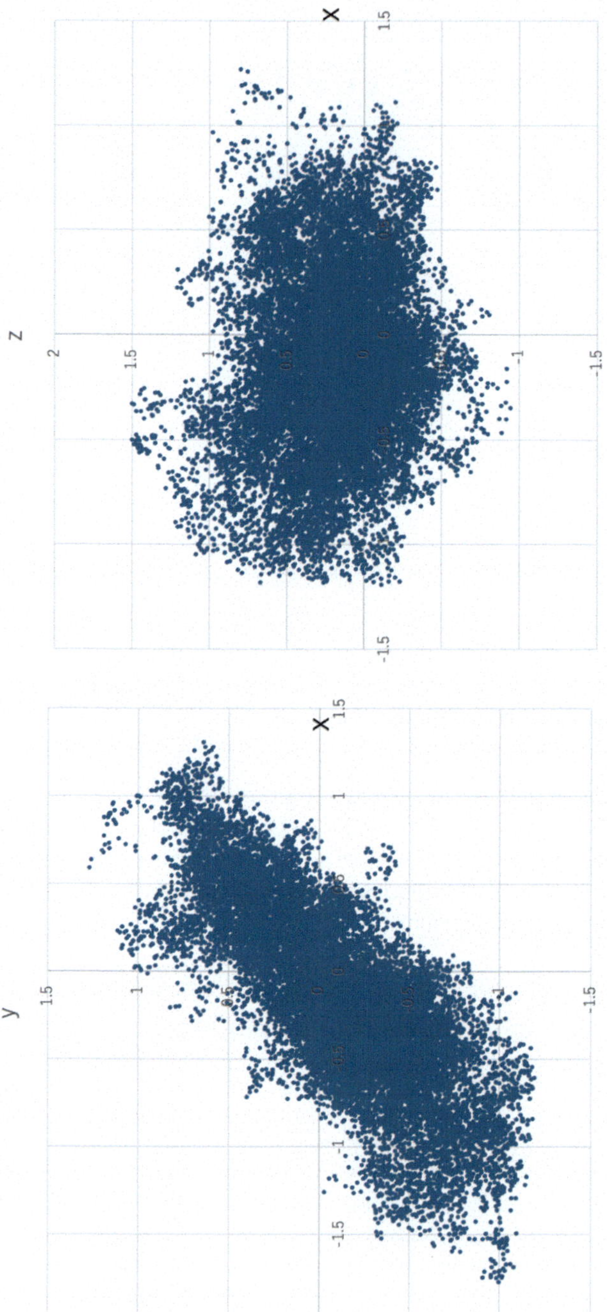

Fig. 5.7 Phase map of the two-dimensional Lorenz dynamic equations (x, y) in left and (x, z) in right

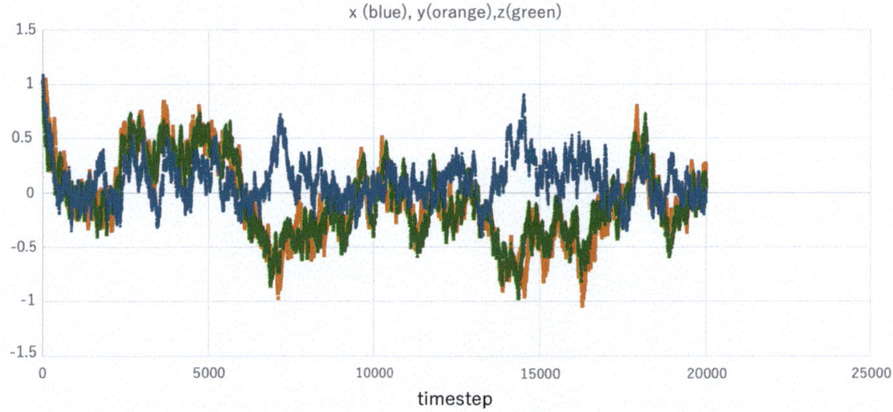

Fig. 5.8 Multivariate timeseries of Lorenz dynamic equations of x (blue), y (orange), and z (green)

a little delay. Furthermore, the 1dCNN predictor from training timeseries of segment lor1-1 can be applicable to the regression of the testing timeseries of lor1-2, thereby showing the plausible prediction of timeseries on the nonlinear dynamics with noisy drift term even in the case of multivariate system as well as one-dimensional system.

Furthermore, it is needed that whether the timeseries governed by the composite nonlinear dynamics system having independently different differential equations is possibly predicted by deep neural network modeling or not. In this investigation, the following composite NLD systems composed of independent two simple two-state dynamics will be discussed,

(1) Random mixture system of two NLD.

$$\begin{aligned} dx/dt &= a\left(1-x^2\right) + wN\left(0, v^2\right), \\ dy/dt &= b\left(y-y^3\right) + wN\left(0, v^2\right), \\ z &= wNx^2 + (1-wN)y^2, \end{aligned} \tag{5.4}$$

in which N is the Gaussian drift term and w is the weight constant of the mixture system.

(2) Constant mixture system of two NLD.

$$\begin{aligned} dx/dt &= a\left(1-x^2\right) + wN\left(0, v^2\right), \\ dy/dt &= b\left(y-y^3\right) + wN\left(0, v^2\right), \\ z &= cx + (1-c)y, \end{aligned} \tag{5.5}$$

in which c is the constant weight of the composite system.

To investigate the natural seismicity phenomena, it needs that the physical process of the solid earth should involve the multivariate linear system combined with various

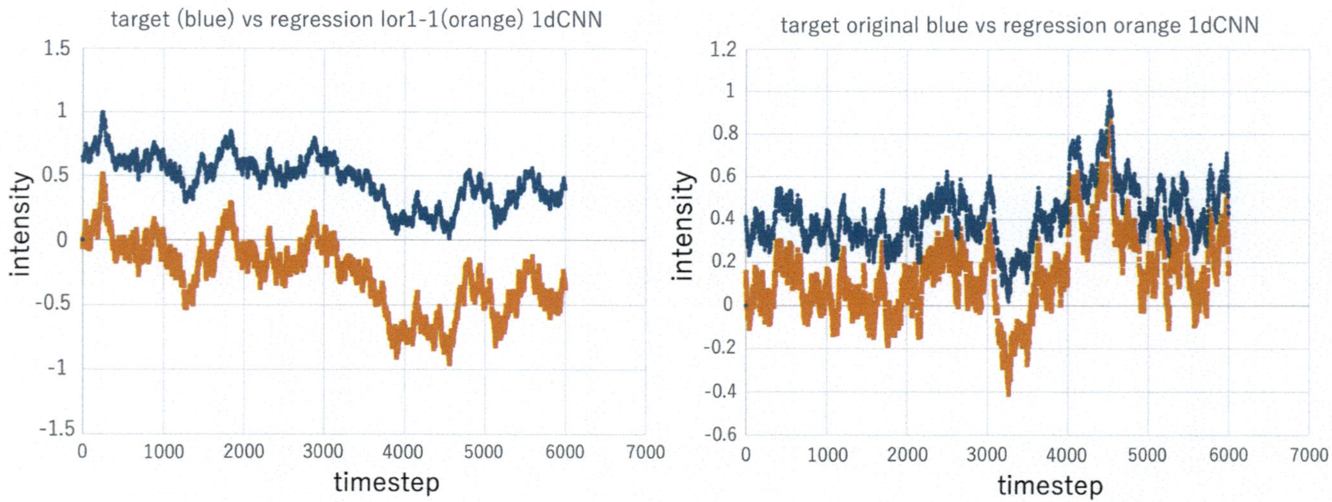

Fig. 5.9 2dCNN regression results of Lorenz chaos from training segment to testing segment using training parameters set. Left; testing timeseries (blue) and regression result (orange) of x, right; testing timeseries and regression result of z

nonlinear dynamical systems, and their coupling mode may have various modes. Above complicated mixed nonlinear dynamical system should be represented as composite nonlinear system. The present types of coupling mode may be plausible to investigate the mode of coupling of the different type of nonlinear dynamical equations as well as the recent developing rate-induced tipping model for nonlinear chaotic dynamics as proposed by Panahi et al. (2023), who studied the complex high-dimensional ecological network.

In the first case of the randomized mixing nonlinear system as displayed in Fig. 5.10, it seems that the self-regression timeseries is quite likely to the simulated one, thereby showing that the 1dCNN modeling of the simulated timeseries is surely available for the regression (Fig. 5.11). In addition, as shown in Fig. 5.11, 1dCNN predictor made from the training segment of the simulated random mixed timeseries is plausible for the regression of the testing timeseries because of clear matching between the original timeseries and regression one.

On the other hand, in the case of solid mixed nonlinear dynamical system (Fig. 5.12), the regression timeseries of the testing segment from the training one is in accordance with the original timeseries of the testing one (Fig. 5.13), thereby manifesting the availability of the 1dCNN modeling as the predictor even in the complex nonlinear dynamics. From these experiments of 1dCNN modeling of the mixing nonlinear dynamical system, it seems that the complicated multivariate timeseries, involving the plural nonlinear dynamics even with chaotic behavior, is possibly predicted by deep neural network modeling as in the simple nonlinear cases mentioned previously.

For both cases of composite nonlinear dynamics, it seems that the feasible basic parameters of 1dCNN modeling are likely to those of time step to be 5 rather than

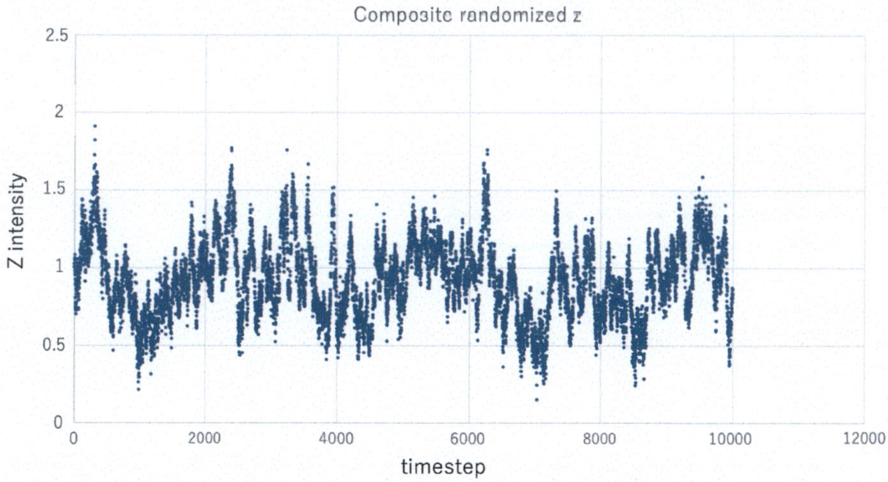

Fig. 5.10 One-dimensional timeseries of X of randomized two nonlinear dynamical equations with noisy Gaussian drift

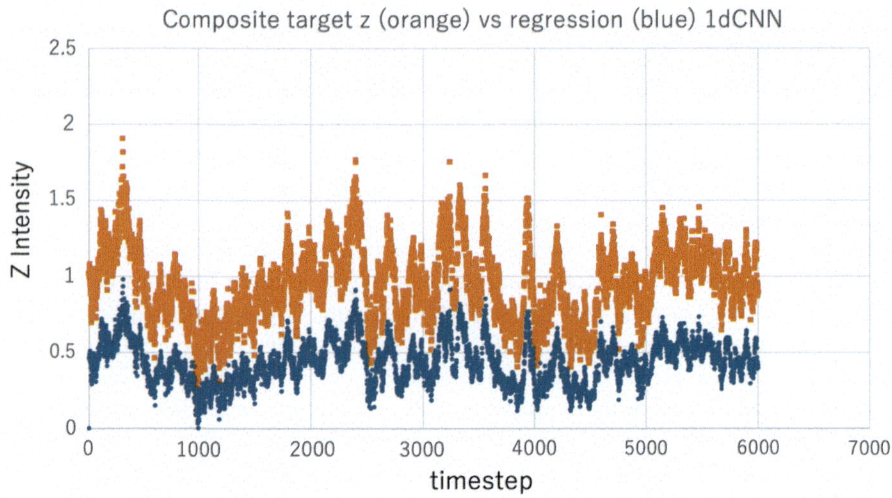

Fig. 5.11 1dCNN regression result of the randomized two nonlinear dynamical equations. Original testing timeseries; orange, regression; blue

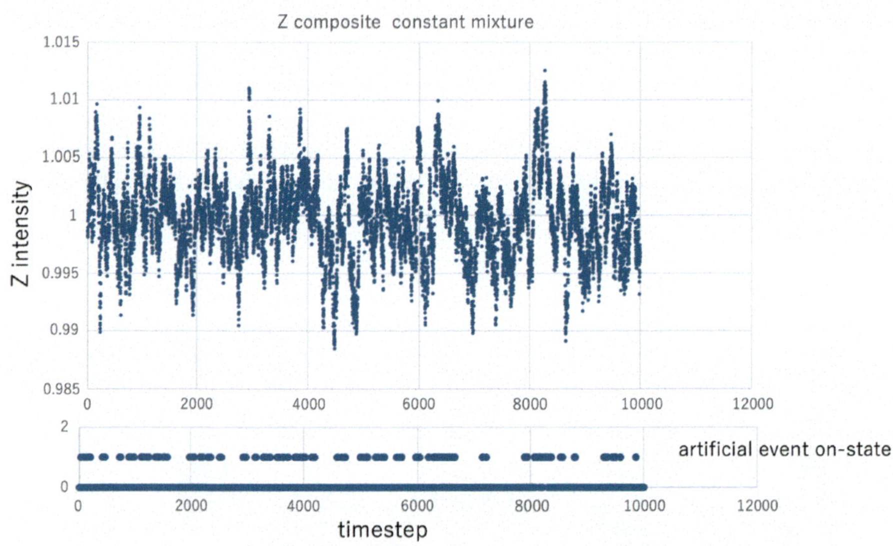

Fig. 5.12 One-dimensional timeseries of solid mixture of two nonlinear dynamical equations with noisy Gaussian drift

3, 4, and 7, and of kernel to be 2 than 3, and the activation function of the final step to be of hyperbolic tangent ('tanh') but not 'relu'.

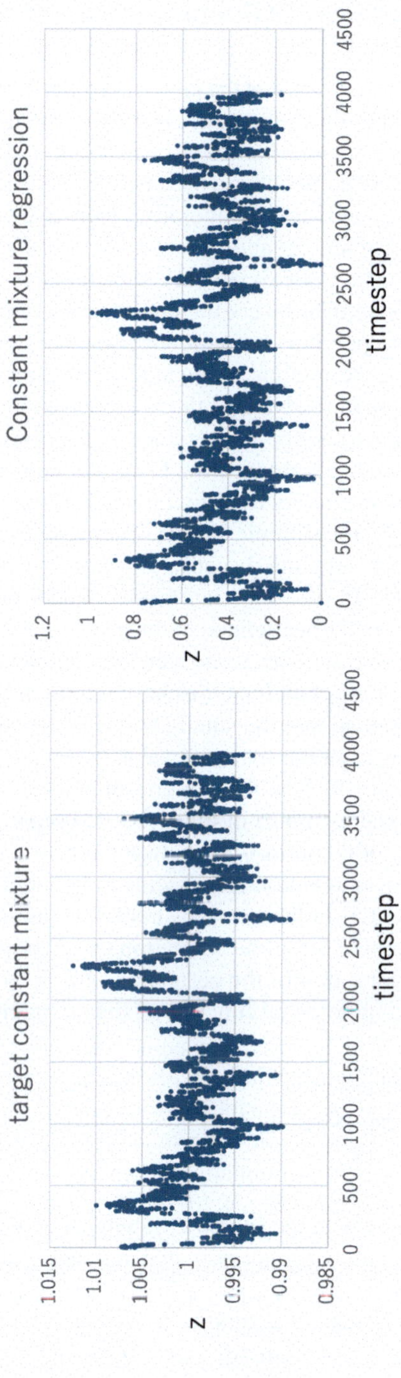

Fig. 5.13 1dCNN regression result of solid mixture of two nonlinear dynamical equations with noisy drift from training segment to testing one. Original testing timeseries; left, regression; right

5.3 Equivalence Between Nonlinear Dynamics and Network Dynamics of Global Correlated Seismicity

In the previous book, it has been clarified that the global correlated seismicity of z_1 and z_3 is governed by the minimal nonlinear dynamics having two dynamical states, and that the correlated seismicity shows the characteristic network of the high seismicity localities along the subduction plate boundary. Meanwhile, a characteristic network of correlated seismicity corresponds to a temporally distinctive stable node in the phase map of the correlated seismicity z_1 and z_3, and in the phase map, the change of nodes manifests the change in the mode of network showing the connection among the gridded localities which is defined by coefficient intensity of eigenvector of the data matrix as discussed in the previous book. It suggests that the network change represents the nodal change in the nonlinear dynamical equation of z_1 and z_3, and thus the hidden physical process of the NLD of the z_1 and z_3 should be derived from the network dynamics with phase change as suggested in previous book.

In the previous book, the author proposed the Gaussian network model for the global and regional correlated seismicity change, in which the potential function is defined by the bilinear form of logarithmic seismicity rate with Kirchhoff matrix defined by geometry of network structure, because of symmetric nature of potential between two grids of network. In the Gaussian model (Rader et al. 2006; Felice and Mancini 2015), the partial potential defined by bilinear combination of network nodes i and j is represented by $x_i x_j$ with the same weight coefficient. However, in the actual seismicity network, the important localities have strong intensity coefficients in the correlated seismicity rate, so that the small change of seismicity rate in their localities represents large change in the potential value presented there. Therefore, the Gaussian network potential using Kirchhoff matrix should be modified to involve the term of intensity of the nodal localities of the correlated seismicity.

It is sure that the symmetry of the interaction between network nodes is required, and thus the minimal model of interaction term seems to be a bilinear form as like as the Gaussian network potential in the previous book. For the reason mentioned previously, the weight coefficient of the bilinear function should be proportional to the intensity coefficient of the eigenvector of the correlated seismicity rates. Therefore, we prefer the following minimal potential function of the seismicity network,

$$
\begin{aligned}
U &= a\left(x'K_1^t x'\right) + b\left(x'K_1^t x'\right)^2 + c\left(x'K_1^t x'\right)^3 + \cdots \\
&= az_1^2 + bz_1^4 + cz_2^2 + dz_1 z_2 + \cdots . \\
x' &= Ax,
\end{aligned}
\tag{5.6}
$$

in which K_i is the Kirchhoff matrix of the network system and A is weight coefficient of eigenvectors of seismicity matrix X, that is z_i, and a, b, and c are constant terms, respectively.

Considering two network systems, the transition between them is interpreted by taking the minimum condition of potential difference between two network systems.

In addition, the transition rate seems to be expressed by the derivative of the above potential difference as,

$$
\begin{aligned}
\mathrm{d}z_1/\mathrm{d}t &= \mathrm{d}\Delta U/\mathrm{d}z_1 + N(m, v),\\
\mathrm{d}z_2/\mathrm{d}t &= \mathrm{d}\Delta U/\mathrm{d}z_2 + N\left(m', v'\right),
\end{aligned}
\tag{5.7}
$$

in which N is the Gaussian noise function for random drift. These equations are to be Langevin type stochastic equations for the correlated seismicity rates of the seismic network systems.

It is obvious that the above Langevin type differential equations have several stable and unstable nodes in the z_1 and z_2 phase map by taking the null clines of two ODEs. In addition, it is important that there should be the tangential relation between null clines. In this type of null cline relation, the behavior of the solution of ODEs should be so-called Millner attractor and it has a chaotic behavior with large variance, suggesting the appearance of very largely correlated seismicity rates following the giant earthquake. In the previous books (Toriumi 2021, 2022), it has been pointed out that the 2004 Sumatra and 2011 Tohoku-Oki giant earthquakes were associated with the very large and global correlated seismicity together with very large variance.

As is formulated previously, the global correlated seismicity dynamics is governed by the seismicity network potential gradient, and it holds,

$$
\begin{aligned}
\mathrm{d}z_1/\mathrm{d}t &= \mathrm{d}\Delta U/\mathrm{d}z_1 + aN(0, v)\\
&= \left(2az_1^1 + 4bz_1^3 + dz_2\right) + C_1,\\
\Delta U &= 4az_2^2 + 12bz_1^4 + dz_1z_2 + C_2.
\end{aligned}
\tag{5.8}
$$

In these equations, a, b, and c are the coefficient terms, and C_1 and C_2 are constant terms with the Gaussian noises for z_1 and z_2, respectively. Therefore, the null clines of them are,

$$
\begin{aligned}
&z_1 + (2b/a)z_1^3 + (d/2a)z_2 + C_1/2a = 0,\\
&z_2 + (d/2c)z_1 + C_2/2c = 0.
\end{aligned}
\tag{5.9}
$$

These equations have two types of solutions: one is three solutions, and another is two solutions as shown in Fig. 5.14. It is noticed that the latter contains a single of double solution. The former solution means that there are two stable nodes and one unstable one, and the latter shows one degenerated node and one stable node in the ODEs presented previously.

Therefore, it is concluded that the seismic network dynamics which has been proposed in the previous book (Toriumi 2022) is just represented by the nonlinear dynamical differential equation concerning the correlated seismicity in the global earth mechanics. The model system in the correlated global seismicity dynamics involves the minimum two-state dynamics composed of the single node to two stable nodes with advancing the parameters of the system, being likely to the pitchfork

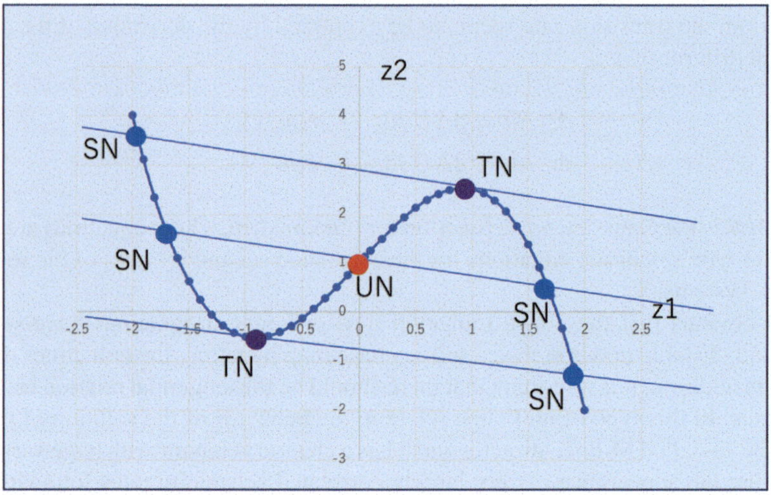

Fig. 5.14 Null clines of nonlinear dynamical equations representing different types of nodes by various parameters of z_2 null cline. SN: stable node, UN: unstable node, TN: tangential node

bifurcation. In this dynamical model, it is expected that the one critical node is the tangential relation between two null clines above, and that critical node is characterized by the large variance of the correlated seismicity triggering the large earthquakes. It is sure that the giant earthquakes of 2004 and 2011 occurred near the critical node of the global correlated seismicity as labeled cluster C in the $z_1 - z_3$ phase diagram (Fig. 5.15).

By the way, in the previous book (Toriumi 2022), the author presented the circular-connected slider-block model of the global seismicity network, and the characteristic correlated motions and strain energy of the many nodes are obtained from the singular value decomposition as like as the global correlated seismicity in nature. In that model, there is a single connection of nodes along the plate boundary: that is the circular connection both of high seismicity subduction boundary and low seismicity ridge boundary. The time evolution of the correlated motions and strain energy seems to be represented by the transition of the characteristic network structures of nodes as well as the global correlated seismicity timeseries.

However, it is further required that the network geometry of the circular subduction plate boundary in the global earth is more complicated and keeping the basic connection of studied nodes of gridded localities along the subduction zone and ridge boundary, and consequently it shows that the network geometry should be composed of three arms of linear array of seismicity nodes as shown in Fig. 5.16. It is important that the three terminal points relate to the global ridge, which are inactive in the seismicity because of dissipation of strain energy by viscous flow of the solid earth in the high temperature condition. Furthermore, these terminal points of the network are to be fixed for the time scale of this investigation because of the triple point of the ridge-ridge-trench.

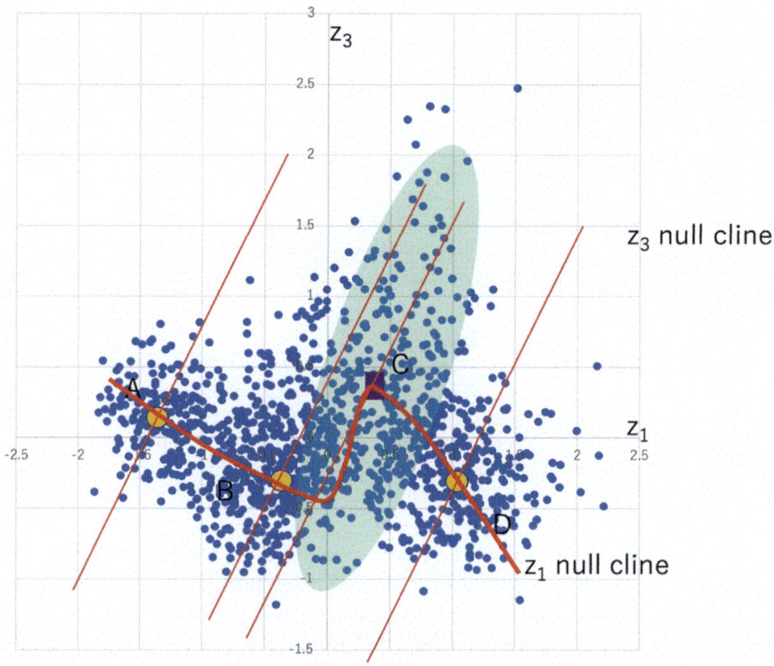

Fig. 5.15 Model of null clines of nonlinear differential equations concerning the global correlated seismicity dynamics. A, B, and D: stable node, C: tangential node with large variance

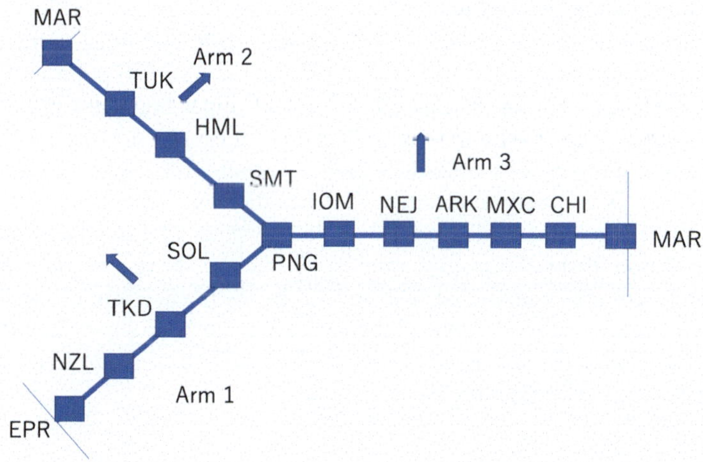

Fig. 5.16 Three-arms slider-block model of global subduction zone system, showing the arm1 connecting from Papua New Guinea (PNG) to New Zealand (NZL) and ended at East Pacific Rise (EPR), arm2 from PNG to Turkey (TUK) ended at Mid-atlantic Ridge (MAR), and arm3 from PNG to Chile (CHI) ended at MAR. Arrows: direction of relative plate velocity

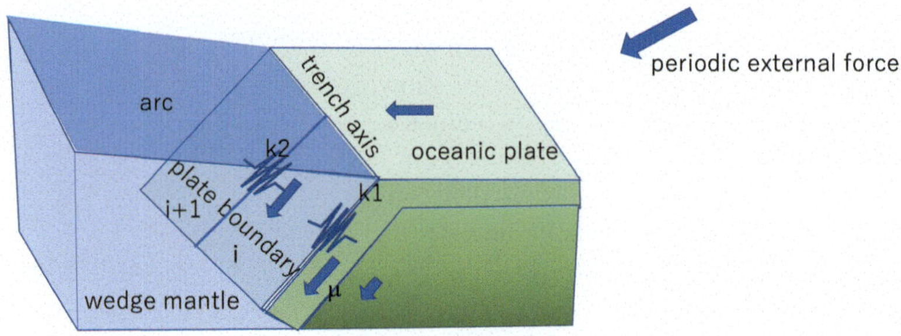

Fig. 5.17 Illustration of mechanical situation of representative nodes i and $i + 1$ in the three-arm network. k_1: elastic modulus between node i and $i + 1$, k_2: elastic modulus between node i and oceanic plate, μ: viscosity between oceanic plate and underling mantle

The one arm of the above network structure is the Tonga-Kermadic-New Zealand arm, the second one is the Sumatra-Himalaya-Turkey arm, and the third is the Izu-Mariana-Northeast Japan-Alaska-Mexico-Chile arm. The triple junction of their arms is just west of Papua New Guinea as displayed in Fig. 5.16. In this study, the slider-spring block model of the above-mentioned three-arm networks will be simulated as like as the previous circular network system, and it will be investigated that whether this simple system with specified geometry behaves like the nonlinear dynamics with double states or not.

The slider-spring block model is constituted with the three arms linearly connecting the many nodes and ended at the immobile terminal point on the ridge boundary. The node is connected side by side and upper boundary by the spring (elastic connection), and the bottom of the node is by dash pod as shown in Fig. 5.17, and furthermore it affected by time-dependent external force (earth rotation change and tidal force). It is greatly concerned that whether the correlated oscillation of nodes displays the nonlinear resonance with double states like stochastic resonance or not. In addition, it should be quite interesting that the above system displays the characteristic networks constituted with several nodes in each arm or inter-arm. In the natural seismicity network of the global system, it has been shown that the nodes of NEJ, SUM, CHI, and TKN are obviously connected with each other, and the mode of network changes with the time advancement.

5.4 Three-Arm Connected Slider-Block Network Model

The geometry of the simple slider-block model connected three arms linearly is shown in Fig. 5.16. It is assumed that the block of i is elastically connected to $i + 1$ block with spring constant k_2, and to the overriding arc crust and mantle with k_1, and furthermore block i is suffered from viscous drag force with viscosity μ and velocity different v_i. As shown in Fig. 5.16, the arm1 and arm2 are the subduction boundary of the oceanic plate and continental plate, but the arm3 is constituted with the subduction boundaries of continent and continent from Himalaya to east of Mediterranean, and oceanic plate and continent from west of Papua New Guinea, and therefore, it is available that the elastic constant between oceanic plate and overriding plate should be smaller than that between continent–continent subduction boundary. In this simulation, k_1 and k_2 are given by 10 and 30, because of simple subduction of oceanic plate and continental collision, respectively.

Each block of the three-arm system is initially given by small random displacement, and initial velocity, and the block is given by external force fa. The boundary condition at the end of the three arms is given by the zero displacement and velocity. The external force on each block is given by the sinusoidal function with some degree of amplitude and wavelength, to investigate the synchronicity between correlated displacement of the blocks and external force, being analogous to periodic force operating on solid earth by secular change of the earth rotation.

The formulation of this three-arm slider-block network motion holds as,

$$
\begin{aligned}
F_i &= \mathrm{fa}(t) + k_1 x_i + k_2 \Delta x_{ij} - m\mathrm{d}x_i/\mathrm{d}t, \\
F_i &= m_i \mathrm{d}^2 x_i/\mathrm{d}t^2,
\end{aligned} \tag{5.10}
$$

in which fa(t) is periodic external force, k_1 and k_2 are the elastic constants, and m is the mass of the block. F_i and x_i are the force on block i and its displacement, respectively, as shown in Fig. 5.17.

Experimental results by simulation are shown in Figs. 5.18 and 5.19, by the spatial–temporal change of displacement of each block along three arms, and the correlated displacement of all blocks. The correlated displacement is estimated from the singular value decomposition (SVD) of the data matrix composed of timeseries of the displacement of all blocks. The data of displacement timeseries in representative blocks are shown in Figs. 5.18 and 5.19. It shows that the displacement timeseries of the tenth block in the arm1 to arm3 appear apparently noisy but slightly periodical, and these display mutually different patterns. In addition, the spatial changes of the displacement of each block are different in each arm, thereby suggesting that the characteristic features of three arms slider-block connection model are likely a little bit to synchronized patterns with periodic external force.

On the other hand, the timeseries of the correlated displacement of total system appears the interesting temporal variation as shown in Fig. 5.20. It is obvious that the timeseries involves temporal variation of short-term periodicity and long-term one, and it is incoherent with the periodic change of the external force. Thus, it is very interesting that the phase map of the correlated displacement vs external force suggests the multiple state modes of them as shown in Fig. 5.20, thereby suggesting that if the external force periodically changes, the correlated displacement timeseries of the three-arm slider-block model displays likely to that of the nonlinear multiple

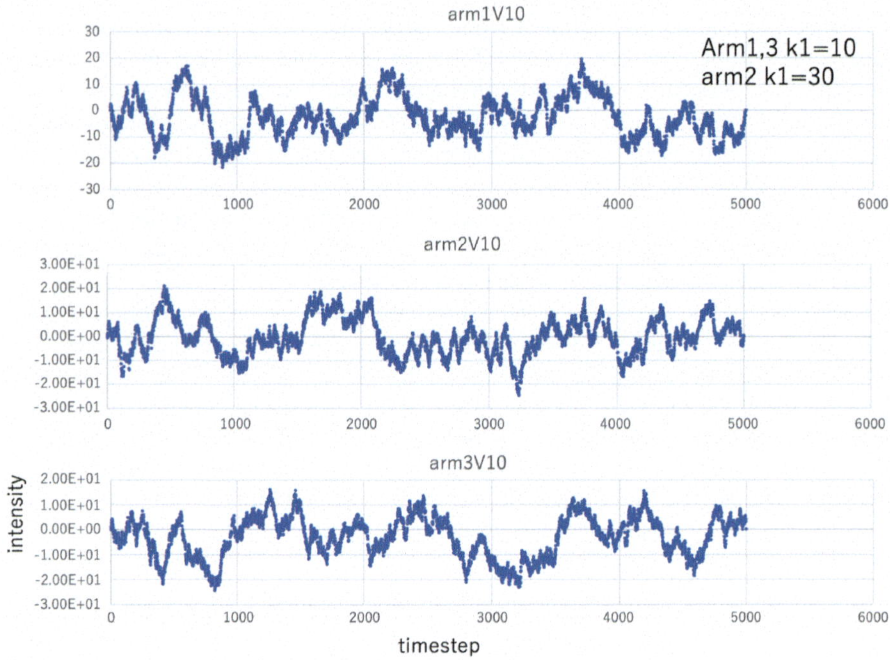

Fig. 5.18 Temporal variation of simulated displacements at position $x = 10$ of arm1, arm2, and arm3, under periodic external stress condition, showing the weak cyclic variation

state dynamics, as being discussed in the global seismicity change against the earth rotation rate.

On the other hand, it shows that the phase map of the z_1 and z_2 of the correlated displacement of each block in total arms suggests characteristic features consisting of three segments around the center as shown in Fig. 5.21. Furthermore, it seems that three segments of them manifest the temporal sequence of the timeseries, thereby suggesting the appearance of the correlation with specified network of three arms. Therefore, it is needed that how connection of the arms and their blocks are strongly connected with each other in the dynamics involving the periodic external force.

By the way, for investigation of the mode of connection in the network dynamics, it is available that the SVD method of the multivariate timeseries matrix can give the eigenmatrix composed of the eigenvectors, which show the major blocks contributable for the eigenvector in all arms, and thus the mode of connection of block in the present network (Fig. 5.21). As shown in Figs. 5.22 and 5.23, it is obvious that there are different compositions of major components of the slider block in the specified network, corresponding to each segment of the correlated displacement of z_1 and z_2 in the phase map. In the primary and second phases of their timeseries (A and B), the blocks of arm 1 and arm 3 are dominantly active and arm 2 is subsidiary, but in the third phase (C), the blocks of arm1 are inactive and arm 2 and arm 3 are less active rather than those of A and B. Therefore, it suggests that the global

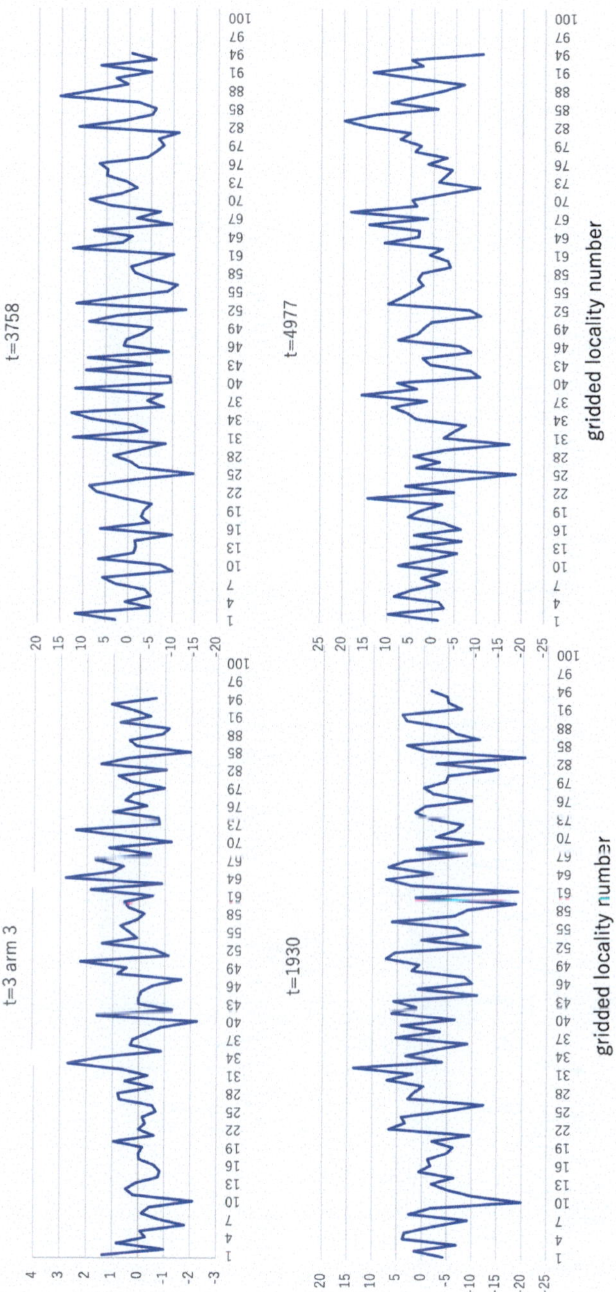

Fig. 5.19 Spatial patterns of node displacement with time step advancement of the arm3 (top) and arm2 (bottom)

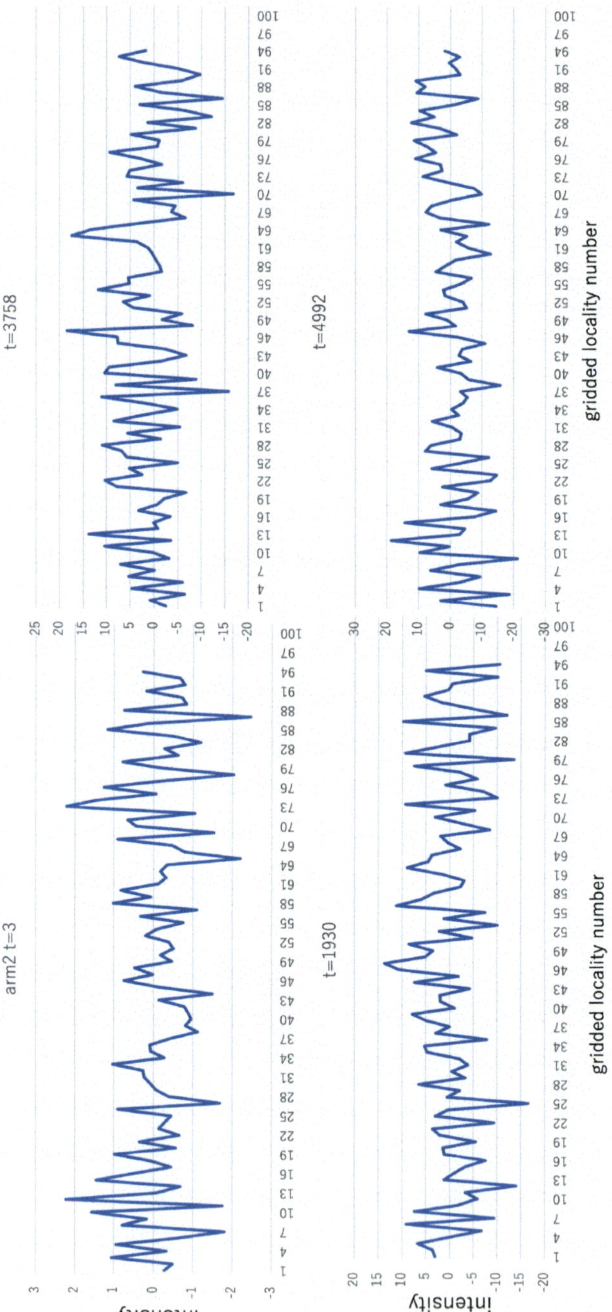

Fig. 5.19 (continued)

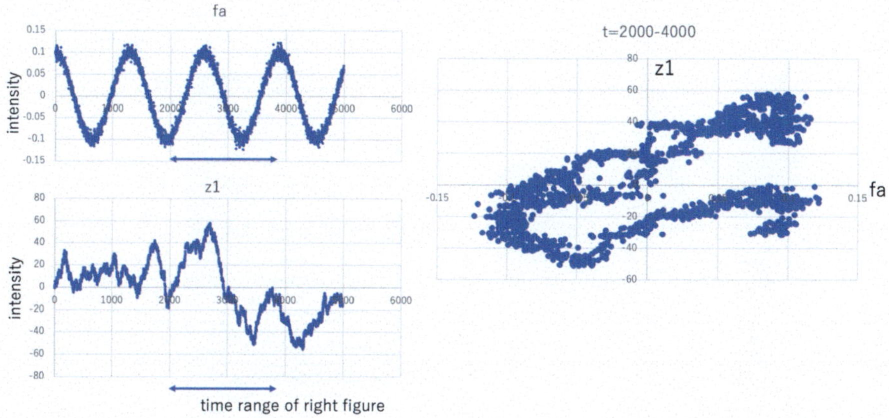

Fig. 5.20 Temporal variation of global correlated displacement z_1 of three-arms system and periodic external force fa in the left figure and phase map of fa and z_1 of the selected time range shown in the right one

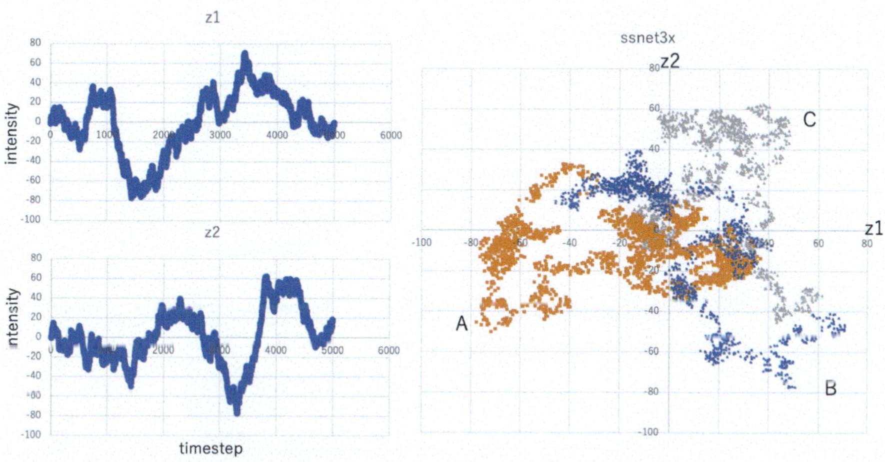

Fig. 5.21 Temporal variations of global correlated displacement of three-arm model z_1 and z_2 (left) and their phase map (right). Phase map appears three cyclic loops A, B, and C with clear symmetry

behavior of the present network dynamics displays the cyclic activity of the different slider blocks belonging to the arms, being driven by the periodic external force, thereby suggesting the rotational itinerancy of the stable nodes in the phase map of the correlated multivariate parameters.

In the seismicity dynamics of Japanese region, it has been clarified that the correlated regional seismicity rates are controlled by the seismicity network dynamics and by the minimal nonlinear dynamical equation of the correlated regional seismicity rates (Toriumi 2022). The key observation is that there is a critical node showing

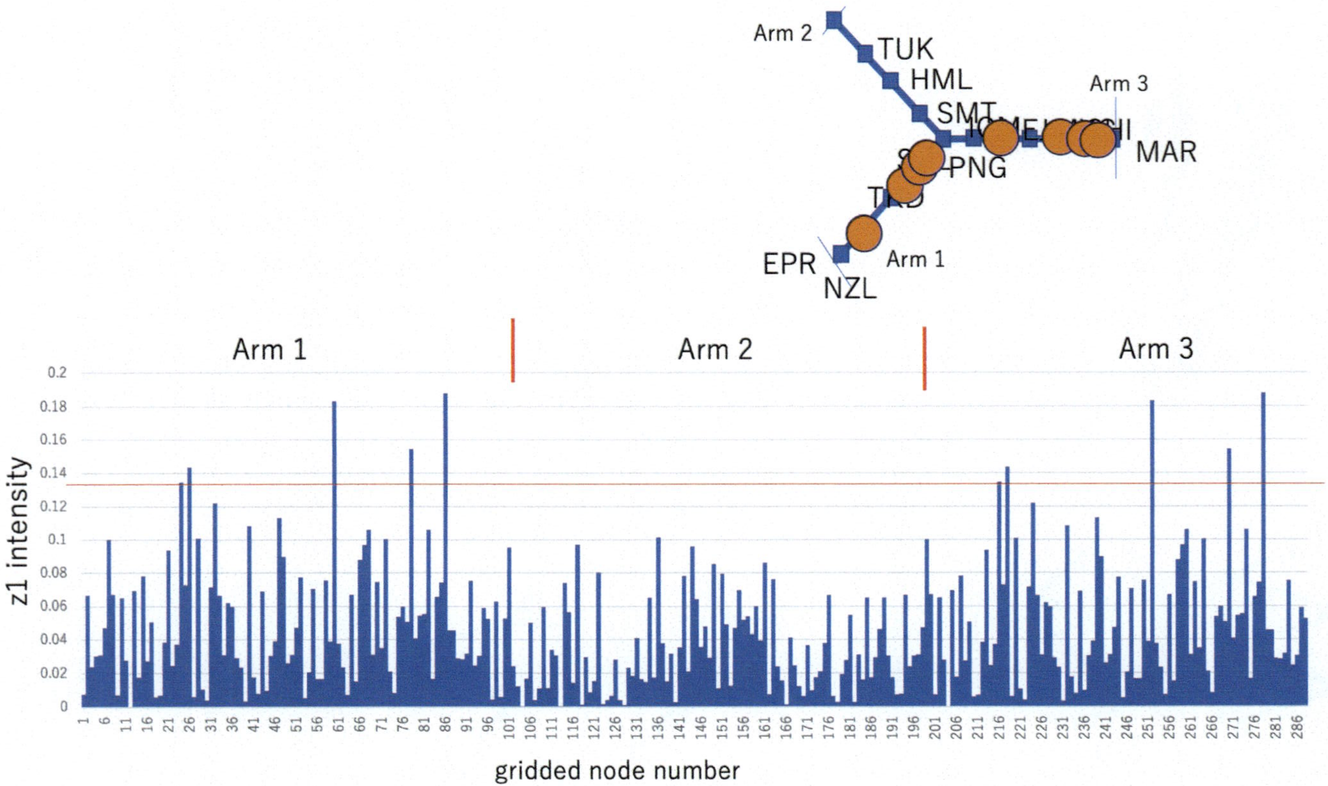

Fig. 5.22 Intensity map of correlated displacement z_1 in the period of A and B in Fig. 5.21 showing active nodes in arm1 and arm3, but inactive in arm2

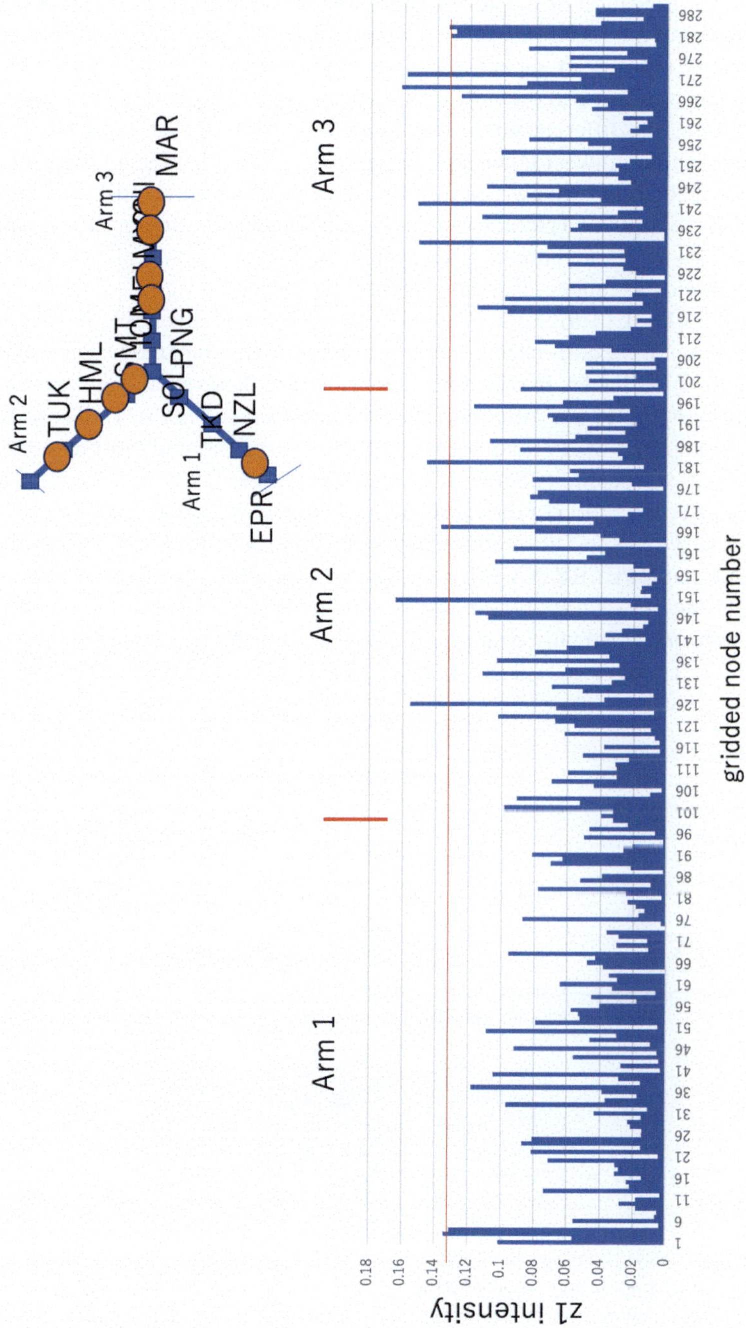

Fig. 5.23 Intensity map of correlated displacement z_1 in the period of C in Fig. 5.21 showing active nodes in arm2 and arm3, but inactive in arm1

the rapid and large variance of the correlated seismicity in the phase map, and that there is a gradual transition of the seismicity network modes near the critical node discussed in the previous section.

The seismicity network of Japanese islands region seems to be likely to that of the global system as discussed previously. The Japanese islands region is composed of three arms along the different subduction boundary: the northeast Japan arm, Izu-Ogasawara arm, and southwest Japan arm. In natural system, it has been clarified that the correlation of nodes of Okinawa, central Kyushu, Kii, Off-Kanto, Off-Fukushima, Off-Sendai, Off-Tokachi, and Nemuro are intense and changes with time. The change of network modes mentioned above is seemingly important for investigation of the seismicity dynamics of Japanese island region.

In the previous book (Toriumi 2022) and in this chapter of this book, it has been clarified that the set of the correlated seismicity z_1 and z_3 of the Japanese island region appears sporadic five clusters in their map from A to E with advancing the time, and they are corresponding to the different seismicity networks (Fig. 5.24). These five networks are composed of Ryukyu segment, Nankai segment, and Northern Japan segment. The present seismicity network does not contain the Izu-Bonin-Mariana arc because of weak contribution of the correlated seismicity of Japanese island region. Network geometry of A to D/E mentioned above changes from A; RY-NA, B; RY-NJ, C; NA-NJ, and D/E; RY-NA-NJ-RY, thereby suggesting that before D stage, the seismicity join between two segments is apparently active, but in the recent D/E stage, the network drastically changes to the whole seismicity connection among RY, NA, and NJ. It is obvious that the Japanese island region changes a single whole seismic active zone.

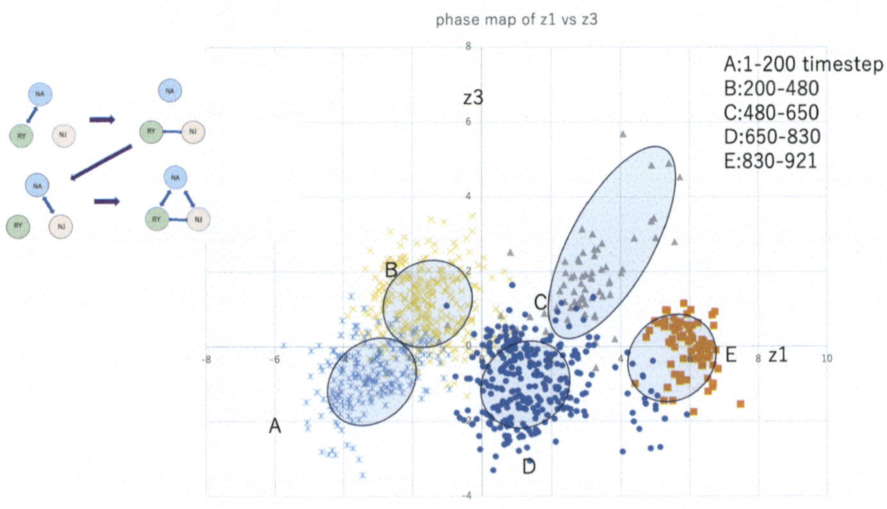

Fig. 5.24 Phase map of the z_1 and z_3 regional correlated seismicity of Japan, showing the discrete clusters of A to E from 1998 to 2023. Time step ranges are shown in right top. Left top: regional network of regional correlated seismicity appeared in previous book (Toriumi 2022)

On the other hand, it is important that the phase map of the correlated seismicity z_1 and z_3 (also z_2) displays the apparent discrete clusters of A to E as shown in Fig. 5.24. Considering the random Gaussian noises contained in the correlated seismicity (Toriumi 2021), it seems that the discrete cluster of correlated seismicity rates manifests the characteristic physical meaning on the seismicity network geometry: the jump of the correlated seismicity set in the phase map may correspond to the geometry change of node connection of the network as discussed previously. In addition, it is probable that the stable set of the correlated seismicity of the Japanese island region keeps during the present seismicity network geometry as shown in Figs. 5.25 and 5.26, even if the external force and related global seismicity change takes place in this geometry of seismicity network as shown in Fig. 5.27.

By the way, it is shown that the present network model displays the characteristic behavior of the displacement timeseries of every node as shown previously, and the correlated displacement timeseries involving all nodes of the three arms are plotted to be formed as several discrete clusters and scattering between these clusters, thereby suggesting that each cluster is formed by the correlated displacement sets of the single network geometry of the coherent displacement of nodes. Accordingly, it may conclude that the dynamical behavior of the multiple network system of the seismicity in the global and regional cases represents nonlinear dynamical equations having multiple stable nodes as discussed in the previous book (Toriumi 2022).

In this section, it has been clarified that the three-arm slider-block model operated by external force, applicable for the global seismicity network, is characterized by the multiple stable states of the correlated displacement of the constituent nodes. The migration of the stable states shown in the phase map of the correlated displacement of nodes is driven by the periodic variation of the external force with random noise, but during the finite period under the external force, the system behavior represented by the present correlated displacement has a stable state with random noise. It is, furthermore, noteworthy that the transition from discrete stable states takes place during the short time in the phase map of the correlated displacement, suggesting that network geometry changes drastically in the short time as like as the phase transition of the crystal.

This type of geometry transition of the network dynamics is, therefore, approximated to be the nonlinear dynamics of the phase correlation parameters operated by the periodic external force. In addition, it suggests that this adjoint nonlinear dynamics has multiple stable nodes in the phase correlation map, which are associated with rapid trails between the stable nodes. Moreover, if the case is ranged within the single cycle of the periodic external force, the phase map displays the pitchfork type pattern of transition from single state to double state, thereby concluding that the dynamics of the correlated parameters in the network dynamics of multiple segments is equivalent to the nonlinear differential dynamics having multiple stable nodes in the phase map.

On the other hand, it has been shown that the global correlated seismicity changed drastically after 2003 just before the 2004 Sumatra giant earthquake in the cluster C of the z_1–z_3 phase map. Furthermore, the global network of the seismicity changed from the Pacific type (TK-CH-NJ) to the global type involving Indian Plate type

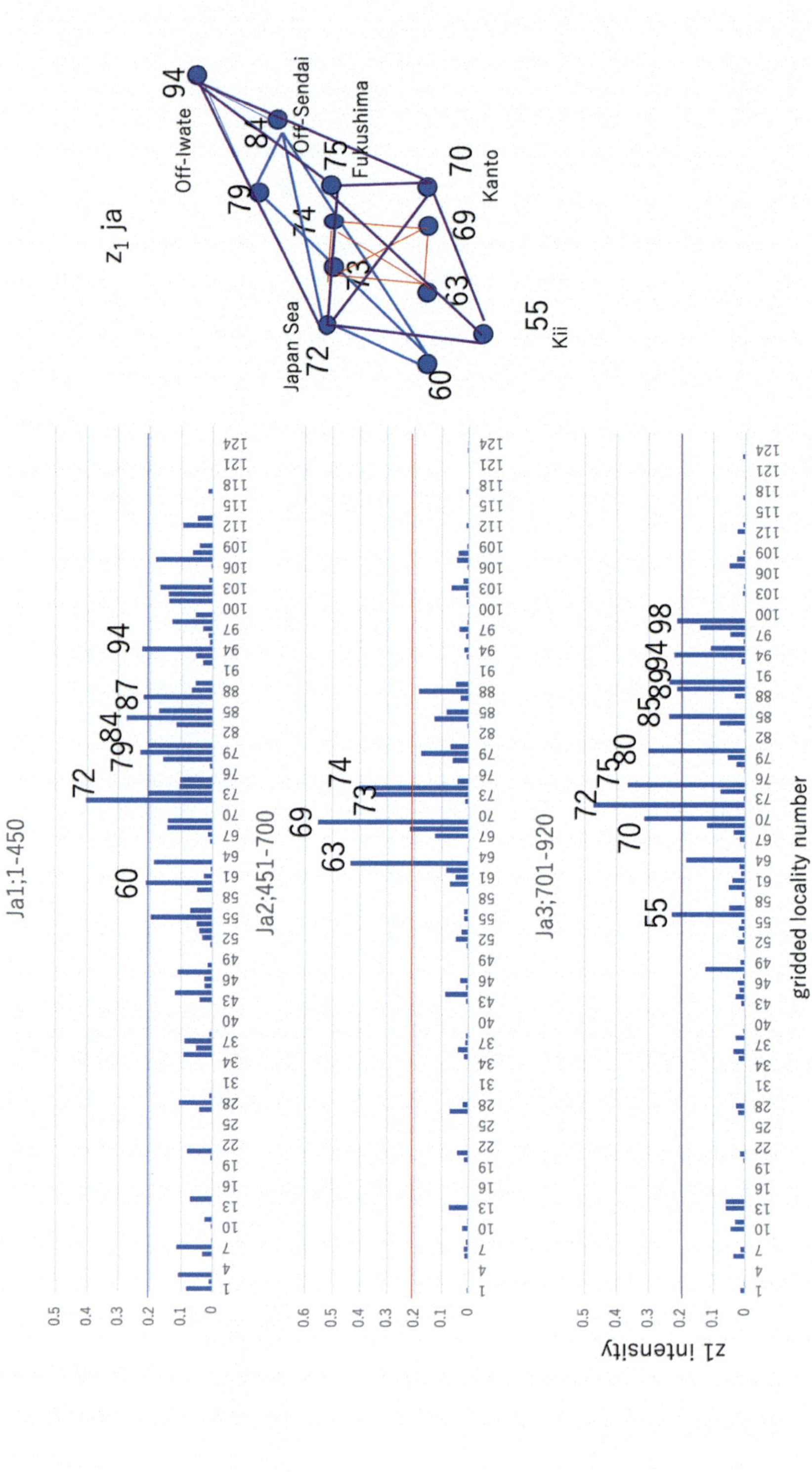

Fig. 5.25 Intensity map of regional correlated seismicity z_1 of Japanese region in the periods of 1–450, 451–700, and 701–920 steps (left) and inferred seismicity network structure from blue > red > violet

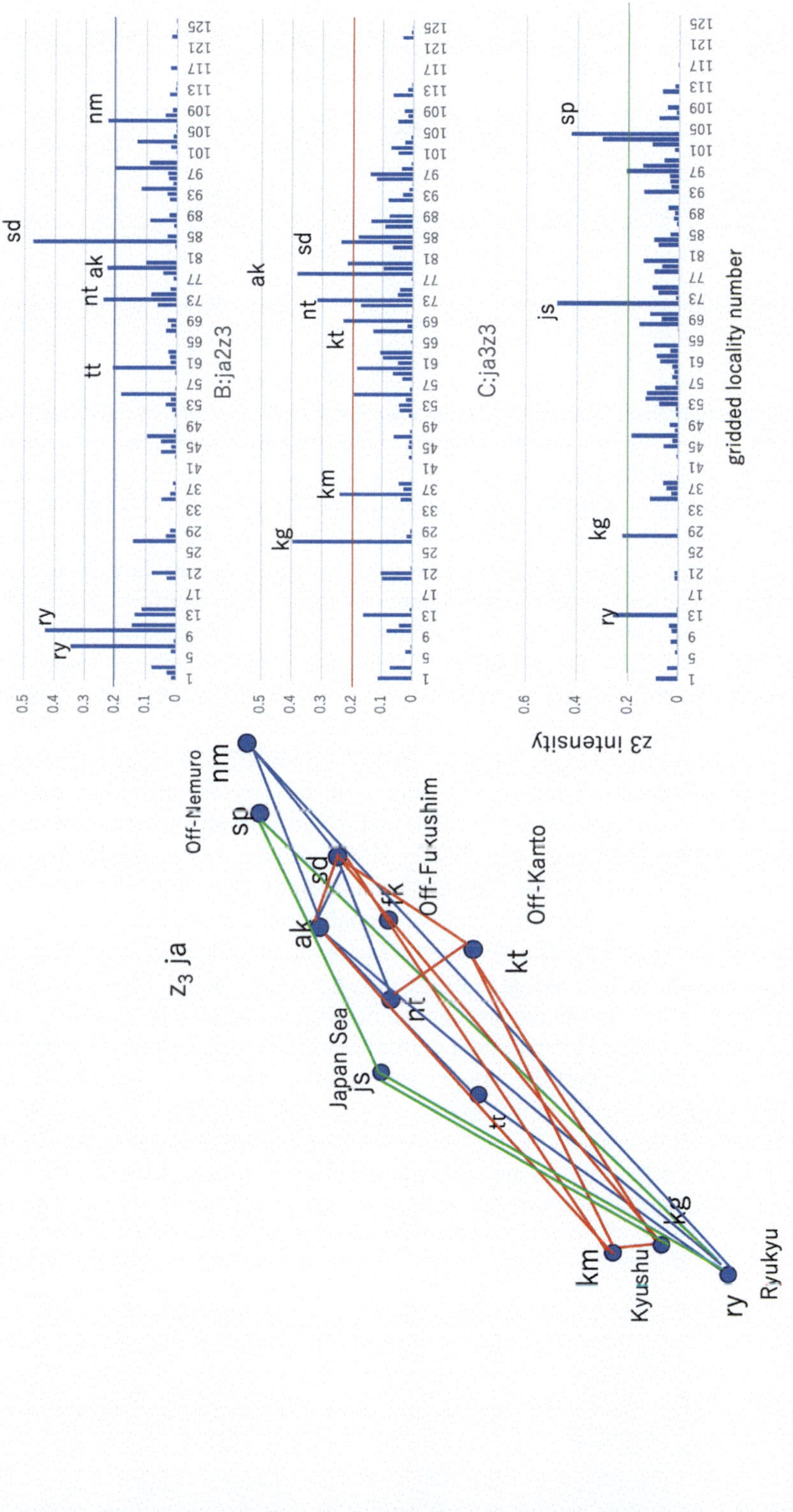

Fig. 5.26 Intensity map of regional correlated seismicity z_3 of Japanese region in the periods of 1–450, 451–700, and 701–920 steps (left) and inferred seismicity network structure from blue > red > green

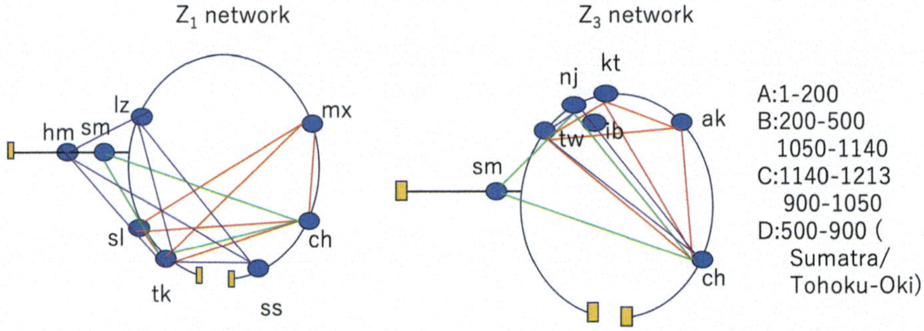

Fig. 5.27 Global correlated seismicity network inferred from intensity map of the global correlated seismicity rates

(SM-NJ-CH), suggesting transition toward the whole earth activity of seismicity as shown in Fig. 5.27. Then, it seems that the important change in seismicity network of the regional system of the Japanese islands took place in connection with the global change of the seismicity network. In the previous book (Toriumi 2021) clarified that the regional correlated seismicity of the northern Californica displays the gradual increase of z_1 and z_2 components, and that the phase map of the their correlated seismicity rates manifests the distinctive change of the clear cluster from A (low z_1) to B (high z_1), thereby showing the evidence that the change in seismicity network of the global seismicity dynamics strongly controls the change of regional seismicity network system of the various areas including northern California and Japanese Islands.

By the way, as discussed in the previous book (Toriumi 2022), it is very important that the global correlated seismicity z_1 is clearly governed by the earth rotation rate, that is the rate of the length of day (LOD), and in the spin-down regime of earth rotation it splits into high level and low level of the correlated seismicity, but in the spin-up regime, it shrinks the medium level, thereby indicating the pitchfork bifurcation pattern. It is noteworthy that the global correlated seismicity z_2 by the decomposition method of non-negative matrix factorization (NMF) against the LOD rate displays the same pattern as that of bifurcation previously noted (Fig. 5.28). This type of pitchfork bifurcation in the nonlinear dynamics is just seen in this three-arm slider-block system applied by external periodic force. It looks like that the external force is the periodic change of the whole earth strain by change in ellipticity of the earth by the changing rate of earth rotation. Furthermore, it seems to be important that the relation between the regional correlated seismicity z_1 and the rate of LOD of Japanese islands region shows the bifurcation pattern with supercritical transition type, showing the middle intensity of the correlated seismicity in the spin-down regime of the LOD (Fig. 5.29).

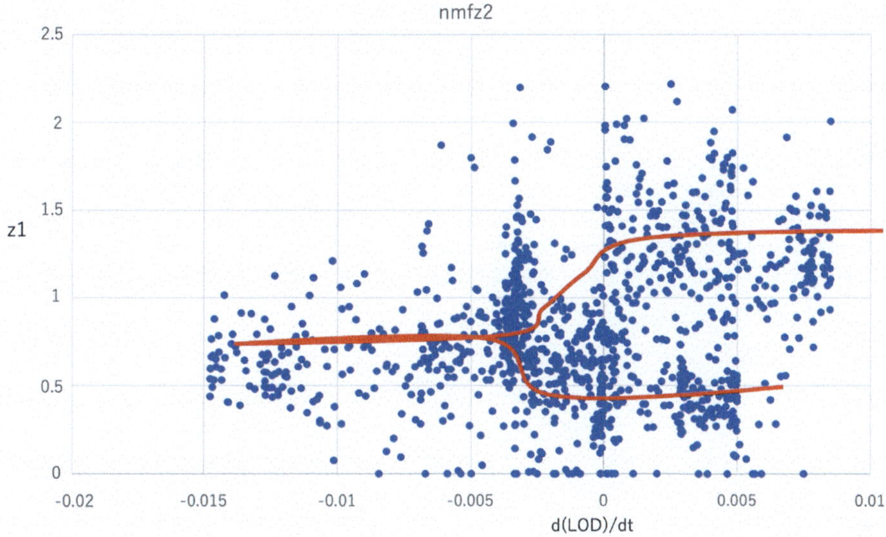

Fig. 5.28 Pitchfork bifurcation pattern of the non-negative factorization decomposition (NNF) of multivariate global seismicity rate timeseries matrix against rate of earth rotation (LOD), showing the same pattern of global correlated seismicity obtained by SVD

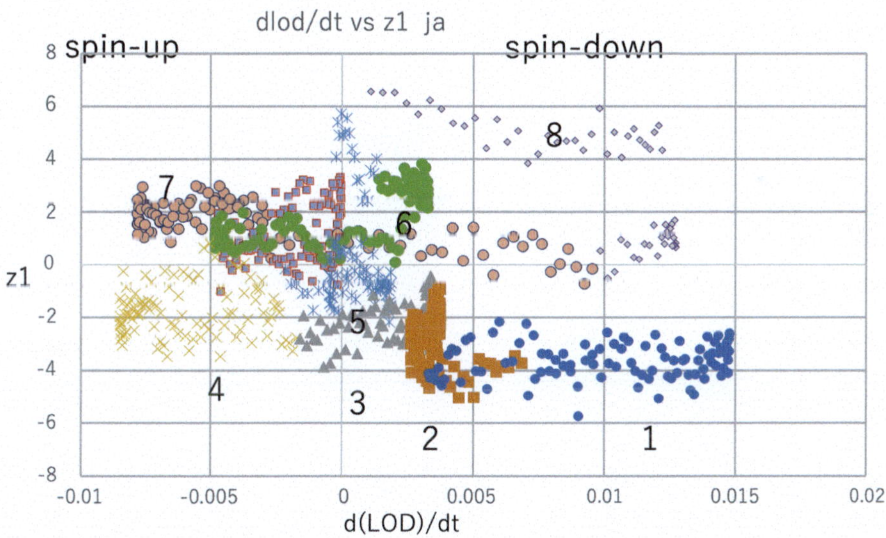

Fig. 5.29 Supercritical bifurcation pattern of the relation between rate of earth rotation and regional correlated seismicity of Japanese islands z_1. Number shows the successive order of the time step from 1998 to 2023

References

Felice D, Mancini S (2015) Gaussian network's dynamics reflected into geometric entropy. Entropy 17:5660–5672. https://doi.org/10.3390/e17085660
Panahi S, Do Y, Hastings A, Lai Y-C (2023) Rate-induced tipping I complex high dimensional ecological networks. PNAS 120(51):e2308820120. https://doi.org/10.1073/pnas.2308820120
Rader AJ, Chennubhotla C, Yang LW, Bahar I (2006) Gaussian network model: application and theory, BICH: 41–64
Toriumi M (2021) Global seismicity dynamics and data-driven sciences. Springer
Toriumi M (2022) Geochemical mechanics and neural network model. Springer

Chapter 6
Augmentation of Timeseries and DNN Modeling of Seismic Activity

Abstract To predict the large earthquakes by deep neural network modeling and more generative AI modeling, it is required that available augmentation of multivariate timeseries should be developed for the sets of large earthquakes and background characteristic features inferred from data-driven science investigation of monitoring big data in the earth science. In this chapter, the new augmentation method of multivariate timeseries of the global and regional correlated seismicity and local seismicity rates are proposed, being based on generative diffusion augmentation using Gaussian mixed modeling. In addition, after one-step ahead prediction of multivariate timeseries of correlated seismicity and local seismicity rates, the full prediction of one-year ahead timeseries of large earthquake is possibly performed by 2dCNN modeling. Furthermore, it suggests that the latent space mapping method of dense layer before final node in 2dCNN model is available for the identification of the clusters belonging large earthquake and no earthquake labels.

Keywords Diffusion augmentation of multivariate timeseries · Full prediction of large earthquake · Latent space mapping of dense layer in 2dCNN modeling

6.1 Augmentation of Timeseries by Diffusion Method

Augmentation of original multivariate timeseries is a powerful tool for generative AI studies with deep machine learning necessary for large dictionary and ground truth in the imaging and language processing fields (Schreiber and Schmitz 2000). In the case of generative AI, the multidimensional latent space is constructed by some characteristic features obtained from various encoder models, and then the present features added various degree of Gaussian noises that are decoded to the image data like to the original image data by the reverse NN sequence used in the encoding. In many cases of this type of generative autoencoder, image data are commonly used to be more than two-dimensional one, and thus the decoding from the latent parameters with Gaussian noises is like the denoising from the diffusive parameter assemblages, being likely to the diffusion model in generative AI (Takanohara 2023).

M. Toriumi, *Physics of Geochemical Mechanics and Deep Neural Network Modeling with Diffusion Augmentation*, Advances in Geological Science,
https://doi.org/10.1007/978-981-97-9376-1_6

On the other hand, considering the rare events arising on the background noisy timeseries, such as large earthquakes on the background seismic activity both in the regional and global system (Toriumi 2021, 2022), the number of events to be used as the ground truth is not enough to deep learning in the almost all neural network modeling for available prediction of the future events. In the previous study (Toriumi 2022), the prediction testing of large earthquake over the magnitude 6 by deep neural network modelings (RNN, LSTM, 1dCNN) was not enough accurate, which has been carried out by using the actual multivariate timeseries of the characteristic seismicity of global and regional cases, being less than 40% accuracy. This seems to be due to the small number of earthquake events during the observation periods in the global and regional cases. Therefore, to perform the available probability prediction of large earthquakes, it should be needed that enough amounts of augmented timeseries from original ones are acquired on the basis with the physical meanings. In addition, to predict the large earthquakes which must be labeled by multivariate background seismicity matrix, many future data sets of the characteristic global and regional correlated seismicity and actual seismicity of the present locality to be investigated, which is correlated with large earthquake, should be required at all by extension from the big data set involving the augmented multivariate timeseries.

Augmentation of the timeseries is commonly carried out by various surrogate methods in the engineering fields, and it is based on the processing by the reversal timeseries, and Fourier transform phase-randomized timeseries, and simple addition of random noise (Schreiber and Schmitz 2000). However, these augmentation processing methods may not be available for the rare event cases on the background noisy seismic timeseries as noted in the previous book (Toriumi 2022). It seems that the physics-informed augmentation method which keeps the local timeseries kinetics and long-range ones should be required for actual rare-event-learning by DNN. In this context, it seems available that the physical model for processing the approximated augmented timeseries is probably based on the Langevin stochastic dynamics, as being widely applied for the diffusion model of generative AI modeling.

In the generative AI modeling such as generative variational autoencoder (GVAE), the generated images are formed by decoding the characteristic features-added Gaussian noises in the latent space, which is encoded by the DNN (CNN, RNN). In the latent space, the characteristic features are successfully classified into several clusters based on the ground truth of the input image data, and then it is probable that the augmented images are possibly made as the different sets approximated by diffusional scattering around the training input data. However, if the timeseries is the present case, because many segments embedded from the input timeseries are encoded with the target ground truth, the feature data sets obtained in the latent space from encoding the multivariate time series are randomly plotted without correlation to the target labels of large earthquake.

In this paper, therefore, the author intends to propose the Langevin–diffusion type augmentation of the timeseries to obtain the enough events data on multivariate seismic activity timeseries.

In the timeseries of the global and regional seismic activity, there is quite large range of moment release at the earthquake event from 10^{15} to 10^{20} Nm, that is the magnitude range from about less than M0 to over M9, but in this study, the background seismicity will be ranged from M1 to M3 in the regional case, and from M4 to M5 in the global case, because of avoiding the observational errors, as denoted in the previous book (Toriumi 2021, 2022).

In this section, it is needed for available augmentation that how the appropriate big data sets of the multivariate timeseries will be made reasonably from the observed multivariate timeseries, being consistent with the physical process of earthquake generation and seismic activity of the solid earth. If the signals of precursor of large earthquakes would be identified in terms of their physical mechanisms, it should be possible to construct the available-augmented large number of data sets for probability prediction of the large earthquake even in the case of large noisy events. Regrettably, it is still unknown that what is the precursor signals of the large earthquake events in the specified localities, although the basic mechanism of the large earthquakes should take place due to the rapid and large unstable shear slip on the fault planes. At this simple model of earthquake, there remains unknown problem whether the earthquakes having vast range of their moment release rate are of the same mechanism as the simple shear slip or not (Ohnaka and Matsu'ura 2002).

In the previous book (Toriumi 2022) and previous chapters of this book, it has been presented that the basic mechanisms involve the various-sized shear crack instability from stable growth (slow slip velocity mode) to explosive propagation (very fast slip velocity mode), and the regulation mechanism of the critical size of shear crack by abundant secondary fluid filled open crack and jog formation as shown in Fig. 6.1. This regulation mechanism explains possibly the vast range of the critical (Griffith) size of stable growing shear crack, and thus the large earthquake is of similar mechanical process as the small one. Therefore, it is concluded that the important factor controlling generation of every earthquake is what is the mechanism governing the critical (Griffith) crack size, which is determined by the density of the microcracks associated with the main shear crack as studied in the previous chapter. Furthermore, it seems that the density of micro-open cracks forming the cluster increases rapidly with increasing number of subsidiary small shear cracks parallel to the main shear slip plane.

The generation of the microcrack and subsidiary small shear cracks near the main shear crack is obviously represented by the stochastic occurrence of micro to small seismicity around the shear plane, thereby showing the stochastic enlargement of noisy seismicity timeseries in the observatory station. Thus, it seems that the probability of the large earthquake attaining the critical size, L_c, should be apparently governed by the following Langevin type stochastic equation following Haken (2006),

$$\begin{aligned} dL_c &= K dt + g dw(t) \\ < dw > &= 0 \\ < dw(t) dw(t') > &= \delta(t - t') dt, \end{aligned} \tag{6.1}$$

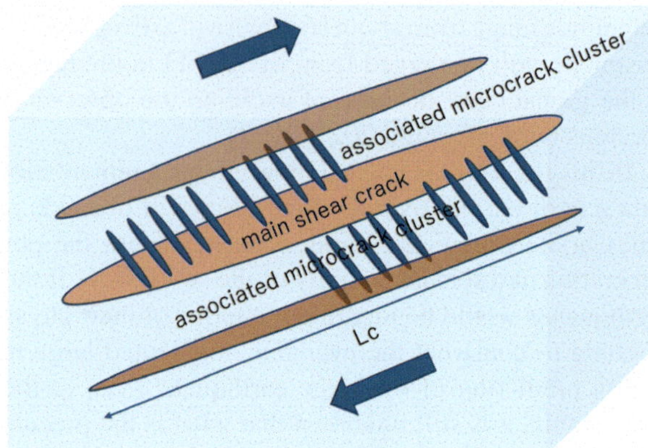

Fig. 6.1 Illustration of main shear crack enlarged by dissipation due to microcrack cluster formation around the main shear crack. L_c; main shear crack size

in which dw is the stochastic drift term and K being assumed to be the transition function of L_c, and in this study, it is assumed to be constant.

Then, the Fokker-Plank stochastic equation of probability of $f(L_c)$, that is the master equation, becomes,

$$\partial f(L_c)/\partial t = \partial K f(L_c)/\partial L_c + D\partial^2 f(L_c)/\partial L_c^2, \tag{6.2}$$

in which D is the diffusion term of L_c.

As the $f(L_c)$ is the probability occurrence of the large earthquake corresponding the magnitude of $M = GuL_c{}^2$, in which G and u are rigidity and displacement of the shear slip, respectively, the above equation should clearly suggest the stochastic equation of the large earthquake to be considered. As clarified previously, the L_c is proportional to the density of the microcrack filled with fluid, and thus, to the probability of the background seismicity rates, which vary in the time and space.

Considering that the relation of L_c and microcrack density, n, becomes as shown in the previous section and in Fig. 6.2,

$$L_c = L_g(1 + n), \tag{6.3}$$

in which L_g is the constant Griffith crack size. Then, above equation can be differentiated as,

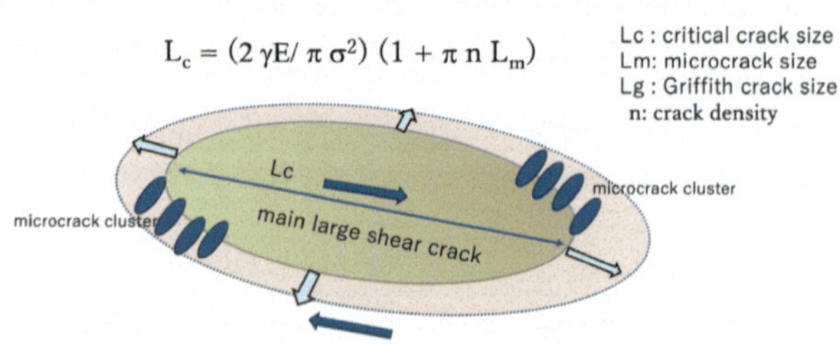

Fig. 6.2 Model of enlarged main shear crack size L_c and associated microcrack cluster with crack density n and microcrack size L_m. L_g; Griffith crack size

$$\begin{aligned} \mathrm{d}f(L_c)/\mathrm{d}L_c &= L_g^{-1}\mathrm{d}f(n)/\mathrm{d}n \\ \mathrm{d}^2f(L_c)/\mathrm{d}L_c^2 &= L_g^{-2}\mathrm{d}^2f(n)/\mathrm{d}n^2, \end{aligned} \tag{6.4}$$

On the other hand, considering the steady state of the probability function $f(L_c)$, the Fokker-Plank equation becomes,

$$\partial Kf(n)/\partial n + (D/L_g)\partial^2 f(n)/\partial n^2 = 0. \tag{6.5}$$

Integrating above equation, it holds,

$$(D/C_g)\partial f(n)/\partial n = -Kf(n). \tag{6.6}$$

If K is proportional to n, then, we have,

$$f(n) = A\exp(-aC_g n^2/D) \tag{6.7}$$

in which A is the normalized factor of the probability of $f(n)$.

This equation suggests that the probability of large earthquake (large n) decreases rapidly in the high concentration of microcracks, and $f(n)$ of the small background crack density showing the Gaussian distribution function, thereby suggesting that the precursor signals may involve the spatial distribution of seismicity rate derived from microcrack generation in the present region, and the possible signals should contain large Gaussian noises dependent on the time and space.

Accordingly, it is reasonable that the augmentation of the observed multivariate seismicity timeseries of various localities and their correlated seismicity, corresponding to appearance and disappearance of large earthquake event should be carried out by uphill -diffusing of the time dependent Gaussian probability function on the observed timeseries. It is possible that the uphill-diffusing as illustrated in Fig. 6.3, that is augmentation by denoising of observed segmented timeseries, is possibly carried out by gradient uphill method of the temporal-variable Gaussian distribution function, called in this book as diffusion augmentation method.

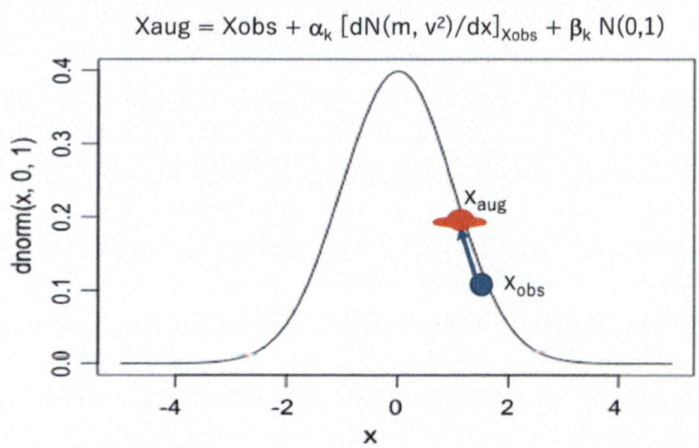

Fig. 6.3 Representation of diffusion augmentation method of timeseries by uphill diffusion of observed data point using Gaussian distribution function of segmented observed timeseries. Abbreviation of the augmentation formulation is denoted in text

In this diffusion augmentation method, it is necessary that from the multivariate timeseries, the time-dependent Gaussian distribution of the data should be calculated by estimating the mean value and the variance of the segmented timeseries. The method for this may be performed by the state-space modeling (Kitagawa 2009), but simply by the successive calculation using Excel code from matrix array obtained by embedding of the timeseries. The author obtained the timeseries of mean value and variance of the successive time intervals by embedding of the correlated seismicity and local logarithmic seismicity timeseries as shown in Figs. 6.4 and 6.5. It seems that the frequency pattern of the regional correlated seismicity of Japanese island clearly displays the multi-Gaussian distribution patterns, which belong to the timeseries with different ranges of the time step as shown in Fig. 6.6.

Gradient uphill method of augmentation is possibly carried out by linear combination of the elements of the timeseries, shift value of the derivative of the Gaussian error function at the point of present elements of the timeseries, and random noise as,

$$u_{\mathrm{a}}^{k}(t) = u_{\mathrm{p}}(t) - a g_k \left(\mathrm{d}N\left(m_{\mathrm{p}}, v_{\mathrm{p}}^{2}\right) / \mathrm{d}u \right)_{up} + b f_k(0, m), \tag{6.8}$$

in which u_{a} and u_{p} are the augmented signal and primary one, respectively, g_k is the Gaussian noise for k-th augmentation from 0 to 1, and $f_k(0, \mu)$ is Gaussian noise by diffusion, $N(m_{\mathrm{p}}, v_{\mathrm{p}}^2)$ is the Gaussian distribution function of the segmented timeseries having mean $\mathrm{m_p}$ and variance v_{p} at u_{p} by embedding of the primary timeseries, but α and β are the controlling parameters. In this augmentation model, it is reasonably assumed that the probability functions of the correlated seismicity of z_1 and z_3 and logarithmic seismicity of the gridded locality are mutually independent Gaussian error function in every range of the timeseries, thereby possible neglecting the cross correlation of them.

In this study, the augmentation is repeated by 10 times for multivariate timeseries of correlated seismicity rate z_1 and z_3 and local logarithmic seismicity rate in the present gridded locality, and these are labeled by 1 (large earthquake event of M6 over in the global case, but of M5 over in the regional case) and 0 (no event of large earthquake) as same as the primary multivariate timeseries as illustrated in Fig. 6.7.

It seems to be important that how the augmented, that is generative, multivariate timeseries matrix is lined in the order, and, for example, that whether random insert into the primary timeseries, or the augmented timeseries is arranged after primary timeseries in the order of augmentation number, k, and or augmented segments is lined after primary segment in the order of k. For the 2dCNN neural network model is used in the work to avoid the sequence error at the connection between augmented timeseries, it seems that the augmented segments line method is possibly applied for the processing enough machine learning of the multivariate timeseries labeled by large earthquake event.

Furthermore, it needs clearly that the parameters of α and β should be determined by evaluation of the predictability of the augmentation CNN modeling proposed above. To obtain the available parameters of α and β, the evaluation of

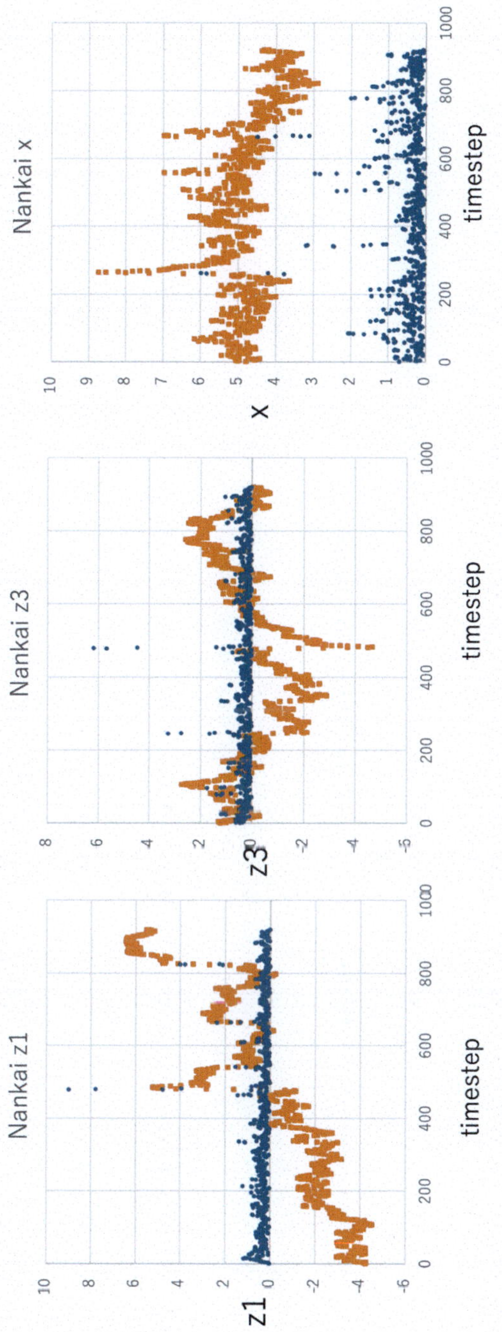

Fig. 6.4 Timeseries of mean values (orange) and variances (blue) of regional correlated seismicity z_1 and z_3 of Japanese islands region, and logarithmic seismicity x of Nankai region from 1998 to 2023

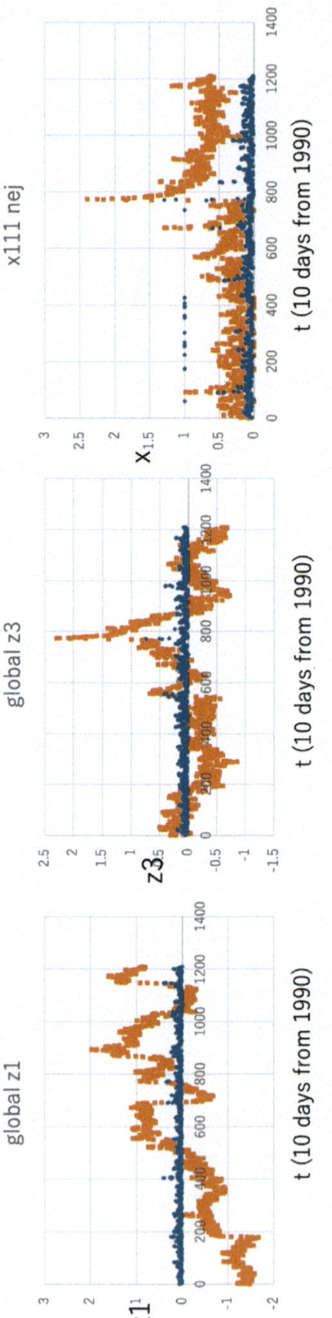

Fig. 6.5 Timeseries of mean values (orange) and variances (blue) of global correlated seismicity z_1 and z_3, and logarithmic seismicity x of northeast Japan region from 1990 to 2023

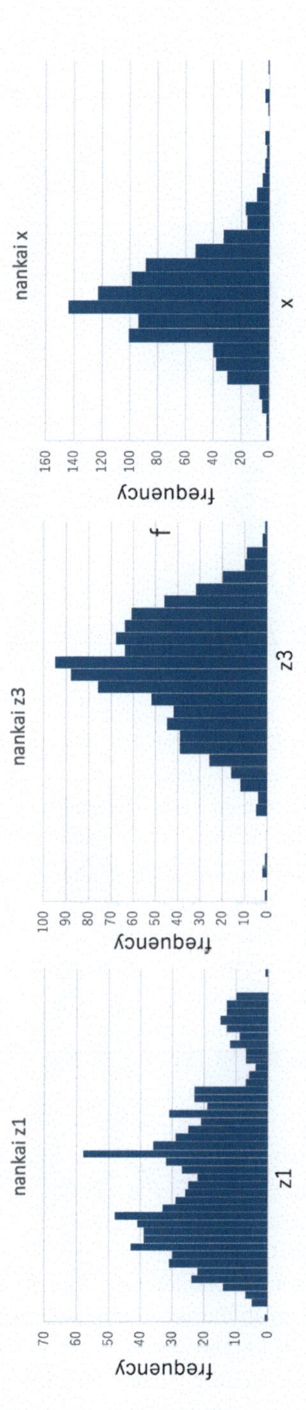

Fig. 6.6 Frequency diagrams of regional correlated seismicity z_1 and z_3 of Japanese islands region and logarithmic seismicity x of Nankai region, showing multiple Gaussian frequency distribution pattern

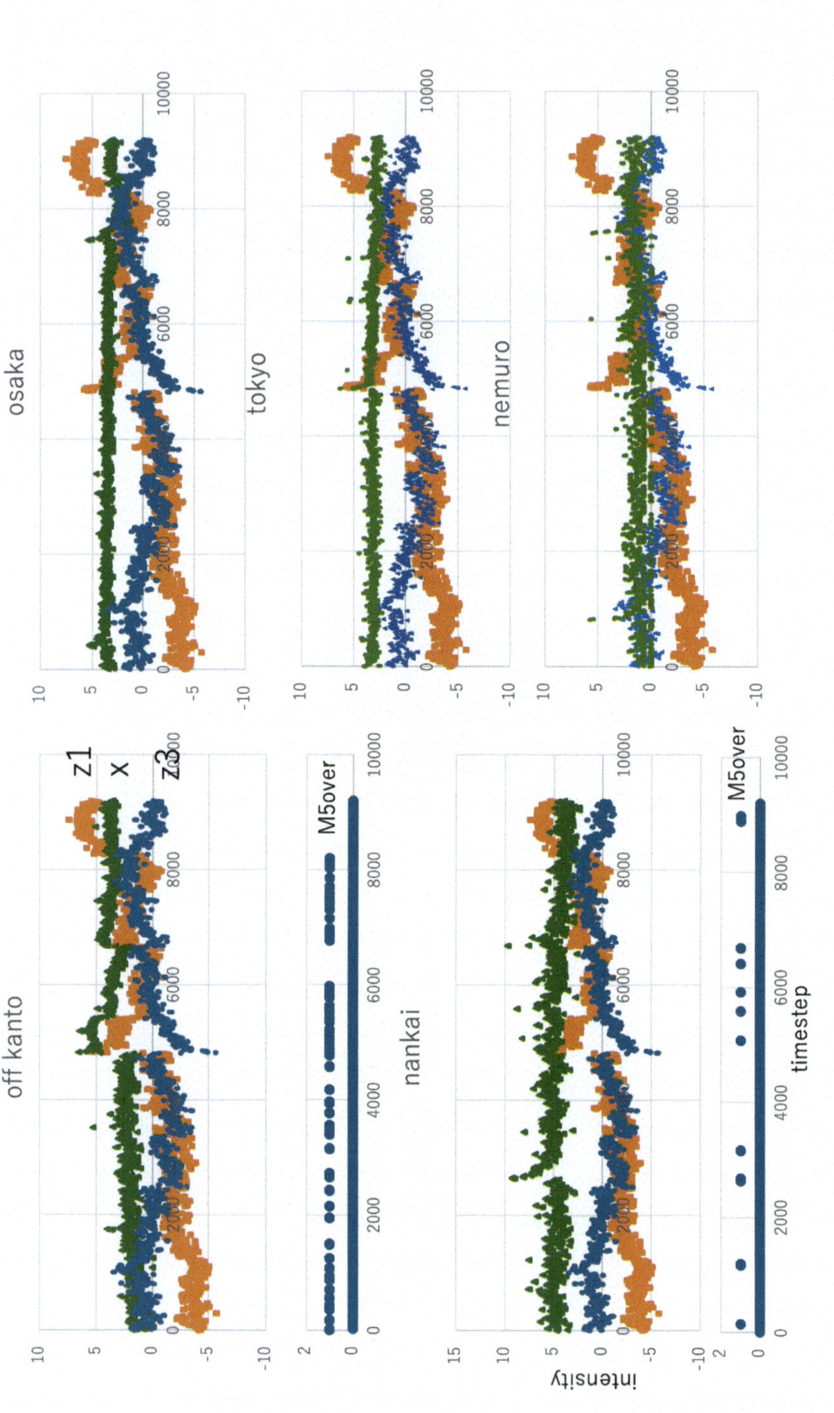

Fig. 6.7 Augmented timeseries of regional correlated seismicity z_1 (orange), z_3 (blue) and logarithmic seismicity rate x (green), and timeseries of M5 over earthquake event (1) and no event (0) in the lower column of Off-Kanto (left top), Nankai (left bottom), Osaka (right top), Tokyo (right medium), and Nemuro (right bottom). In the right figures, earthquake of M5 over is not shown

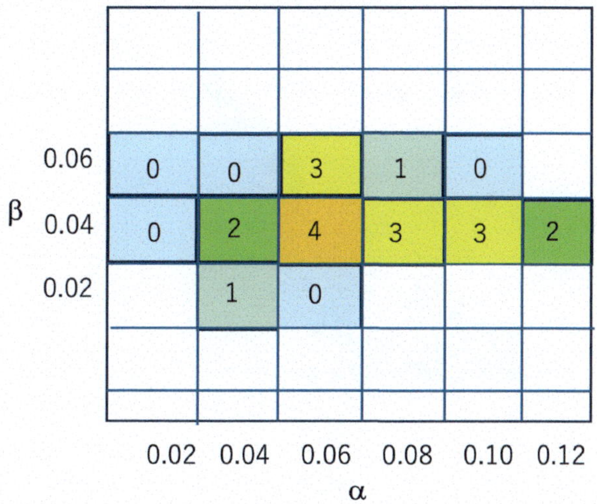

Fig. 6.8 Representation of maximum augmentation effort in various set of (α, β) of the augmentation parameters. See text

the predictability of earthquake labels has been performed by the accuracy of the prediction using augmentation–2dCNN modeling with the grid search on parameter set (α and β) as shown in Fig. 6.8. The results of large earthquake prediction after training by deep learning of the augmentation timeseries for the testing timeseries in the range of 500 to 920 step of the Japanese regional seismicity case are shown for various parameters in the above augmentation parameters, thereby suggesting the available parameters of α to be 0.06 and β to be 0.04. The grid search is based on the accuracy in the time step range from 250 to 350 step shown in Fig. 6.7, and the central parameter set is just at the (0.06, 0.04) of (α and β). Therefore, in the following, the augmentation parameters are set on the above values for the practical prediction study.

6.2 Latent Space Mapping of Augmented Timeseries Labeled by Large Earthquakes

To evaluate the availability of the present augmentation method for the prediction of the large earthquake, it is required that the labeling data sets of the augmented timeseries should be investigated by the latent space mapping of the generative variational autoencoder modeling (GVAE) with CNN as discussed in the previous book (Toriumi 2022). In the previous book, it concluded that the latent space maps of GVAE from the primary timeseries labeled by event of large earthquake of several gridded localities show the random mixing between characteristic parameters with different labels, thereby suggesting too small amount of training data sets.

The augmentation data sets attain several ten thousand data sets labeled by 0 and 1, and thus it is possible to check that whether the latent space mapping of them

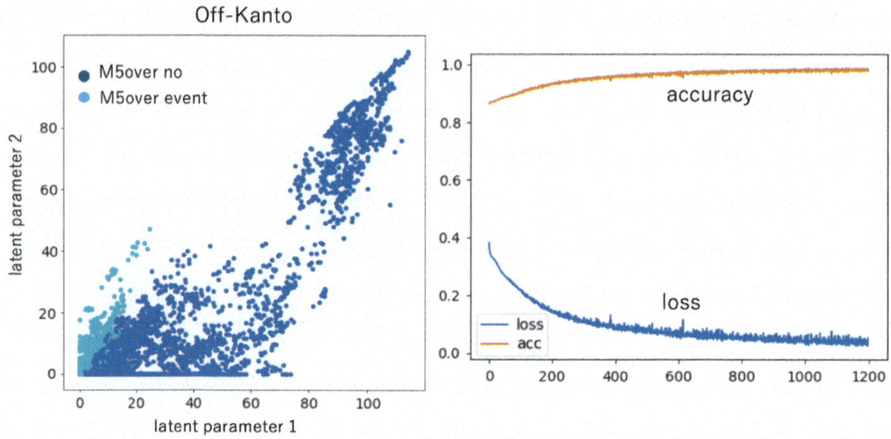

Fig. 6.9 Latent space mapping of parameters in dense layer (16 nodes) before final node, showing sharp separation of the clusters of M5 over earthquakes event and no event of Off-Kanto region. The right figure shows the accuracy and loss function curves against repeating number of optimization procedure of 2dCNN modeling. See text

shows the distinctive clustering of the data sets with labeling or not. In this work, it shows that the latent multidimensional data are acquired from the final dense layer before single node in front of target label. In the present dense layer of the 2dCNN modeling, the multivariate timeseries with 16 dimension of feature data are stored, but many of components are seemingly singular and some of them involve important information for identifying the clusters of different labels. To search the assemblage of components available for the classification of clusters, it is possible that the widely scattered data sets in the present latent space of the dense layer should be acquired by the latent space mapping of them, as shown in Fig. 6.9. The diagram of the accuracy and loss function at the 2dCNN execution appears in the same figures, suggesting the good performance of the augmentation proposed here.

On the other hand, it is obvious that the case of the global seismicity prediction has the same difficulty as the regional seismicity discussed previously, that there is very scarce occurrence of the large earthquakes to deep machine learning for seismicity prediction, although it will benefit for human society avoiding the natural disaster. Therefore, the augmentation of timeseries of the global correlated seismicity and target local logarithmic seismicity have been carried out in the diffusion augmentation method proposed in this chapter.

Firstly, it should be manifested that the diffusion uphill augmentation is feasible to apply to the multivariate timeseries of the global correlated seismicity and target local logarithmic seismicity with large earthquake labeling. As shown in Fig. 6.10, it is noteworthy that the frequency of global correlated seismicity z_1 shows the multiple Gaussian distribution function about three modes, but that of the z_3 displays nearly Gaussian pattern with rather long-term tail.

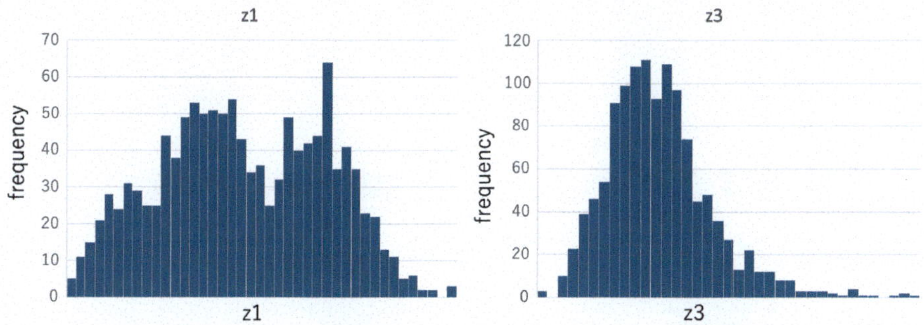

Fig. 6.10 Frequency diagrams of global correlated seismicity z_1 and z_3, showing multiple Gaussian frequency distribution pattern

Furthermore, as shown in Fig. 6.11, the augmentation timeseries of the z_1, z_3, and x of the target locality appear for the case of northeastern Japan (NEJ), Himalaya (HIM), and Chile (CHI) regions. The representative set of the augmented timeseries of global correlated seismicity z_1, z_3, and local logarithmic seismicity x is shown together with the timeseries of labeled (event:1, no event:0) large earthquakes of M6 over.

The latent space maps of the results of 2dCNN modeling for augmented timeseries with large earthquake labeling are shown in Fig. 6.12, thereby suggesting that the maps manifest distinctively two clusters of labels 1 (earthquake) and 0 (no earthquake) in the diagram. The parameters in these diagrams are chosen carefully from the 16 variables stored in the final dense layer of the 2dCNN model for enough scattering of the distinctive labels in the diagram. The accuracy and loss function of the execution of DNN are displayed in the same figure as the latent space mapping, showing the availability of the model. From these figures, it suggests that the cluster of large earthquakes is rather sharp in the case of northeastern Japan than those of the Himalaya and Chile, and this is probably derived from the frequent occurrence of the large earthquakes in the northeastern Japan.

6.3 Neural Network Testing of Artificial Earthquake Events on Minimal Nonlinear Dynamics

In the previous books of the global and regional correlated seismicity dynamics (Toriumi 2021, 2022), it has been clarified that the present dynamics has stable two modes and one unstable mode of the behavior, and thus it should be represented by the minimal nonlinear dynamical equation.

The ordinary differential equations of nonlinear dynamics hold in the previous chapter as,

$$dx/dt = x - kx^3 + y$$

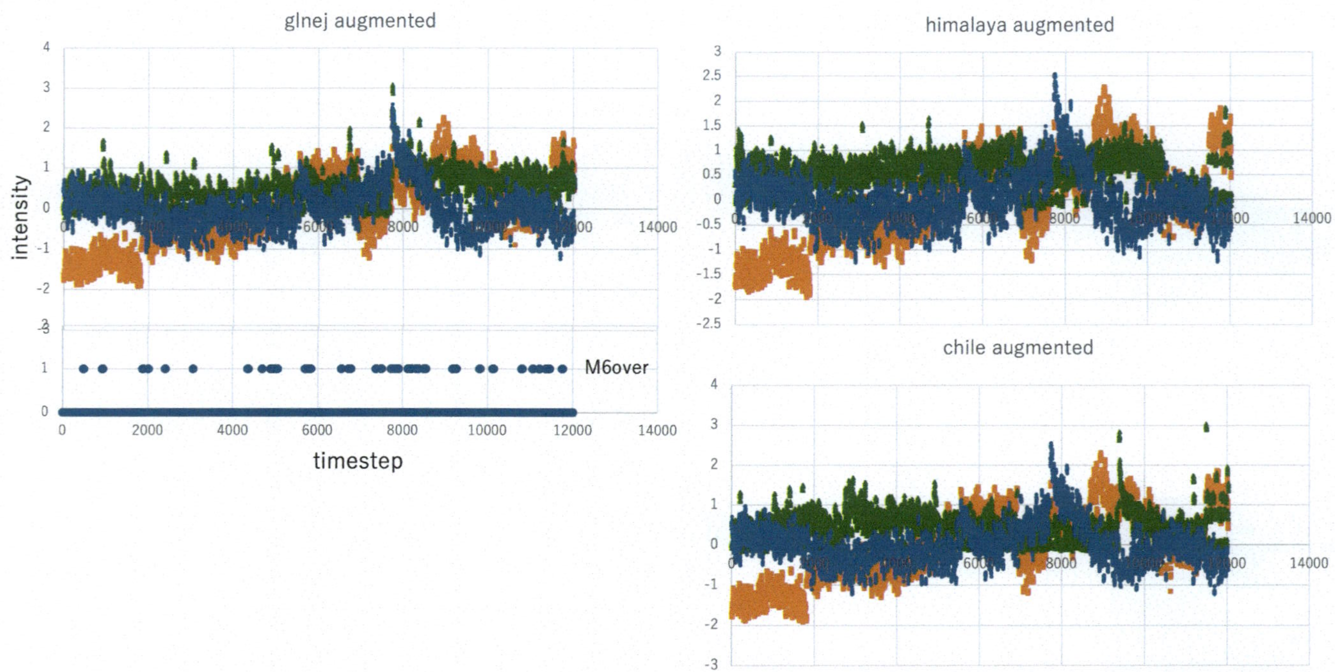

Fig. 6.11 Augmented timeseries of global correlated seismicity z_1 (orange), z_3 (blue) and logarithmic seismicity rate x (green), and timeseries of M6 over earthquake event (1) and no event (0) in the lower column of northeast Japan (NEJ; left), Himalaya (right top), and Chile (right bottom). In the right figures, earthquake of M6 over is not shown

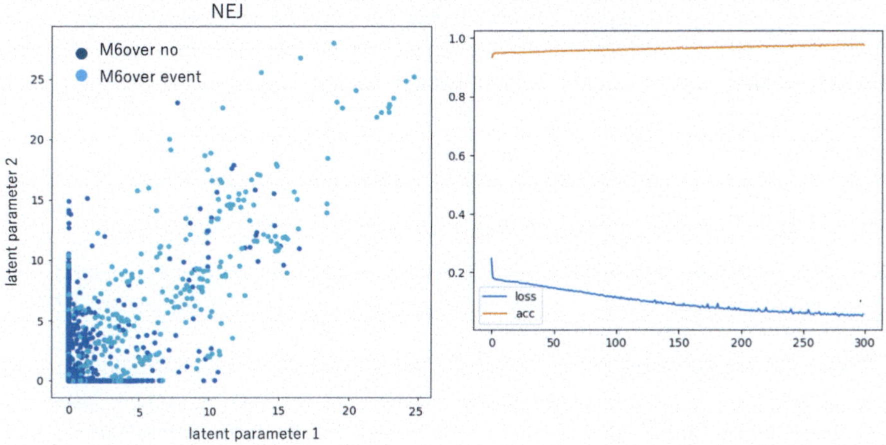

Fig. 6.12 Latent space mapping of parameters in dense layer (16 nodes) before final node, showing sharp separation of the clusters of M6 over earthquakes event and no event of northeast Japan (NEJ) region. The right figure shows the accuracy and loss function curves against repeating number of optimization procedure of 2dCNN modeling. See text

$$dy/dt = h\cos(wt). \tag{6.9}$$

In these equations, the y means a periodic drift term for x, and thus the two-state model equation has various stable and unstable nodes, as inferred by their null cline equations. It is sure that sets of stable nodes of these null clines are possibly classified into three types: one is single stable node, second is two stable nodes and one unstable node, and third is one stable node and one degenerate node as described in the previous chapter.

On the other hand, in the previous book, it was pointed out that the deep learning testing of the dynamical behavior of the global and regional correlated seismicity dynamics has not succeeded in approximation from the training, because of the multi-state nonlinear dynamics controlling the multivariate timeseries of them. This seems to be due to the reason why the training by deep machine learning was carried out within the single process, but the testing was far from the interpolation of that single process. If it is possible to be carried out in both regimes of two or more modes of dynamical behavior, it seems to be possible that two or more dynamics should mathematically connect them to make the available nonlinear dynamical equations controlling the present system.

By the way, the variance of the NLD solution near the tangential point becomes very large because of very large value of the inverse second derivatives of the potential function of the above nonlinear kinetics, thereby showing the existence of the large random drift of the dynamical states. In the case of global correlated seismicity dynamics, the author has pointed out the possibility of the prediction by the deep learning from the one of the time series to another range of the time series having secular variation and drift term.

In the previous books, it has been concluded that the global and regional correlated seismicity of the micro-and small-seismic activity of the plate boundary regions displays the different series of stochastic stable nodes in the state-space of several correlated seismicity parameters (Toriumi 2021, 2022), thereby suggesting that there are several different types of dynamical states in the 1990–2023 time series of global seismicity, and the transition between their nodes should take place near 2003–2004 and 2015–2016. In the case of Japanese regional correlated seismicity dynamics, the similar transition occurred near 2010–2011 in the timesseries of 1998 to 2021 (Toriumi 2022). The transition of the correlated seismicity dynamics seems to be coeval and overlapped within the periods of the giant earthquakes of Sumatra (2004) and Tohoku-Oki Earthquakes (2011).

Thus, in the section, the author intends to study the training feasibility and testing of the artificial earthquake event on the timeseries from minimal nonlinear dynamics with Gaussian random drift by recurrent and convolution neural network modeling.

The abnormal events on the timeseries of the variable in the simple NLD are simulated by,

1. If sum of $x(t-j)$ to $x(t)$ in $j = 2$ is larger than L_c, abnormal event as artificial earthquake occurs.
2. The labeling is to be 1 for on-state of the artificial earthquake and 0 for off-state along on the simple nonlinear dynamics as shown in Fig. 6.13.

It shows that many events are concentrated in the period of the high background state (high x regime), but some of events are found in the short period of the high-level x during the low-level regime, thereby suggesting the occurrence of artificial earthquake event even in both of bifurcated states of the present NLD.

The regression results by 1dCNN and RNN modeling in the training segment of the timeseries obtained from above NLD with Gaussian drift are shown in Figs. 6.14 and 6.15, respectively. It seems that there is no distinctive difference in results between RNN and 1dCNN modeling for the prediction efficiency for the testing of the different segment of the timeseries. It is obvious that the DNN regression results are good in accordance with the simulated timeseries of nonlinear two-state dynamics.

The results of testing for artificial event-on and -off states are evaluated from the consistency between the predicted event signals and the ground truth events simulated presented above. To obtain the graphical representation of the evaluation, the plotting of predicted and ground truth events are shown in Figs. 6.16 and 6.17, indicating that the length of timesteps about 3 is preferable for actual prediction of testing. It seems that the event dynamics in the NLD should depend partly on the deterministic process still even in the chaotic dynamics with Gaussian noise. Furthermore, it suggests that the deep learning with 1dCNN is potential to infer the event sequence on the chaotic dynamics.

It should be noted that the score of the predicted events, that is the event probability of prediction, is scattered from 0 to 1 around the enough predicted events. These patterns of predicted events timeseries are characteristic for the NN deep learning results of two-state nonlinear dynamics, suggesting that the segments of timeseries in the NL dynamics appear compactly the similar sequences with ground truth events.

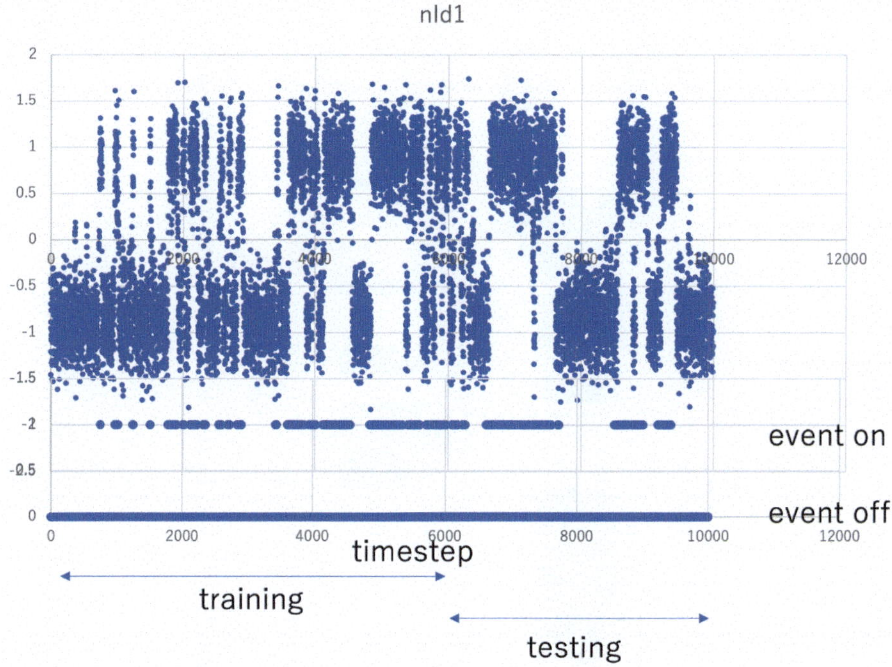

Fig. 6.13 Timeseries of nonlinear logistic dynamics with noisy Gaussian drift and labeling of artificial event-on (1) and event-off (0) in the bottom column. The segments of training and testing of the studied timeseries are shown in the lower corner

To infer the availability of DNN for the machine learning of the abnormal event in both regimes, it should be required that the training experiment on the segments of the present timeseries containing the two states governed by simple nonlinear dynamics will be carried out for artificial large earthquake event on that timeseries. Therefore, the author intends to study the available method of the deep learning prediction of episodic events related to the background timeseries controlled by the nonlinear dynamics.

In this study, the drift term of the Gaussian noises is added on the time derivative of x in the present NLEs, producing the random two-state timeseries, and the deep learning with the available time step to be 3, corresponding to the ground truth is carried out to obtain the 1dCNN regression for another segment of the timeseries. After this validation method of the background timeseries of x was conducted, and the no important discrepancy between actual timeseries and NN predicted timeseries was observed in the different stage of the early timeseries.

The timeseries of the different modes of NLD solution of x mentioned above is shown in Figs. 6.16 and 6.17, and the deep learning of the single and multiple segments of the timeseries will be performed to obtain the NN predictors for inferring the late stage timeseries. In this study, the simple RNN model for timeseries was applied for evaluation of the complex system controlled by nonlinear differential

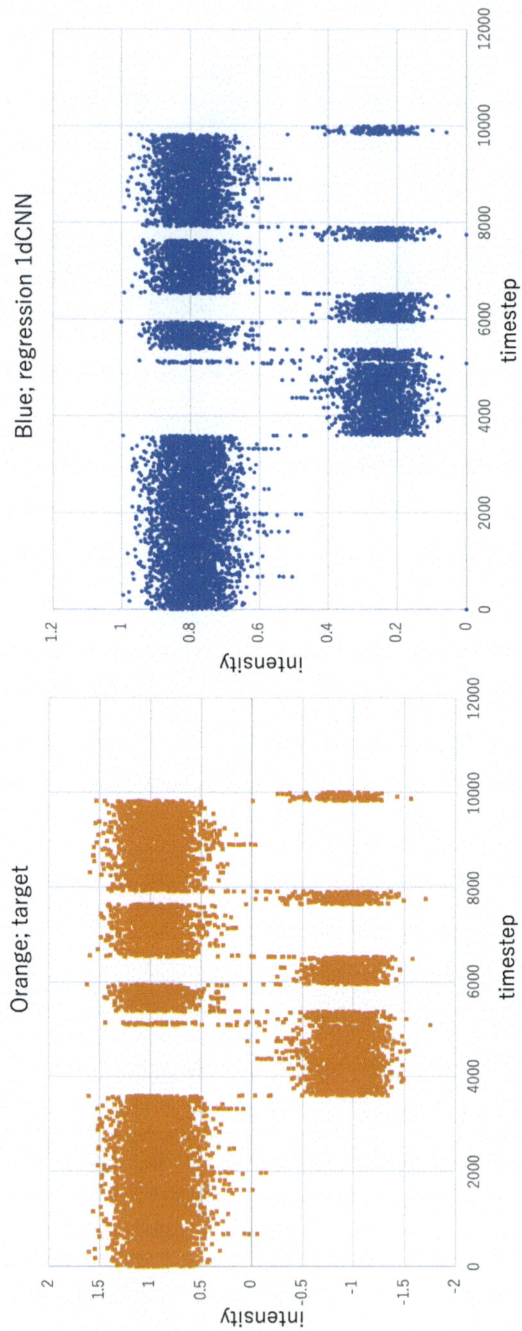

Fig. 6.14 1dCNN regression results of testing segment of the nonlinear two-state dynamics in Fig. 6.13. Orange: testing segment, blue: regression result

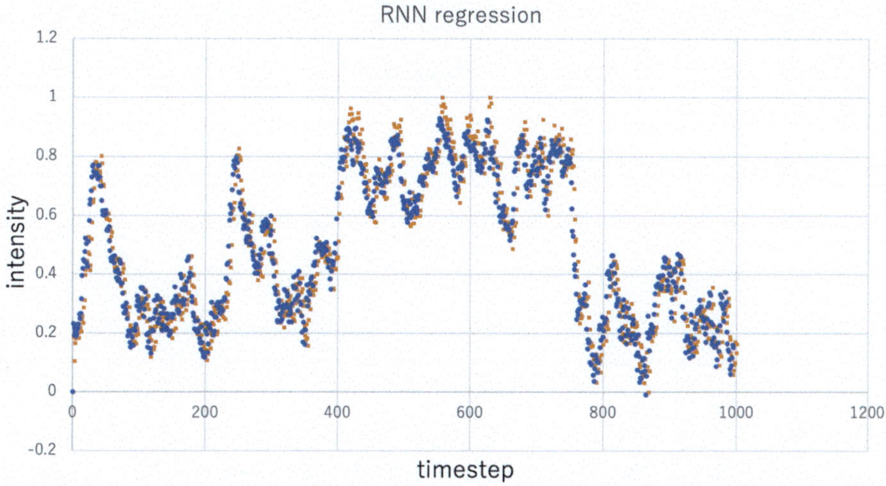

Fig. 6.15 RNN regression results of testing segment of the nonlinear two-state dynamics in Fig. 6.13. Orange: testing segment, blue: regression result

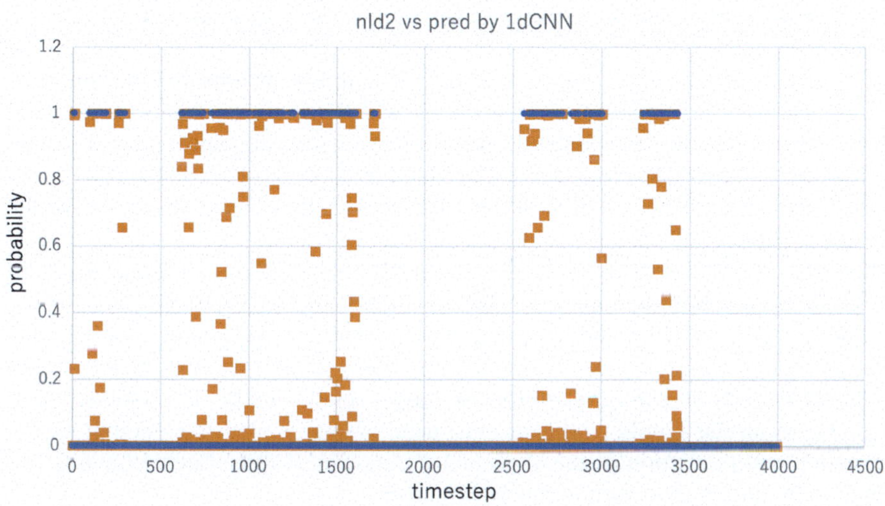

Fig. 6.16 Testing results of 1dCNN model from training of early segment of the timeseries. Orange dots: predicted probability, blue dots: target data of artificial event-on state

equations. Furthermore, it is required to prediction study that the advancing timeseries after given timeseries simulated by NLD will be possibly estimated in the enough length.

The testing study of the artificial earthquake event on the target timeseries of the background is carried out from the training by 1dCNN and RNN modeling as shown in Figs. 6.14 and 6.15, thereby showing that there are target events on the high- and

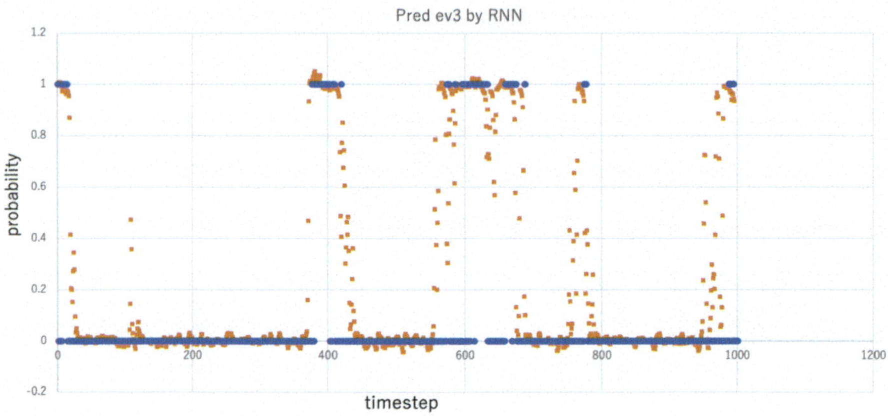

Fig. 6.17 Testing results of RNN model from training of early segment of the timeseries. Orange dots: predicted probability, blue dots: target data of artificial event-on state

low-level modes of the background timeseries of the simple nonlinear dynamics. The training timeseries are in the range from 1 to 6000 steps, and the testing from 6001 to 10,000 steps, but the abnormal events appear in both modes of background variation of x.

The results are shown in Figs. 6.16 and 6.17, and it is obvious that the abnormal events on the testing segment are in consistence with the predicted events having good probability, though near the predicted events there are several points with rather high probability. It seems that these false events, but low probability may be derived from using the finite time step to be 3 in the 1dCNN modeling on the timeseries of the nonlinear differential equation with random noisy drift term.

On the other hand, it is likely that the RNN modeling after training learn of the former segment of the timeseries may give the available prediction points of the probability diagram. However, it seems that near the ground truth points of label 1, the false prediction points are scattered with rather low probability, showing the similar results as 1dCNN modeling. Therefore, it may conclude that the DNN modeling of the behavior in the simple nonlinear dynamical system is plausible enough for the regression and prediction of the abnormal event, even if that system contains two stable states and chaotic variation between them.

6.4 Case of Multivariate NLD; Lorenz Dynamical Equations

The nonlinear dynamic differential equations are denoted in the previous chapter. In Fig. 6.18, the timeseries of the Lorenz chaos with Gaussian noisy drift are shown, indicating that the early behavior appears short-term variations and later one does

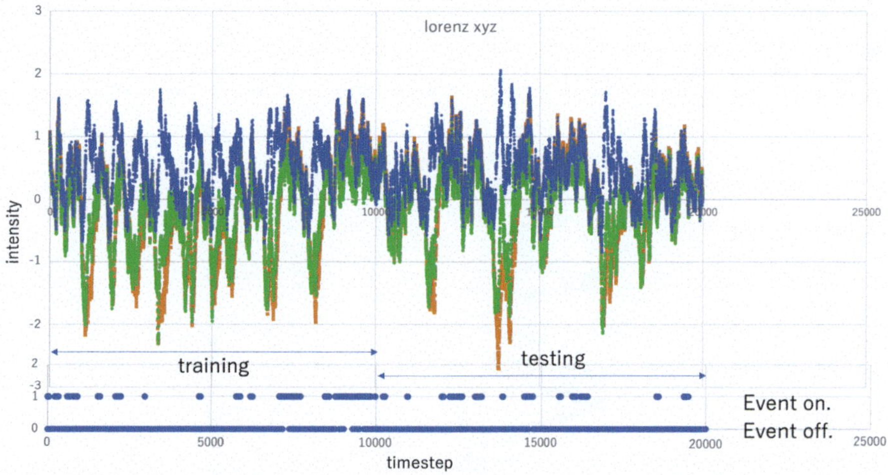

Fig. 6.18 Timeseries of nonlinear Lorenz dynamics with noisy drift, x (orange), y (green), and z (blue), and artificial event-on (1) and -off (0) states in the lower diagram. In the diagram, training and testing segments are shown

rather long-term variations of (x, y, z) variables. In the same figure, the artificial earthquakes on the Lorenz dynamics, which are given by the following procedure, are displayed as labels 1 (on) and 0 (off).

To check the availability of the prediction by 1dCNN, the regression study of the DNN modeling after training the early segment (1–6000 step) is applied for the testing of the late segment (6001–4000) as shown in Fig. 6.19, thereby indicating the plausible result of the reduction of Lorenz chaotic behavior.

Next, let us advance the prediction feasibility of the artificial earthquake on the Lorenz chaos with Gaussian noisy drift.

The artificial events on the multivariate nonlinear chaotic timeseries of the noisy Lorenz dynamics are simulated by,

1. when sum of x, y and z at a time is larger than the critical value K, seismic event is on, but smaller than K, seismic event is off.

The labeling of the seismic events is to be 1 for on-state, and 0 for off-state along on the Lorenz track, to indicate the training range (1–10,000) and testing range (10,001–20,000) as shown in Fig. 6.18.

The deep learning of these events on the Lorenz tracks with Gaussian noise was performed by 1dCNN modeling as like as the previous single variable NLD case. The input data are the multivariate timeseries of variables x, y, and z, and the number of the time step is taken to be 3 because of balancing the number of input timeseries. The number of artificial seismic events in this study is enough for the deep machine learning, and then the prediction is also expected to be enough accurate.

The results are evaluated from the consistency between the predicted event signals and the ground truth events simulated above. To obtain the graphical representation

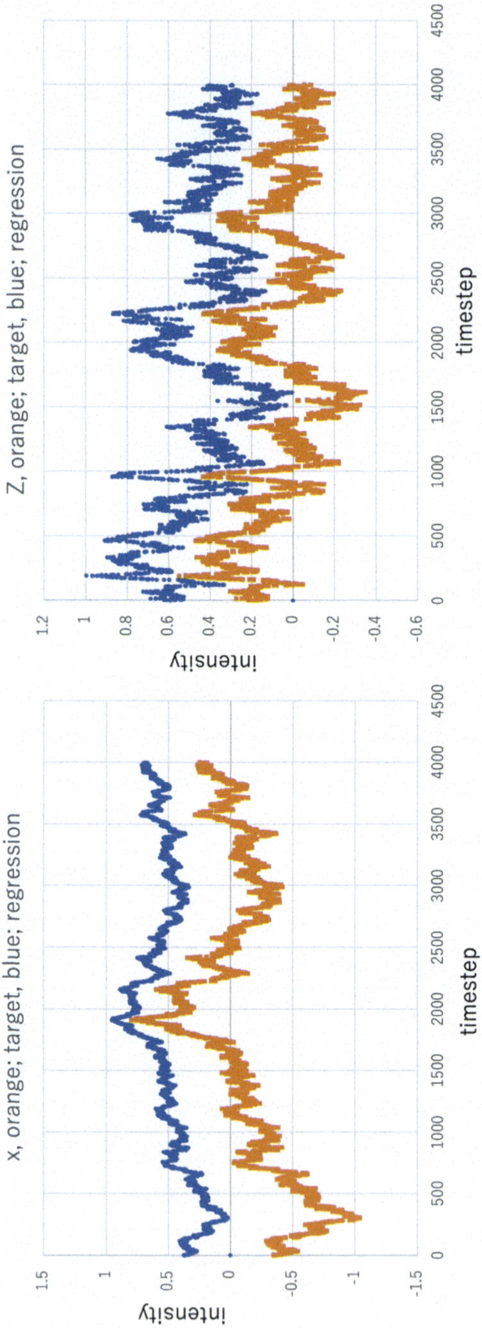

Fig. 6.19 1dCNN regression results of Lorenz dynamics x (left) and z (right). Orange: testing timeseries, blue: regression timeseries

of the evaluation, the plotting of predicted and ground truth events is shown in Fig. 6.20, indicating that the length of time steps about 3 is preferable for actual prediction of testing. It seems that the event dynamics on the NLD should depend partly on the deterministic process, still even in the chaotic dynamics with Gaussian noisy drift. Furthermore, it suggests that the deep learning with 1dCNN is potential to infer the event sequence on the chaotic dynamics. On the other hand, to compare the 1dCNN and RNN modeling for the prediction of artificial events on chaotic noisy timeseries, the prediction results of the testing segment of (10,001–20,000) are shown in Fig. 6.20, thereby suggesting the no significant difference between them. It should be noted that the score of the predicted events, that is the event probability of prediction, is scattered from 0 to 1 around the enough predicted events. These patterns of predicted events timeseries look like those of the NN deep learning results of monovarietal timeseries of simple nonlinear dynamics, suggesting that a universal relationship between the simulated event and predicted one by DNN modeling in the case of various nonlinear dynamics with Gaussian noisy drift. Therefore, it is likely that the composite nonlinear dynamical system with Gaussian noisy drift should be modeled by the similar 1dCNN in the case of multivariate timeseries. In the natural phenomena such as global and regional seismicity dynamics as proposed in the previous books (Toriumi 2021, 2022), there are commonly complicated dynamical system with noisy observables, and these are seemingly constituted partly by some varieties of nonlinear dynamics. Therefore, it will be discussed that whether the composite nonlinear dynamics is possible to be approximated by DNN modeling with labeling method or not. Furthermore, it will be discussed that how we possibly infer the signals of the transition from lower activity level to higher one in the case of multiple stable modes of the nonlinear dynamics with noisy drift.

In the composite nonlinear system to be discussed, the following variables are simulated,

$$\begin{aligned} &1.\ \text{logistic dynamics;}\ \ dx/dt = a\left(1 - x^2\right) + w_1 N(0,1) \\ &2.\ \text{two - state dynamics;}\ \ dy/dt = b\left(y - y^3\right) + w_2 N(0,1) \\ &3.\ \text{output of the composite dynamics;}\ \ z = k_1 x + k_2 y. \end{aligned} \tag{6.10}$$

In these differential equations, a and b are constant term, and w_1 and w_2 are drift term. N is the Gaussian noise. The term k_1 and k_2 are two cases: one is the randomized coefficients contributing the x and y terms, and the other is the constant contribution case in this study.

First, the random mixture composite dynamics will be discussed. The representative regression of the testing segment (10,001–20,000) of the timeseries z is displayed in Fig. 6.21, after the training in the timeseries of earlier segment (1–10,000). The comparison between simulated timeseries of z and the regression timeseries appears a good coincidence with each other. In addition, it seems that the testing results of the artificial earthquake event after training is so matched with the simulated events in the testing segment as the ground truth. It looks like that the artificial seismic event on randomized mixture of the timeseries by independent nonlinear dynamics

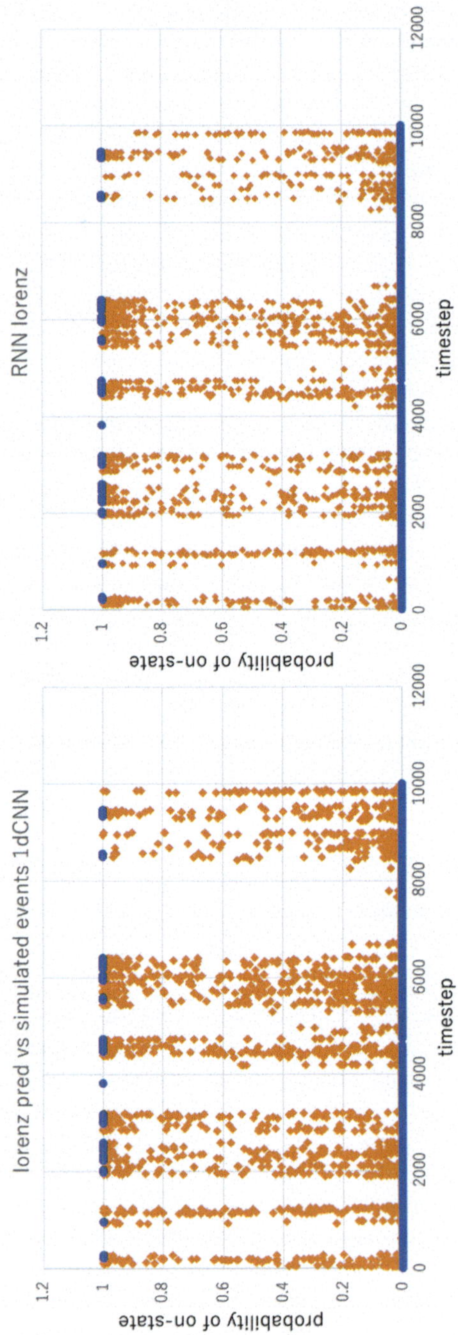

Fig. 6.20 Predicted probability of artificial event-on state (orange dots) and target (blue) by 1dCNN (left) and RNN (right)

is feasibly predicted by the 1dCNN modeling with enough training of the earlier segment of the timeseries as shown in Figs. 6.22 and 6.23, which appears as some scattered prediction points with low probability before and after the simulated point of event. It seems that there are a few simulated events that the DCNN did not give enough probability of prediction.

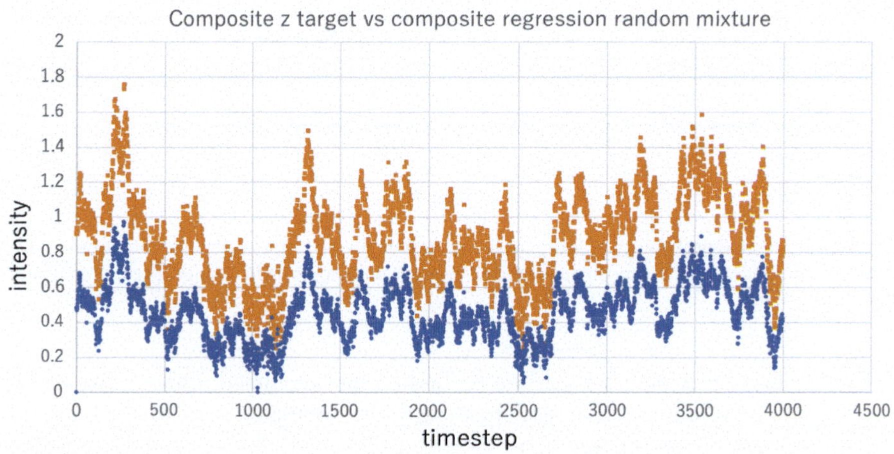

Fig. 6.21 Regression result of randomized composite nonlinear dynamics. Orange: target data, blue: regression data

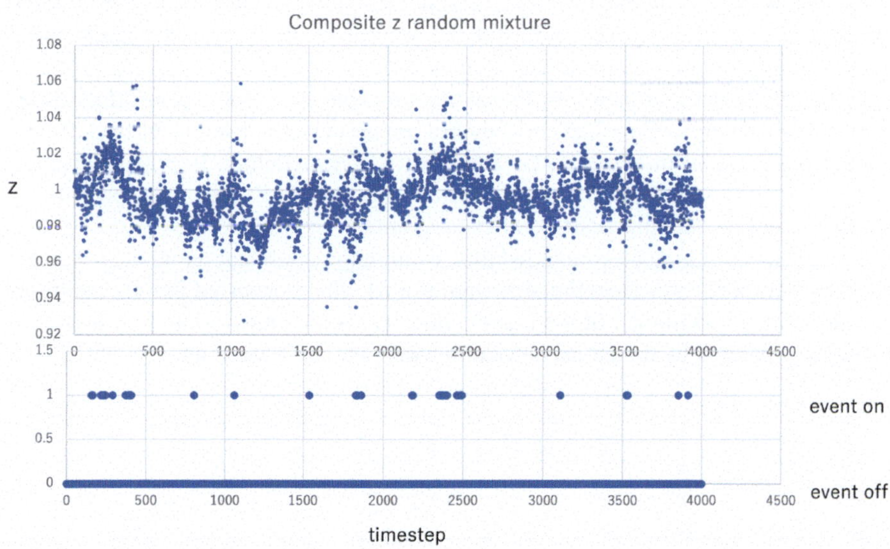

Fig. 6.22 Timeseries of target segment in the randomized composite nonlinear dynamics z and artificial event-on state (1) in the lower diagram

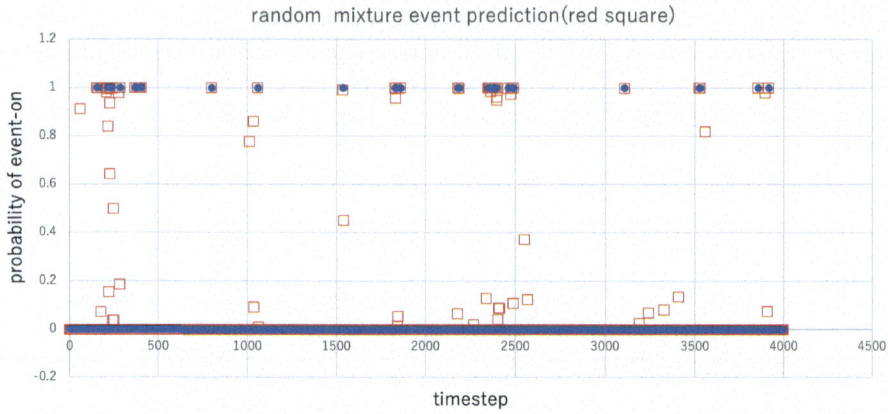

Fig. 6.23 Predicted probability of artificial event-on state on the testing segment from training one in the case of randomized composite nonlinear dynamics with noisy drift. Orange: prediction result, blue: target point

Furthermore, it is obvious that the regression and predicted artificial events by 1dCNN modeling are much in agreement with the simulated ones of the range of testing segment of the timeseries as shown in Fig. 6.23, thereby manifesting that the mixture model of the independent nonlinear dynamics is fairly discussed on the artificial events predicted by DCNN modeling with labeling by enough amounts of training data.

By the way, it is interesting that, before the predicted artificial events, there appear a few predicted points with rather low probability in both the randomized and constantly mixed nonlinear dynamics as shown in Fig. 6.24, but it is probable that either these prediction points display the transient state from event-off to event-on states as shown in Fig. 6.25, or they are apparent but not transient ones, being derived from size of time step to be 3 in the 1dCNN modeling. In practice, for the case of the time step to be 7, much prediction points with rather low probability appear around the simulated artificial event points.

Conclusively, it may be concluded that, even in the nonlinear dynamical systems and their composite systems, the abnormal events which occurred in the conditions over the critical magnitude of the signal intensity on the multivariate timeseries are plausibly predicted by the DCNN modeling with available labeling.

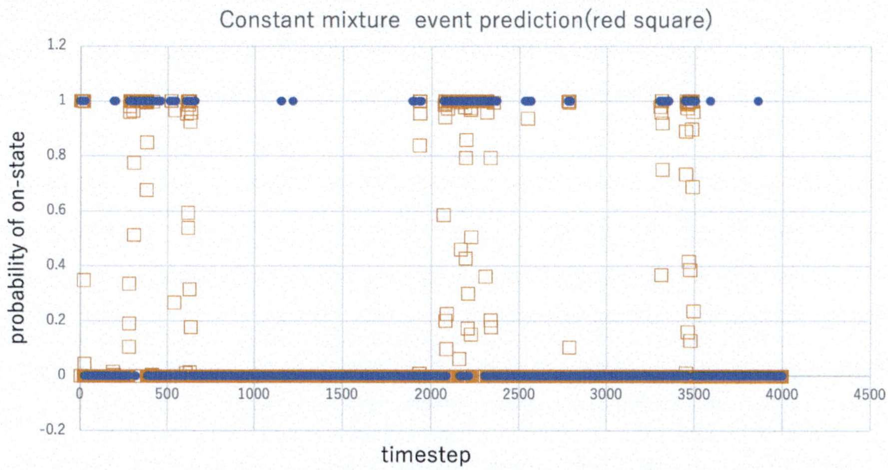

Fig. 6.24 Predicted probability of artificial event-on state on the testing segment from training one in the case of constant mixed nonlinear dynamics with noisy drift. Orange: prediction result, blue: target point, red square: predicted point, blue: target data

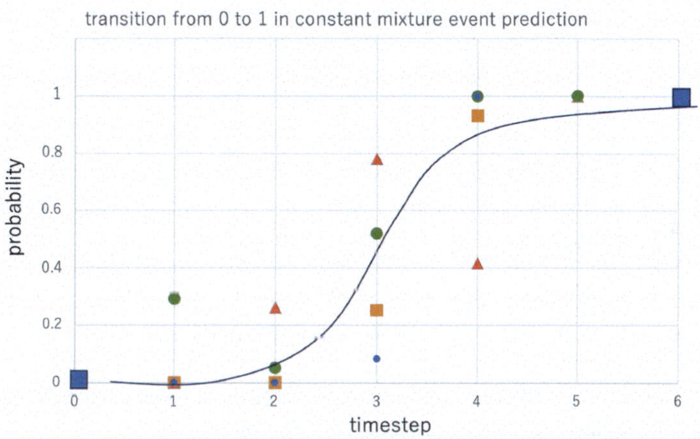

Fig. 6.25 Diagram of temporal variation showing the transition from artificial event off-state to on-state in the composite nonlinear dynamics system with noisy drift

6.5 Probability Prediction of Large Earthquake by Neural Network Modeling of Correlated Seismicity

In the previous book (Toriumi 2021), it has been clarified that the global seismicity and the regional ones are possibly investigated by characteristic features of the global and regional correlated seismicity rates obtained from the singular value decomposition method of the logarithmic seismicity rate matrix. The seismicity matrix is defined by the logarithmic data matrix of the density of seismic events ranged in M4–M5 in the global case but in M1–M2 and M2–M3 in the regional case of the gridded locality of the earth and the present region. The gridded locality is defined by depth range in 0–700 km and longitudinal range of 5° and 2.5° of the latitude range in the global case, but in the regional case it is the depth range in 0–30 km (crust) and 30–700 km (mantle) in the regional case as defined by the previous book.

The characteristic features of the seismic data matrix which shows the dimension of 174 × 1213 in the global case, but in the regional case of Japan it does 125 × 930, and the time window is the 10 days in both matrices. In the previous work, the author clarified that for the acquisition of the characteristic parameters, various time windows of data matrix have been investigated in the global and regional seismic activity from one day to one year, but the 10 days window is available to investigate the dynamics of the global and regional seismicity (Toriumi 2021). Furthermore, it has been clarified that the timeseries of the correlated seismicity rates such as z_1 to z_4 of both regional and global seismicity dynamics manifest several stable nodes in the phase map among these correlated seismicity rates during the period from 1990 to 2021 in the global case and from 1998 to 2021 in the regional case, thereby suggesting that the several isolated stable nodes represent those of the nonlinear dynamical system of their characteristic parameters.

In this book, it has been shown that the timeseries of the global correlated seismicity from 1990 to 2023 displays similar pattern during 1990–2021 as that of the 1990 to 2021 timeseries in the previous book (Toriumi 2022), but the trend after 2021 is somewhat different in the z_3 of the period of 2016–2018 as shown in Fig. 6.26. The strange pattern of this period appears the rapid increase and sudden decrease of z_3, and it seems to be associated with giant eruption of submarine volcano at Hunga-Tonga-Hunga as shown in the previous section. Thus, it should be carefully discussed that the characteristic features are used to be potential indicators for probability prediction of large earthquake.

In the next sections, the author intends to try the large earthquake prediction by means of the deep neural network modeling with input data matrix of correlated seismicity and logarithmic seismicity rate of the present gridded area.

A general structure of the full-prediction procedures of large earthquake as shown in Figs. 6.27, 6.28 and 6.29 is composed of five steps of

(1) To obtain the global and regional seismicity density of small earthquake and large earthquake of gridded locality (along the subduction plate boundary),
(2) To obtain the global and regional correlated seismicity rates,
(3) To obtain the augmented timeseries of combined correlated seismicity and local logarithmic seismicity timeseries with labeling of large earthquake,
(4) To obtain the predicted correlated seismicity and logarithmic local seismicity matrix by means of one-step ahead prediction by 2dCNN, and
(5) To obtain the predicted probability of large earthquake by 2dCNN with deep machine learning of augmented seismicity timeseries to test the predicted matrix obtained (4).

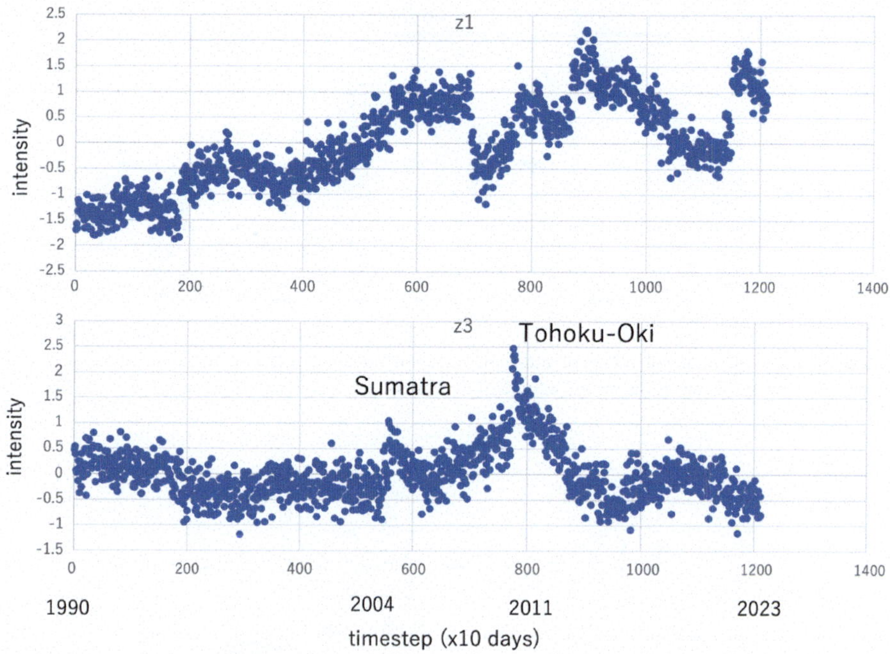

Fig. 6.26 Timeseries of global correlated seismicity z_1 and z_3 from 1990 to 2023

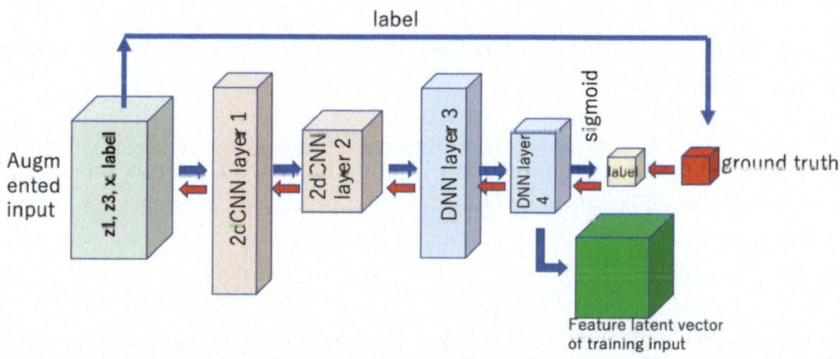

Fig. 6.27 Illustration of full-prediction procedure of 2dCNN machine learning of augmented data array. Final layer output is referred to ground truth (large earthquake event on/off). Latent parameters are taken from DNN4

Therefore, two full-prediction procedures are needed for estimation of the probability of large earthquake prediction: one is the prediction of characteristic feature and local seismicity matrixes in the future, and the other is the prediction of future large earthquake.

By the way, there are several models in the future prediction of the multivariate timeseries using the machine learning, being likely to seismicity and climate

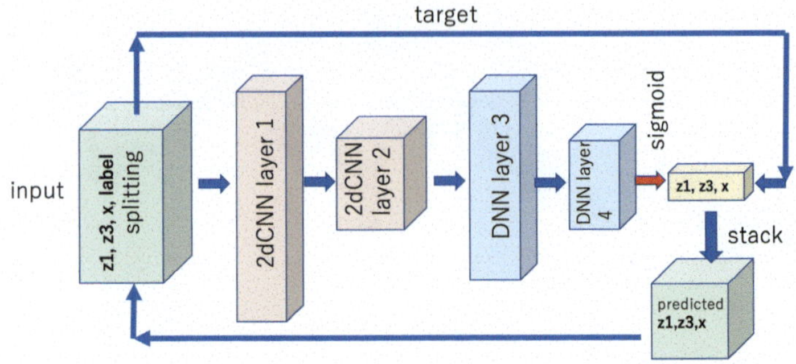

Fig. 6.28 Illustration of 2dCNN prediction procedure of one-step ahead future of global correlated seismicity z_1 and z_3, and logarithmic seismicity of the target locality grid x

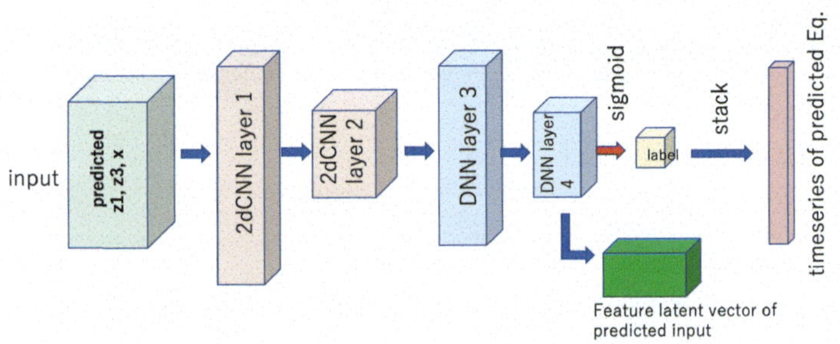

Fig. 6.29 Illustration of 2dCNN prediction procedure of timeseries of predicted large earthquake events during 2023 to 2024/3/30 from predicted multivariate timeseries of (z_1, z_3, x). Latent parameters are taken from DNN4

change in the global and regional cases of the earth. One of them is the state-space modeling (SSM) which is composed of stochastic processes of observable timeseries and hidden state vector timeseries proposed by Kitagawa (1990), as studied in the previous book (Toriumi 2021). On the other hand, neural network modeling (NNM) is a powerful method for prediction problem, it is divided into several models: recurrent neural network modeling (RNN), long- and short-term recurrent neural network modeling (LSTM), convolution neural network modeling (CNN), and transformer/attention modeling (TAM).

In the previous books (Toriumi 2021, 2022), the author tried to investigate the prediction study that contains the accuracy for estimation from deep machine learning of training data to compare the testing data matrix. The multivariate timeseries of the global (1990–2021) and regional (1998–2021) correlated seismicity matrices investigated in the previous book (Toriumi 2022) are extended to 2023 in this book.

First, let us describe the procedure of data processing in the full-prediction investigation of the future earthquake event. The input data are composed of the correlated seismicity timeseries and local logarithmic seismicity ones of the present gridded locality to predict the near-future earthquake event. The former data have been described in detail in the previous books, but it should be necessary to denote that the seismicity density matrix in the time window of 10 days and the space window of 2.5 degree of longitude and 5 degrees of latitude with 0–30 km (in crust) and 30–700 km depth (mantle) ranges in the regional case of the subduction zones is decomposed into the correlated seismicity and its multivariate timeseries. In the global case, the seismicity density is taken in the depth range of 0–700 km, because of various epicenter distribution in the global subduction zone. In addition, it should be noted that the density of epicenters of small earthquakes (M3 and M4) in the global case and micro earthquakes (M1, M2, and M3) in the regional case were chosen because those are very sensitive in the change of stress conditions as like as the acoustic emission from the brittle-deformed rock sample in the deformation experiments. The gridded localities of the regional cases of Japanese islands and northern California are shown in the previous book (Toriumi 2021). In this book, the gridded localities for prediction study are shown in Fig. 6.30.

The shape of data matrix X composed of logarithmic epicentral density of small/micro earthquakes in the gridded localities is then 920×125 from 1998/1/1 to 2023/3/29 by the time window of 10 days in the case of Japanese islands region, and in the global subduction zone, it is 1210×174 from 1990/1/1 to 2023/3/29. These data matrices are decomposed to characteristic features matrix Z and characteristic Eigen vectors array W which is the orthogonal rotation matrix by singular-valued decomposition (SVD).

The formulation holds,

$$X = ZW + N\left(\mu, \sigma^2\right). \tag{6.11}$$

In this equation, N is the Gaussian noise containing observed logarithmic density of small and micro earthquakes, but availability of Gaussian noise of the logarithmic density was already shown in the previous book (Toriumi 2021). Then, the characteristic features of multivariate timeseries contain the Gaussian noises.

Therefore, Z holds,

$$Z = W^t X - N\left(\mu', \sigma'^2\right). \tag{6.12}$$

As shown in the previous book (Toriumi 2021), characteristic features of the correlated seismicity multivariate timeseries contain rather large Gaussian noises.

Secondly, it follows that the single box corresponding to the large earthquake event is labeled 1 (yes) and 0 (no event) in the timeseries of gridded localities of the global and regional cases. If there are several numbers of events, the label is characterized by 1, because of machine learning of symbolic target of the event corresponding to the correlated seismicity and local seismicity arrays in the multivariate timeseries. The i-th row of input multivariate timeseries matrix is composed of major correlated seismicity rates $z_{i,1}$ and $z_{i,3}$, and logarithmic local seismicity density $x_{i,k}$ of k-th grid and large earthquake event label l_i at the i-th time step. In this study, the reason why the characteristic features z_1 and z_3 with local logarithmic seismicity rate x is chosen

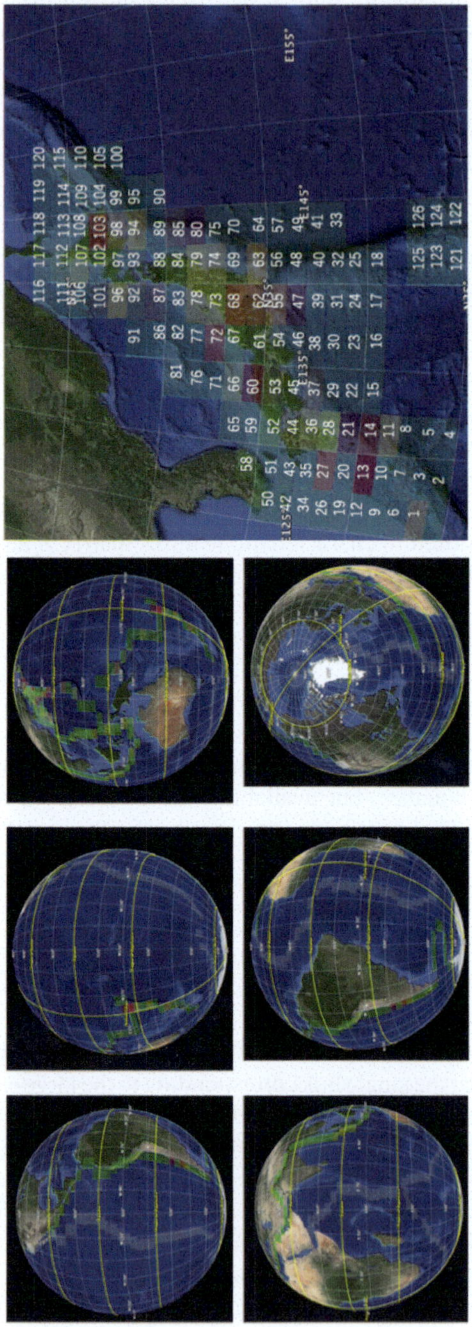

Fig. 6.30 Map showing the gridded localities of global subduction and ridge zones (left) and of Japanese region (right) on Google Earth map

is due to the available efficiency of the accuracy and cost performance of CPU time, using the augmentation of input data matrix described later.

Thirdly, processing of the data matrix is the augmentation routine that is performed in the RStudio platform installed in MacBook Pro, through preprocessing calculation of local mean and variance values on Excel platform. The embedding of the timeseries for above processing is taken to be 3 steps in this study. Number of augmentations of the input timeseries is 10, because of the available amount of data for machine learning of the seismicity timeseries, as described in the augmentation section.

The example of the augmented timeseries of the correlated seismicity rates z_1 and z_3, and logarithmic seismicity of the target localities are shown in Fig. 6.31, thereby indicating the available range of the fluctuation in the augmented timeseries. After augmentation by several ten times, the number of input data is surely available in the machine training of the characteristic features matrices corresponding the labels. The multivariate timeseries of the augmented data matrix and occurrence of large earthquakes labeled by 1 and 0 are shown in the same figures, thereby indicating the fluctuated timeseries of the present data around the primary multivariate ones. It, furthermore, shows that the synthesized timeseries displays the sequence repeated in the same time range as the primary timeseries of z_1, z_3, and x from 1990 to 2023 in the case of global seismicity and from 1998 to 2023 in the regional case of Japanese islands.

As stated previously, the present full-prediction model requires both of predicted multivariate timeseries involving the correlated seismicity rates z_1 and z_3 and logarithmic local seismicity rate, which are estimated from the one-step ahead prediction method using 2dCNN modeling, from the observed timeseries, as an example shown in Figs. 6.32. In this work, the time range of the prediction is performed extending to about 1 year from 3/29/2023. Therefore, the full prediction of the large earthquakes

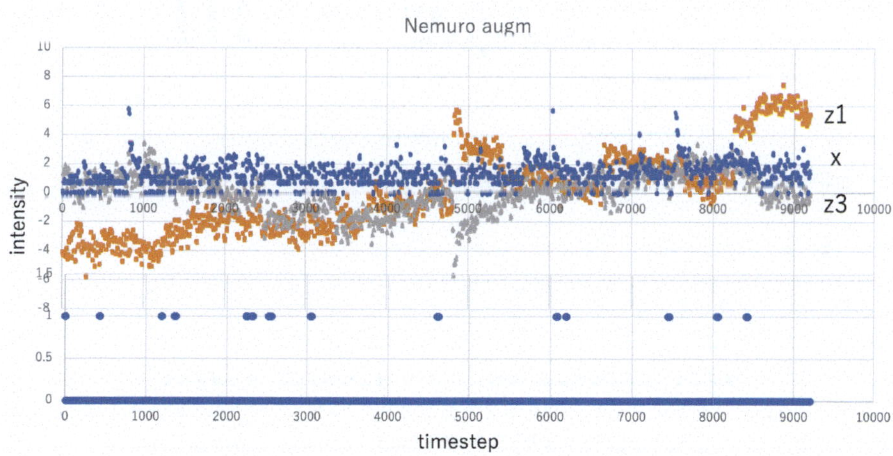

Fig. 6.31 Multivariate timeseries of augmented regional correlated seismicity z_1 and z_3, and logarithmic seismicity rate x of Nemuro region, Hokkaido, Japan. The labeling of large earthquake over M5 is shown in the lower column

by 2dCNN has been carried out for one year after 3/29/2023 in the case of Off-Kanto region as shown in Fig. 6.33.

In the present work, the author intends to refer the intermediate data matrices stored in the later dense layers of the 2dCNN modeling for application of the more

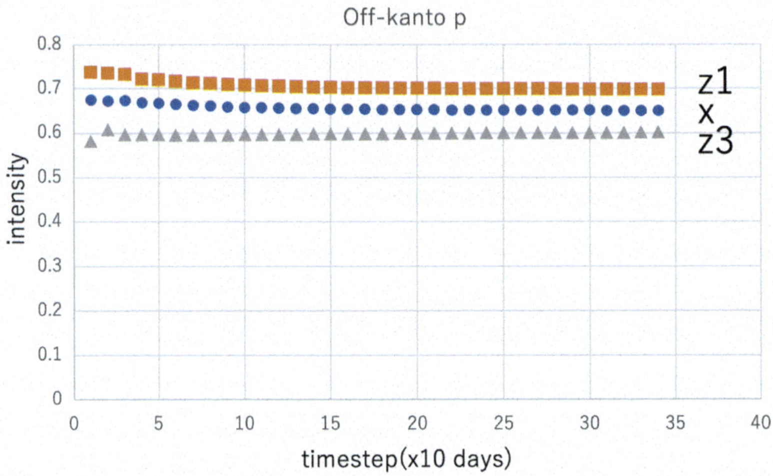

Fig. 6.32 Predicted multivariate timeseries of regional correlated seismicity z_1 and z_3, and logarithmic seismicity x of Off-Kanto region

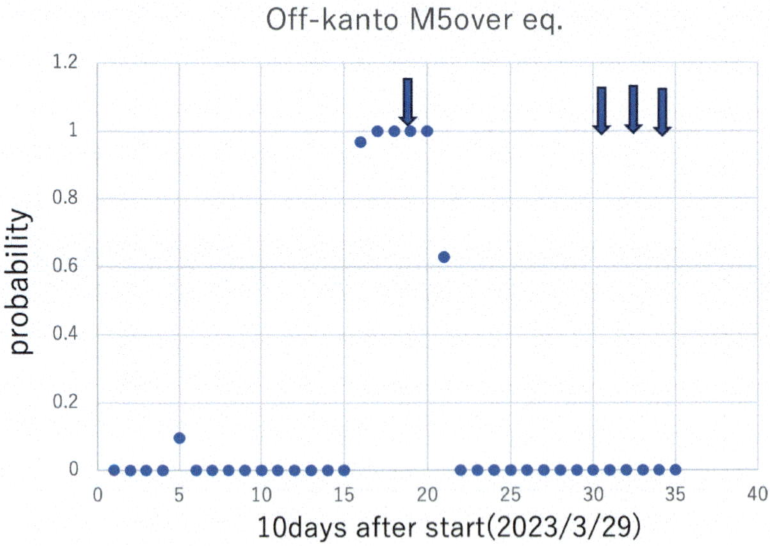

Fig. 6.33 Result of full prediction of large earthquakes over M5 during the period of 2023/3/30–2024/3/29 in Off-Kanto region. Arrows indicate occurrence of large earthquake in this time range

available labeling of the earthquake data to the augmented input data. The dimension of the data matrix stored in the intermediate layers varies depending on the number of the nodes of the present layer, but in the study of try and error, it seems that the dimension of the final dense layer before single node connected by sigmoid transfer function is availably 16, thereby that the dimension of the matrix of the intermediate dense layer is (time step number) * (16 + 1) (label). The present matrices having large number of dimensions represent the latent space expression of the augmented input data.

As discussed in the previous chapter (augmentation method), the latent space mapping is very important, for the reason whether the present augmentation of the input multivariate time series is available for the actual machine learning of the present study or not, and which is the most effective partial space in the latent high-dimensional space for the distinctive classification by each label, and how the partial space can be applied for the practical prediction of the large earthquake as the main object of the book.

In this investigation of DNN study of earthquake timeseries, the input data is now selected as 3 × 3 embedded matrix timeseries, and therefore, the usage of the 2dCNN modeling of the matrix is taken as filters of 2 × 2 and initial layer has 128 nodes. The two layers of dense connection are composed of 64 nodes and 16 nodes before final node (1 node). The acquisition of latent parameters is from secondary dense layer with 16 nodes, thereby the latent space has 16 dimensions with sample size.

To investigate the ability of classification by labeling in latent space mapping, two dimensions mapping of the latent parameters was carried out, and as a result, it concludes that a few maps of them show obviously two clusters of labels '0' (no large earthquake) and '1' (large earthquake occurs). Furthermore, it is noteworthy that the result by the latent spacing prediction is somewhat different with those by the CNN final output, thereby suggesting that the final transformation by sigmoid connection from dense layer to final 1 node contains somewhat noisy components from the indifferent parameters. In this work, therefore, both prediction results are shown for comparison of them.

From the results of the present full predictions by normal 2dCNN with labeling model (Toriumi 2022) in the testing region shown in Fig. 6.32, it suggests that several events of large earthquake are predicted as a large probability but the actual event occurred a little bit of later, and that earthquakes in the recent days of 2024 are not predicted, thereby suggesting that whether the augmentation proposed here is not enough potential to deep training of the present DNN modeling, or the prediction of the characteristic correlated seismicity and logarithmic seismicity of the target locality is not available for transferring to the earthquake 'on and off' label.

In the previous section of augmentation of the timeseries, it shows that the latent space map of two parameters in the dense layer of the 2dCNN modeling for augmented multivariate timeseries represents a clear pattern of distinctive two clusters: one '1' is large earthquake event, and the other '0' is no event. It is obvious that the augmentation proposed here is surely available for the deep machine learning of rather rare event of the earthquake, and therefore it leads that several latent space parameters are useful to apply to DNN modeling, which gives the available target values of label (1 or 0).

It seems that the latent space parameters of which the map displays the distinctive clusters, corresponding to labels exhibiting large earthquake and no event, are capable for the application to prediction of the large earthquake, and it is possible to be compared with the predictable output data of the 2dCNN modeling as shown in Fig. 6.34. To obtain the available set of the latent space parameters, it needs for the random search that many sets of two latent parameters will be tested in 16 parameters of the second dense layer of the 2dCNN model. In this study, the judgment of the availability of the latent parameters is carried out by appearance of obvious two clusters and dispersive plotting of data within the single cluster. The author will intend to combine the present selection procedure of the latent space in second dense layer with the 2dCNN modeling.

The implication of the latent space mapping of the intermediate dense layer parameters is to be considered that the input multivariate vector timeseries, x_1 (sample number × 3 dimension) should be transformed into s_1 multivariate vector timeseries of sample number × 16 dimension by the 2dCNN function, F, with optimal parameters fitted through one-dimensional output in the final dense layer against the target label as,

$$s_1 = F(x_1). \tag{6.13}$$

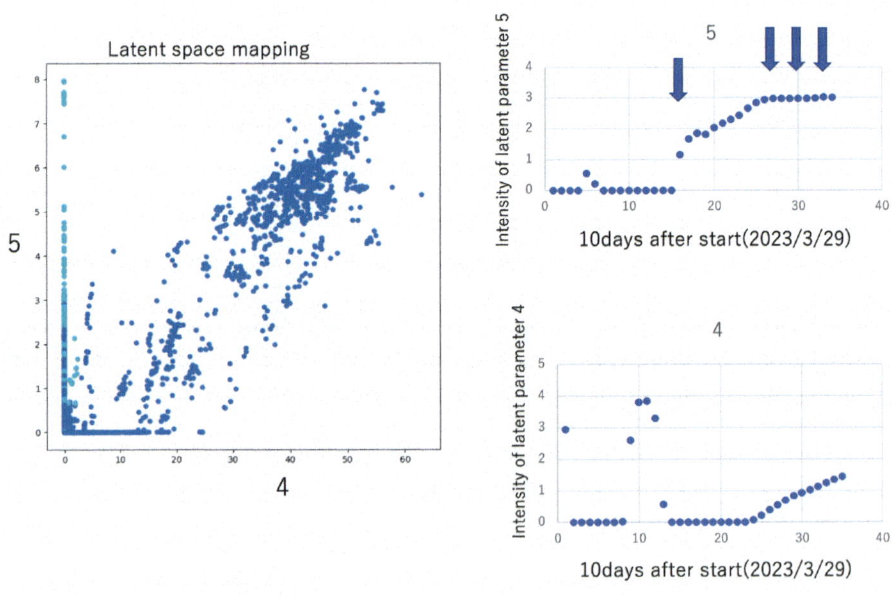

Fig. 6.34 Results of full prediction by latent space mapping from DNN4 showing the probability of prediction. Arrows indicate occurrence of large earthquake in this time range

Furthermore, for there is optimal solution of the output against the target data, some of components of s_1 should contain distinguishable parameters sets for labeling '0' and '1', that is the possible index between the large earthquake event or no event. In other words, there should be the signal of the precursor of the large earthquake in some parameters of the latent space of second dense layer. After transformation of such signals to the final node by sigmoidal function, it seems that the possible precursor signals may change vague by mixing the noisy parameters stored in the different parameters in the same dense layer, and thus, it seems to be appropriate that the output data matrix in the dense layer should be considered for looking for available precursor signals of large earthquake transformed in the latent space of 2dCNN modeling (e.g., Chollet and Allaire 2018).

It is still unknown that whether the seismic precursor signals exit before the large earthquake or not, and what is observable signal of possible seismic or geodesic precursor in the plate boundary zone, although there are abundant post-seismic aftershocks in the area of the large earthquake. However, as discussed in the previous chapters, it has been clarified that the large earthquake has the large high-speed slip area, embedded in the earth interior, and therefore, the boundary between the slip area and the surrounding forms physically the crack tip.

By the way, it is well-known that the large earthquake is represented by large slip area and nearly constant applied shear stress, thereby showing that the large shear crack is generated without explosive propagation, that is stable growth stage before earthquake generation. Considering the earthquake generation is responsible for the transition from stable growth to explosive propagation of shear crack, the reason why the large shear crack grows stably is due to the large dissipation of the stress concentration around the crack tip by generation of microcrack cluster or by plastic deformation, as clarified in the previous chapter. Therefore, it is expected that, before explosive propagation, stable growth of the large shear crack, which is potential large earthquake, should be associated with large amount of microcrack generation around the main shear crack, thereby representing the continuous micro-seismic clusters around the large earthquake epicenter. And this conclusion means the seismic precursor of the large earthquake. Judging from the microcrack size is about 0.1 m with displacement of 0.1 mm, micro earthquake is less than magnitude of about -3, being much less than background seismicity about M1 to M2 in this study. However, it should be considered that the probability distribution of the earthquake magnitude in the seismic region displays a power law (Gutenberg-Richter law), thereby suggesting the seismicity changes during the stable growth of the large shear crack before transition to large earthquake propagation mode.

Otherwise, it is possible that the viscous dissipation by plastic deformation with dissolution and precipitation of metamorphic minerals around advancing crack tip strongly enlarges the crack size without transition to propagation mode. In this case, the seismic precursor before transition to propagation mode is not expected enough, thereby suggesting the abrupt generation of large earthquake without micro-seismic activity. In the characteristic correlated seismicity and local logarithmic seismicity density, transformed parameters of latent space in the 2dCNN modeling should manifest the characteristic pattern of the precursor signals.

From the training data of the augmentation of the input set of timeseries, as discussed in the next section, it appears that the augmented multivariate timeseries of the correlated seismicity rates z_1 and z_3, and the logarithmic seismicity rate of the target locality are associated with the timeseries of labels of large earthquakes. In the following, the target localities are selected as the Northeast Japan (NEJ), Chile (CHI), and Himalaya (HMA) in the global case, and in the regional Japanese islands case, as Off-Kanto (KT), Tokyo (TK), Osaka (OS), Nankai Trough (NT), and Off-Nemuro (NM). The primary multivariate timeseries ranges from 1990/1/1 to 2023/3/29 in the global case, and from 1998/1/1 to 2023/3/29 in the regional case as mentioned previously. The labeling of '1' is denoted by large earthquake single and multiple events over M5 in the regional case but M6 in the global case. The correlated seismicity rates are obtained by the singular value decomposition method on the logarithmic seismicity rate of the gridded localities in the subduction zone as defined in the previous chapter.

6.6 Prediction of the Large Earthquake in the Japanese Island Region and in the World

As shown in Figs. 6.35 and 6.36, the augmented timeseries of z_1, z_3, and x together with label (1, 0) are well-plotted near the primary, their timeseries, and it shows that they are fluctuated in some degree of intensity but denoised by the present up-diffusion processing. The labeling, however, is performed by copying of primary label (1, 0) corresponding the primary input 3×3 matrix of each time step.

Studied gridded localities are Off-Nemuro, Off-Kanto, Tokyo, Osaka, and Nankai, and the augmented timeseries are shown in above figures, thereby showing large difference in the logarithmic seismicity of these localities from 1998 to 2023. The timeseries of the correlated seismicity commonly shown in the diagram are from the regional correlated seismicity rates of the Japanese islands region from 1998/1/1 to 2023/3/29, and it is of unique point of this study that the characteristic features of the local seismicity should be strongly dependent on the regional seismic process and regional mechanical behavior manifesting the local seismicity because of the continuum mechanical process in the regional and global solid earth environment.

Next, it will be shown that the inferred timeseries of the multivariate seismicity vector extrapolated by 2dCNN modeling of primary multivariate timeseries of z_1, z_3, and x are drawn in the range 350 days from 2023/3/30 as in Figs. 6.37 and 6.38. It seems that the predicted multivariate timeseries vary before 30 to 60 days, and then changes slowly and monotonously, thereby suggesting the early stage of the extrapolation controlled by the primary observed data, and later they are rounded by the predicted matrix of z_1, z_3, and x. In addition, it will be denoted that the regional correlated seismicity z_1 and z_3 may have the same timeseries within different localities, but in this study the 2dCNN modeling for extrapolation of them are carried

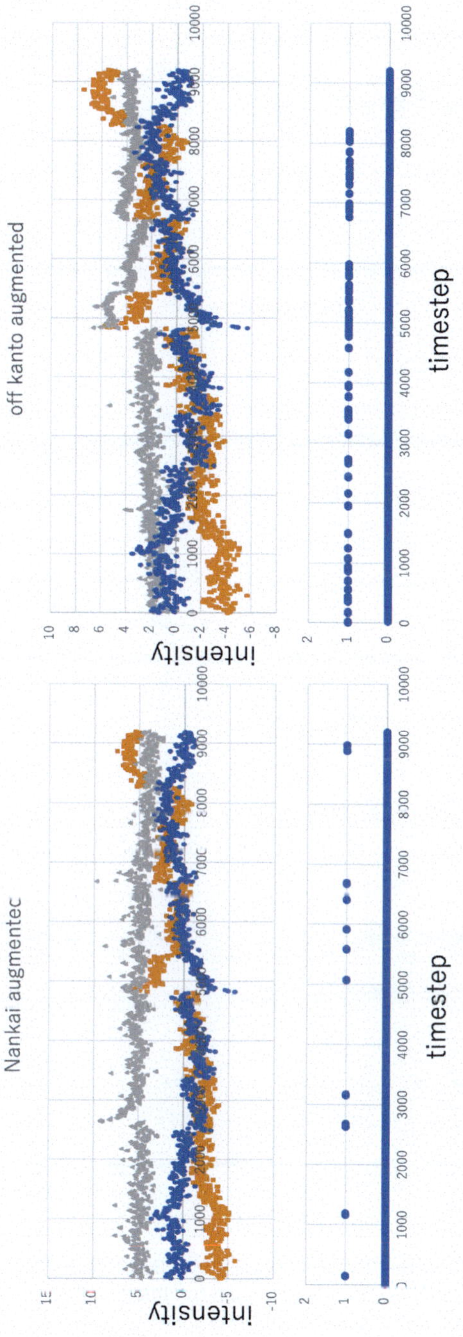

Fig. 6.35 Augmented multivariate timeseries of regional correlated seismicity z_1 and z_3, and logarithmic seismicity x of Nankai region (left) and Off-Kanto region (right). Timeseries of large earthquake over M5 is shown in lower figure

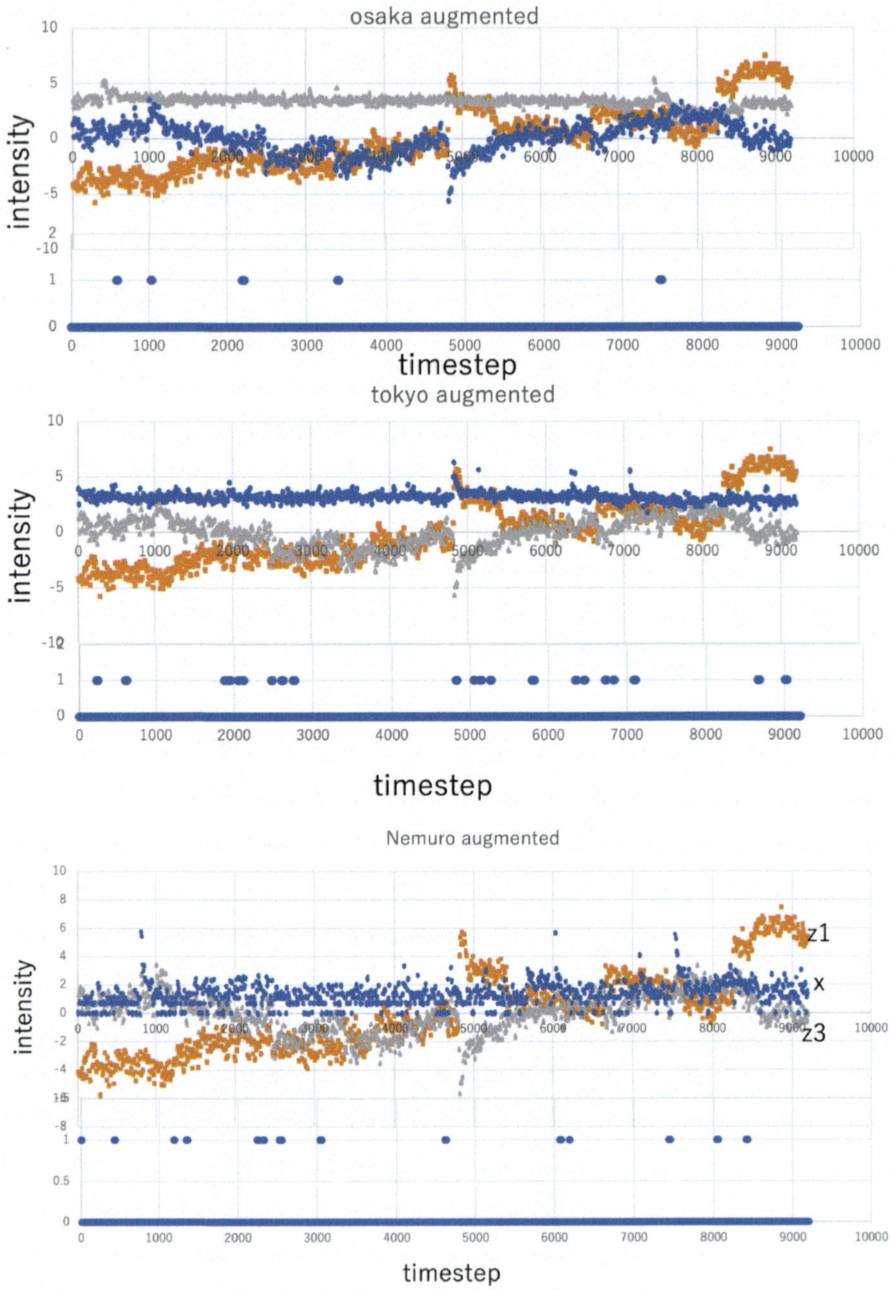

Fig. 6.36 Augmented multivariate timeseries of regional correlated seismicity z_1 and z_3, and logarithmic seismicity x of Osaka region (top), Tokyo region (middle), and Nemuro region (bottom). Timeseries of large earthquake over M5 is shown in lower figure

out by incorporation of coupling of them because of including the possible effect from local seismic activity.

As shown in Figs. 6.37 and 6.38, the predicted timeseries are well-extrapolated by 2dCNN modeling during about one year, and therefore, it will be possible to practical prediction of large earthquake, that means when and where it will occur, that the 2dCNN model after training of the augmented multivariate timeseries of z_1, z_3, and x with target label $(1, 0)$ can be applied to the testing on predicted multivariate (z_1, z_3, x) timeseries from 2023/3/30 to 2024/3/15. The results are shown in Figs. 6.39, 6.40 and 6.41. It seems that the occurrence of large earthquake over M5 in Japanese island region does not all coincide with predicted one during last one year, but some of them are seemingly approximated to the predicted one.

In the case of areas of Tokyo and Osaka, the present full-prediction method gives the possibility of large earthquakes over M5 such that no event in the underneath of Tokyo will occur in the range from 2023/3/30 to 2024/3/15, and that several events underneath of Osaka will take place in the same range of time, but in the early period of the time range, earthquakes less than M4 occurred in the predicted range of the time, as shown in Fig. 6.39. On the other hand, in the case of Off-Kanto, the predicted large earthquakes over M5 are concentrated in the middle period of the studied time range in the Off-Kanto, but the actual occurrence of the large earthquake is concentrated in the later period, although one of large earthquake took place just in the predicted period, as displayed in Fig. 6.40.

It seems to be noteworthy that the predicted earthquakes over M5 in the Nankai area, along the Nankai Trough, are concentrated in the period of rather early stage, and the actual ones but over M3 not M5 were during the period earlier than the predicted times, though one of them took place just in the predicted times, as seen in Fig. 6.41. Moreover, it displays that no earthquake over M5 of Off-Nemuro in northeastern Hokkaido was predicted, and it is in accordance with no large earthquake, thereby suggesting availability for large earthquake prediction by this method.

On the other hand, it may be possible that the mapping of the latent space parameters stored in the intermediate dense layer is available for the classification of the labels $(1, 0)$ by separable clusters in the map during the training stage of augmented input data. However, there are 16 parameters of latent space in the final dense which connected with final single node layer sigmoid transfer function, and thereby it needs that the representative two parameters of the latent space should be selected from the 16 independent latent parameters. The criterions of the present selection of parameters are based on the ability of separation between the clusters of labels 1 (large earthquake event) and labels 0 (no large earthquake). In addition, it should be required for that parameters of the latent space are scattered widely in the map because of no degeneracy of data in dense layers.

In Figs. 6.42, 6.43, 6.44, 6.45, 6.46 and 6.47, the available latent space maps are shown, thereby displaying distinctively separated clusters of labels 0 and 1. In the survey of the available parameter sets for separate clustering of the latent space, the maps of them are almost always unfavorable for separation of clusters between two labels, thereby suggesting that the final sigmoidal output in the single node is seemingly mixed with a few parameters in the final dense layer. Therefore, it seems

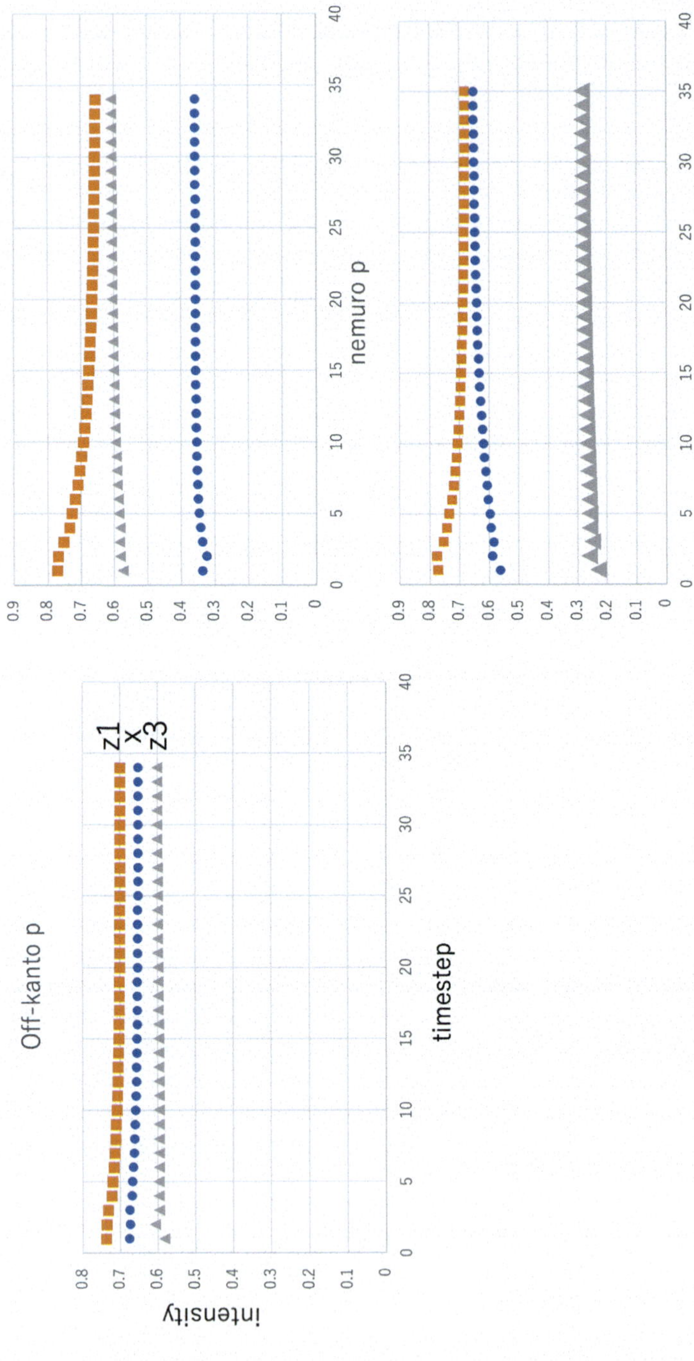

Fig. 6.37 Predicted multivariate timeseries of z_1, z_3, and x from 2023 to 2024 of Off-Kanto (left), Nankai (right top), and Nemuro (right bottom)

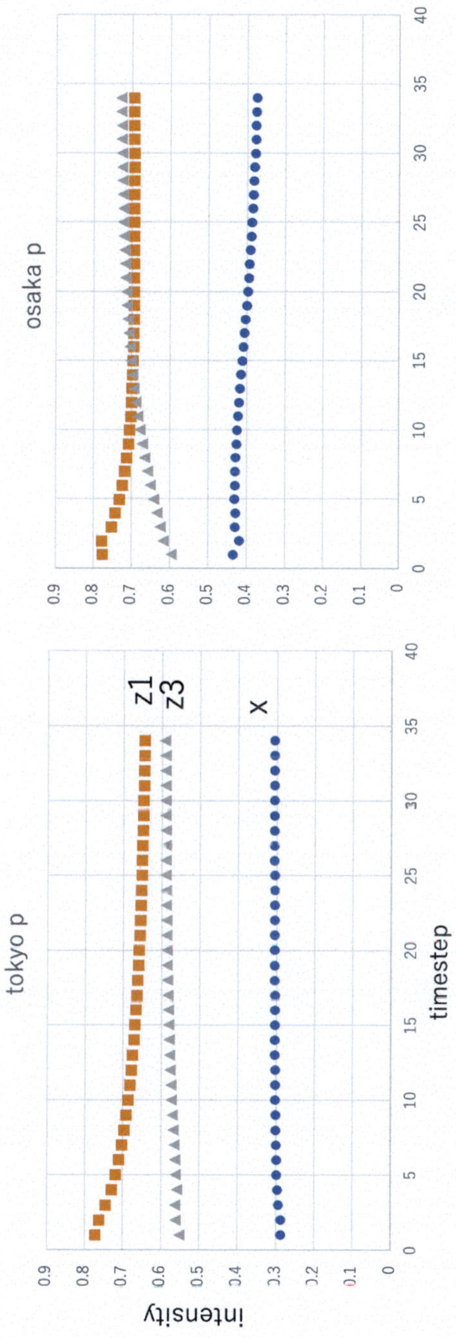

Fig. 6.38 Predicted multivariate timeseries of z_1, z_3, and x from 2023 to 2024 of Tokyo (left) and Osaka (right)

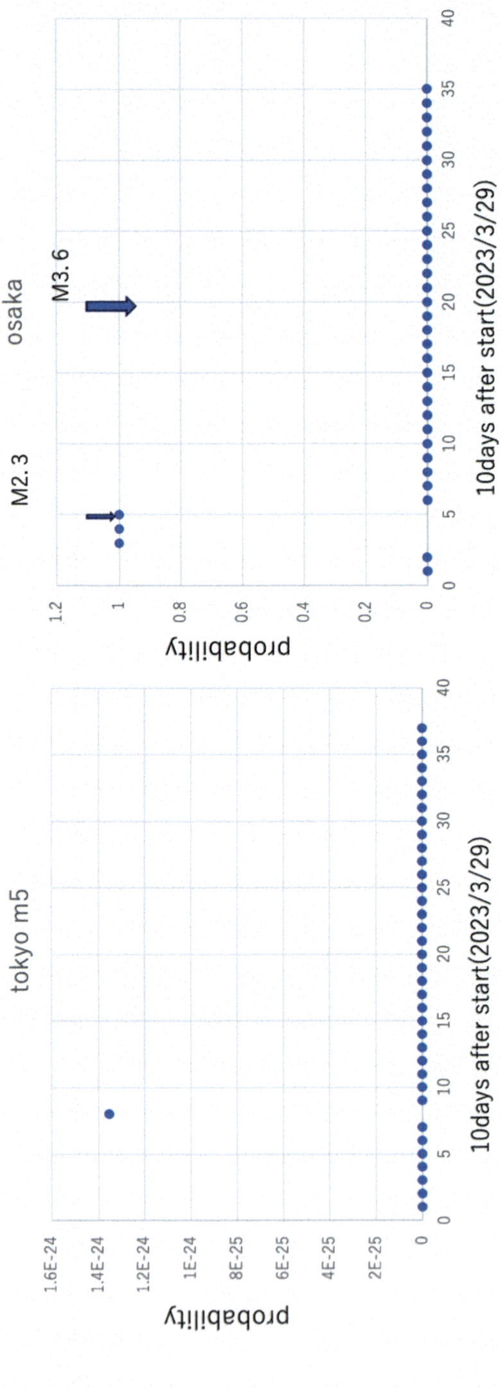

Fig. 6.39 Full prediction of large earthquake over M5 from 2023 to 2024 of Tokyo (left) and Osaka (right). In Tokyo and Osaka regions, no M5 over occurred

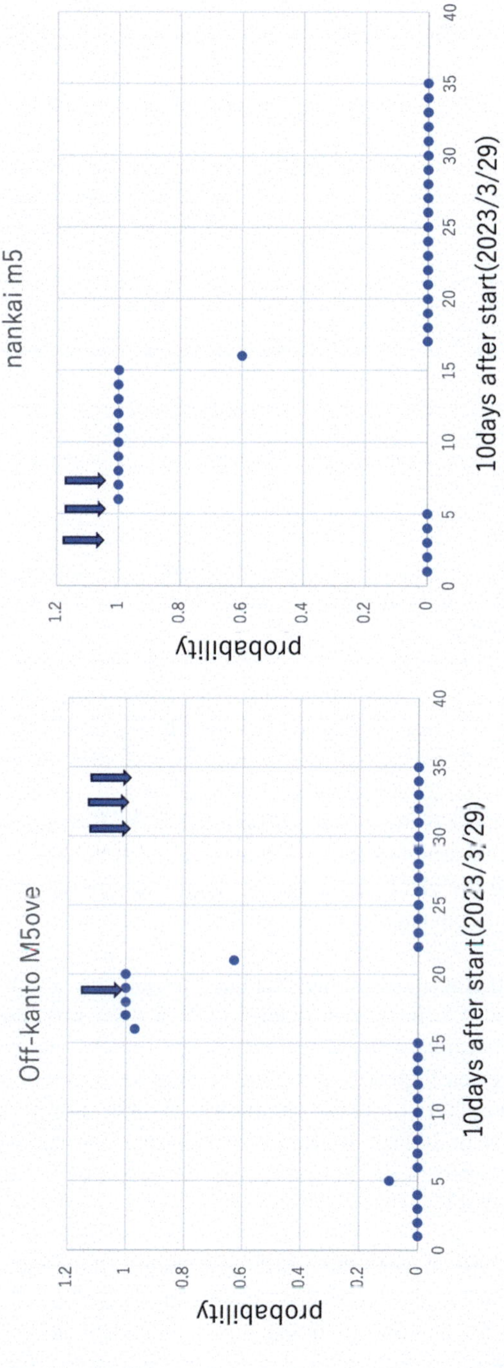

Fig. 6.40 Full prediction of large earthquake over M5 from 2023 to 2024 of Off-Kanto region (left) and Nankai region (right). Arrows indicate the occurrence of M5 over earthquakes

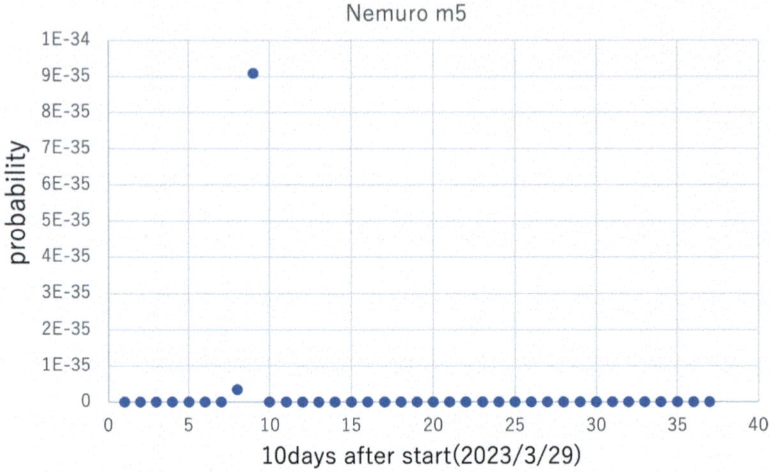

Fig. 6.41 Full prediction of large earthquake over M5 of Nemuro region from 2023 to 2024. No earthquake occurred in this region

that the latent parameters in the final dense layer contain the signals of precursors before the large earthquake in the target area, and these signals are contaminated by noisy signals and become low *S*/*N* ratio.

Let us look on the latent space mapping of the case of the Off-Kanto region (Fig. 6.42), in which rather large earthquakes take place frequently. As shown in the above figure, the cluster of large earthquakes is clearly separated by the scattered cluster of the no large earthquake, and therefore, it seems that the full prediction of the large earthquake after 2023/4/1 to 2024/3/30 is possibly available for. In the present figure (Fig. 6.42), the results of the full prediction by the present method are shown by the inferred timeseries of the characteristic latent parameters in the case of Off-Kanto, suggesting the clear occurrence of large earthquakes in the period of late time steps, which are consistent with the real recent seismic events as shown in the arrows.

On the other hand, in the case of the Nankai Trough, that is Off-Kii to Off-Shikoku, there are distinctive clusters of labels 1 (M5 over eq. on) and 0 (no eq.) in the characteristic latent parameter mapping (Fig. 6.43), thereby suggesting that the full prediction of M5 over earthquakes along the Nankai Trough will be successful. The timeseries of the characteristic latent parameters displays multiple peaks in the early stage after 2023/4/1 to 2024/3/30. It is surprising that the earthquakes a little bit smaller than M5 occurred in the period just in the range of the full prediction shown by arrows in Fig. 6.43.

Moreover, in the case of Tokyo area, it shows that the earthquake in the early stage of the studied period was predicted with a little probability in the latent parameters mapping and timeseries in Fig. 6.44, but there was no large earthquake in this period. In the full prediction by 2dCNN output, it was a very low probability of occurrence. On the other hand, in the region of Osaka, it seems that the successful prediction of

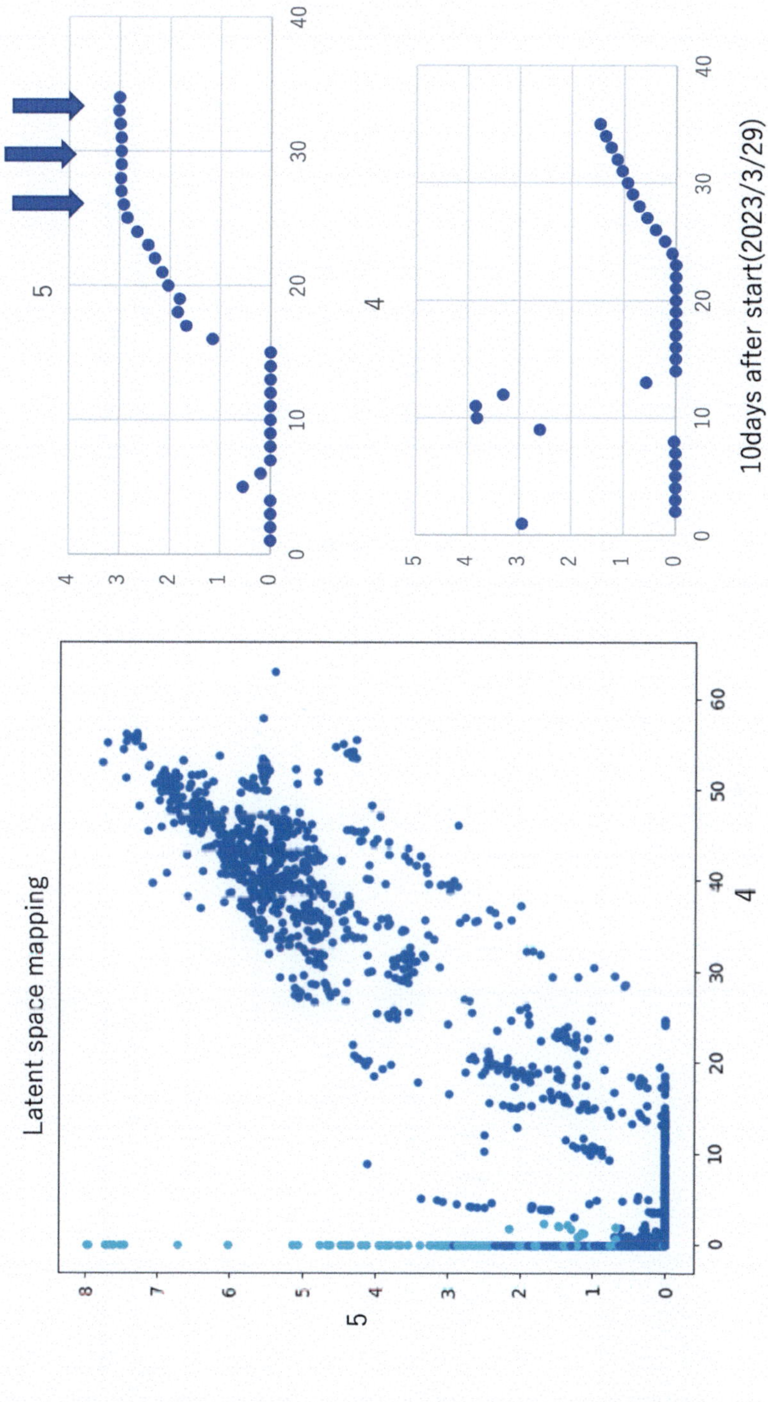

Fig. 6.42 Prediction of large earthquake over M5 of Off-Kanto region by latent space mapping method from 2023 to 2024. Left: latent space map showing large earthquake cluster (light blue) and no earthquake cluster (blue). Arrows show occurrence of M5 over earthquakes predicted in the figure

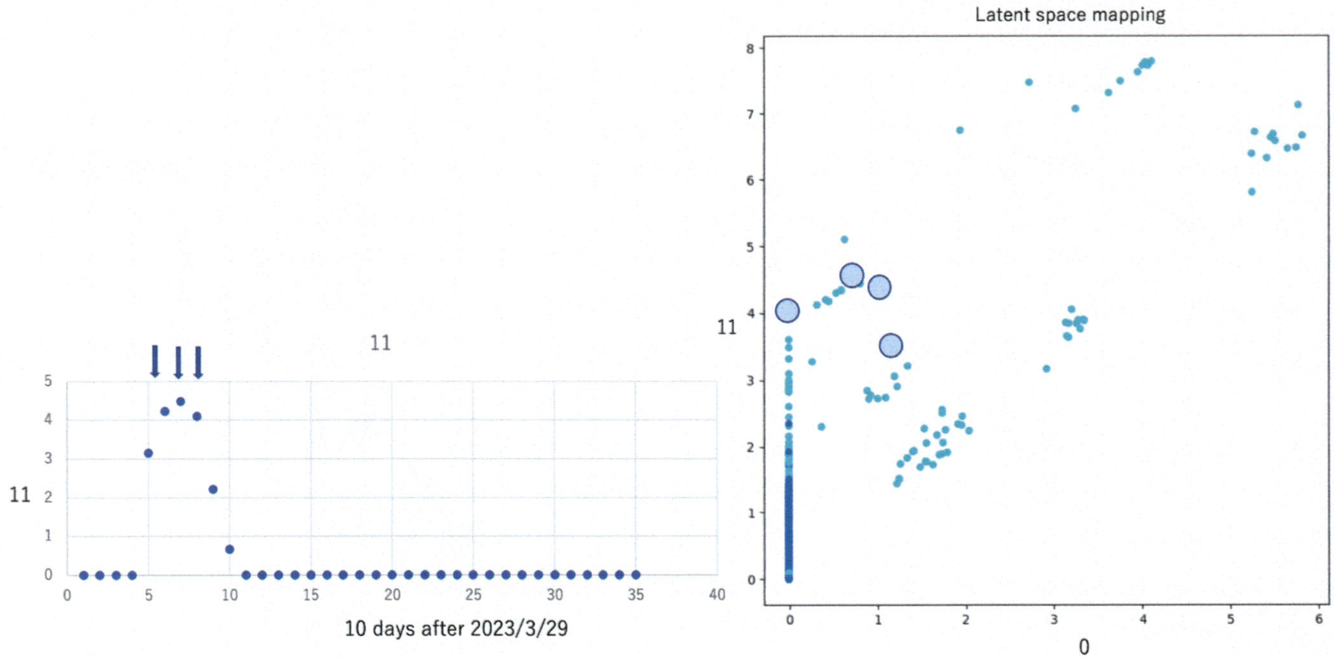

Fig. 6.43 Prediction of large earthquake over M5 of Nankai region by latent space mapping method from 2023 to 2024. Right, latent space map showing large earthquake over M5 cluster (light blue) and no earthquake cluster (blue). Arrows show occurrence of M5 under earthquakes predicted in the figure

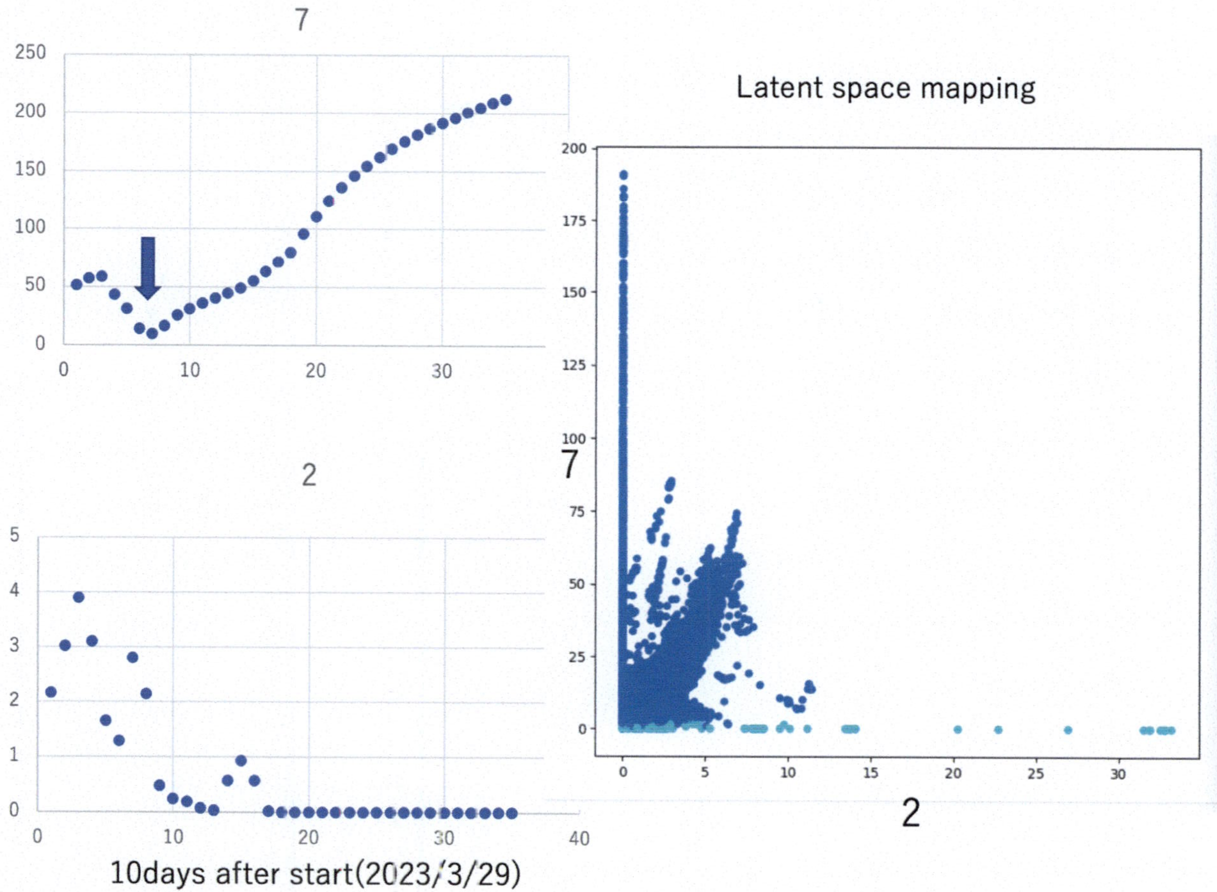

Fig. 6.44 Prediction of large earthquake over M5 of Tokyo region by latent space mapping method from 2023 to 2024. Right, latent space map showing large earthquake cluster (light blue) and no earthquake cluster (blue). Arrows show occurrence of M5 under earthquakes predicted in the figure

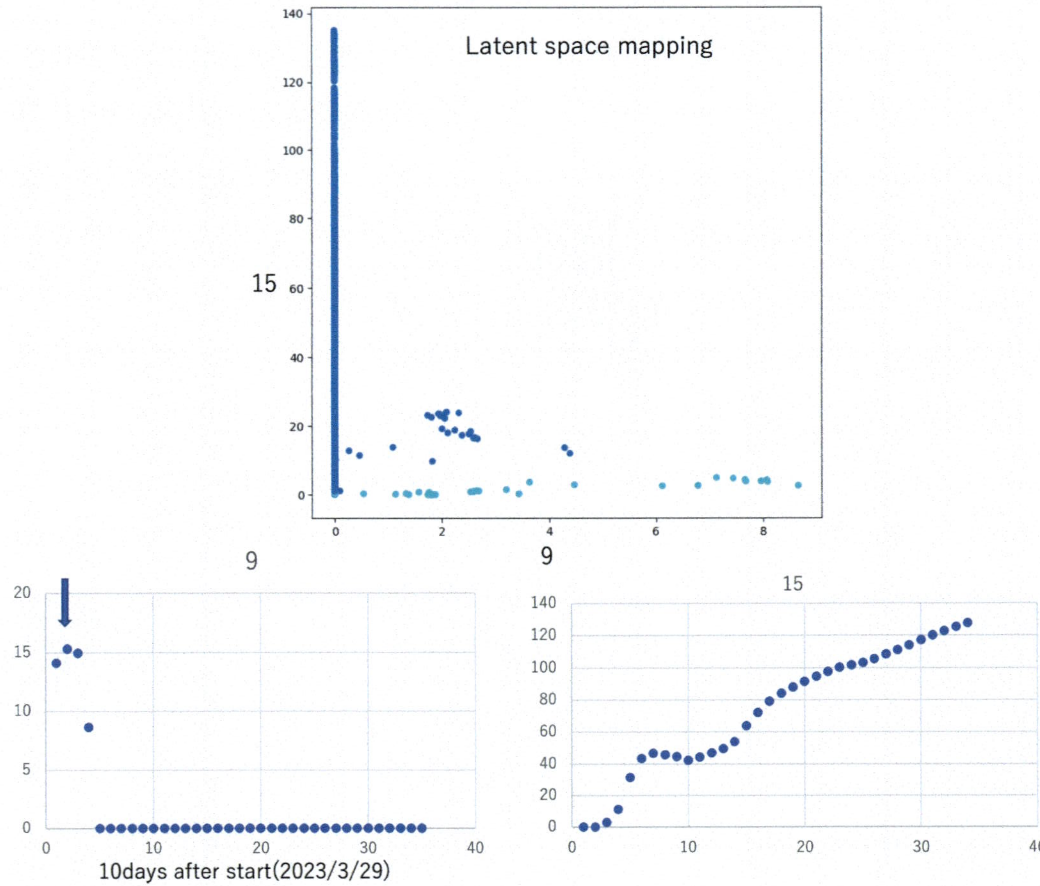

Fig. 6.45 Prediction of large earthquake over M5 of Osaka region by latent space mapping method from 2023 to 2024. Top, latent space map showing large earthquake cluster (light blue) and no earthquake cluster (blue). Arrows show occurrence of M5 under earthquakes predicted in the figure

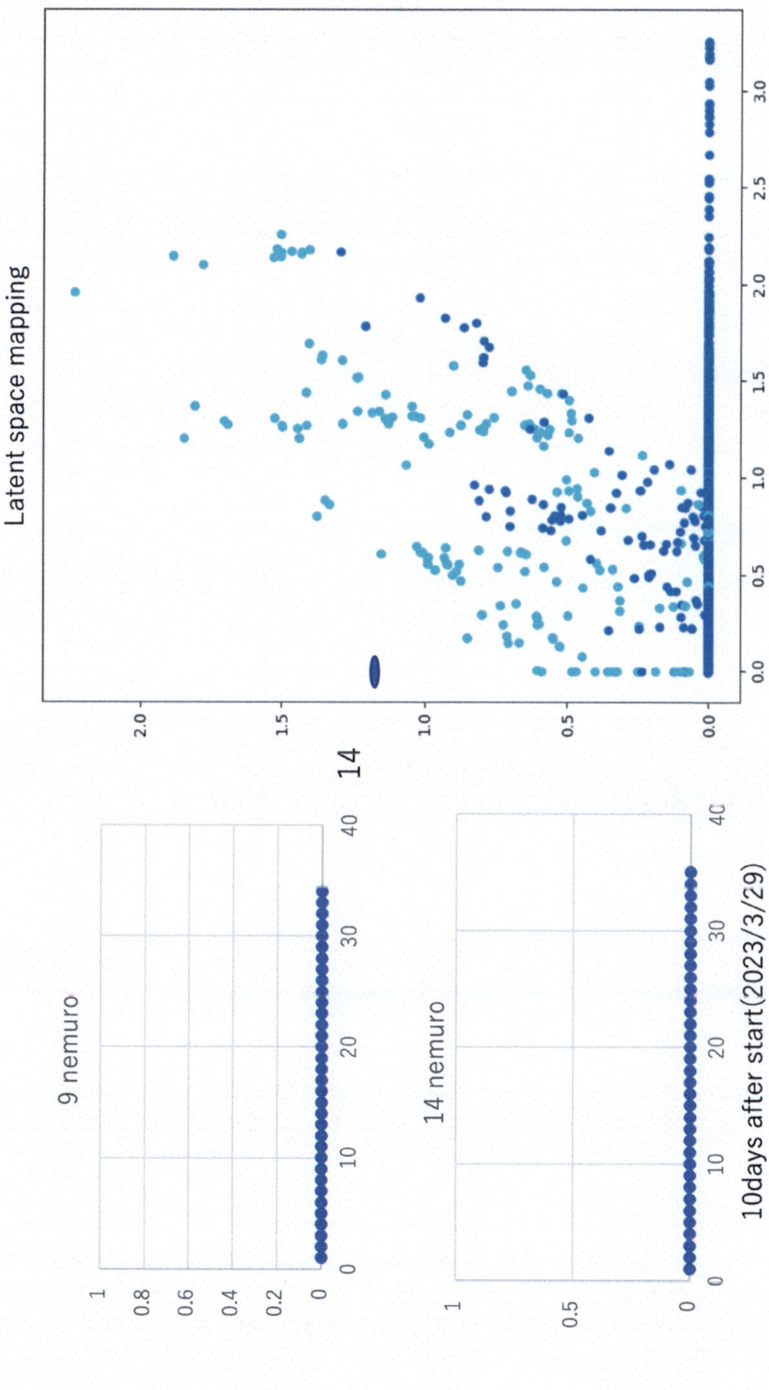

Fig. 6.46 Prediction of large earthquake over M5 of Nemuro region by latent space mapping method from 2023 to 2024. Right, latent space map showing large earthquake cluster (light blue) and no earthquake cluster (blue). No large earthquake in this region

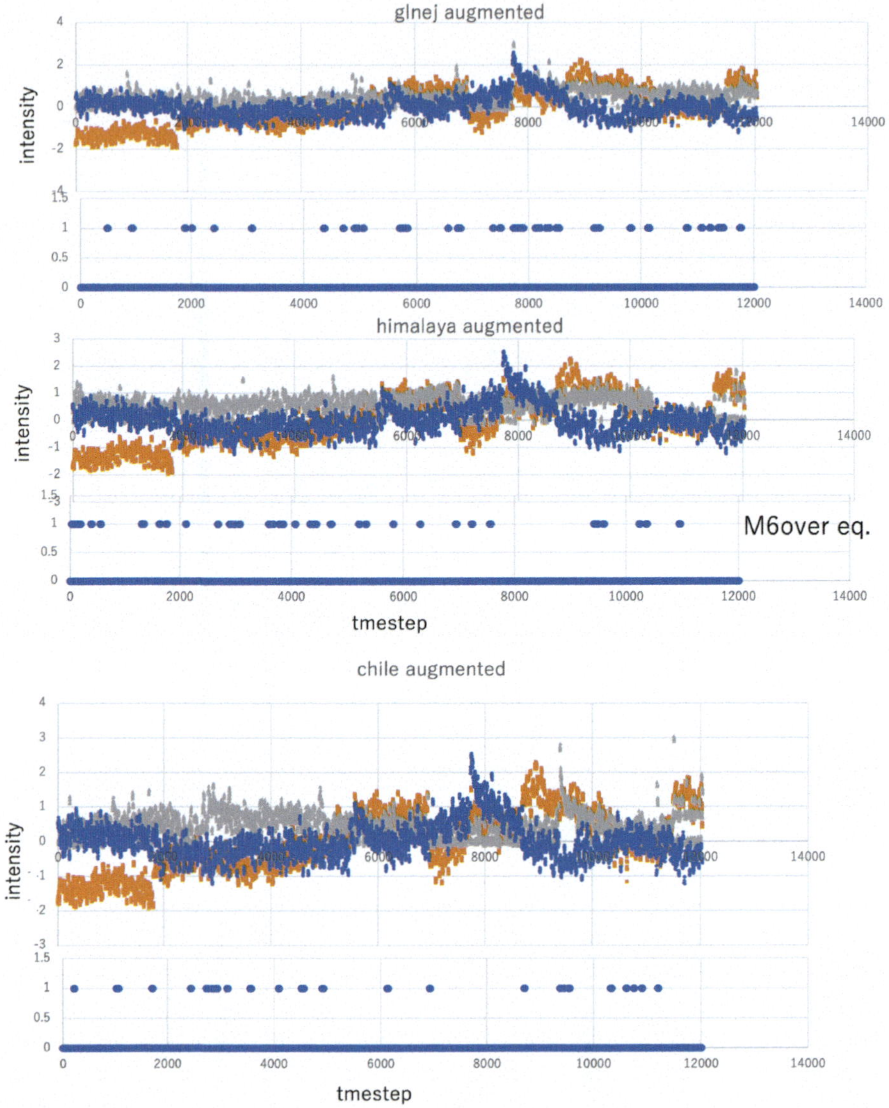

Fig. 6.47 Augmented multivariate timeseries of global correlated seismicity z_1 and z_3, and logarithmic seismicity x of northeast Japan, Himalaya, and Chile regions. Timeseries of large earthquakes over M6 are shown in the lower figures

rather large earthquakes of near M3, but not M5, is performed as shown in the latent parameters plotting of Fig. 6.45, and in the middle earthquake of the studied period, no prediction was performed. On the other hand, in the Off-Nemuro, the clusters of event-on and -off are sharply displayed in Fig. 6.46, but no prediction of the large earthquake event does not appear in the timeseries of the studied range. This result is consistent with the output prediction of 2dCNN as described previously.

Next, let us discuss the results of the global seismicity prediction. The augmented global correlated seismicity and the logarithmic seismicity of the target localities are shown in Fig. 6.47. The full prediction of the northeast Japan (NEJ), Chile (Chi), and Himalaya (Him) regions was carried out by the same methods as those in the Japanese region. As shown in Fig. 6.48, the global correlated seismicity z_1 and z_3, and local logarithmic seismicity x is predicted in the period from 2023/3/30 to 2024/3/30 by 2dCNN modeling from the training of the timeseries from 1990/1/1 to 2023/3/29. The method of the present prediction is such that the input matrix 3×3 is composed of features z_1, z_3, and x with their embedded three components of the timeseries, and then the set of next of these features is inferred from previous 3×3 matrix by one-step ahead method as described in the previous section. It, therefore, shows that the change of the predicted z_1, z_3, and x appears smooth curves (Fig. 6.48), suggesting a little difference in timeseries of the global correlated seismicity z_1 and z_3, because of interaction of them at the prediction by 2dCNN.

The results of full prediction on the case of NEJ, Chi, and Him are shown in Fig. 6.49. It seems that the accuracy of prediction of the large earthquakes over M6 is not satisfied yet, except for the case of Chile. It seems that the reason why the prediction is inadequate is derived from the prediction of the feature timeseries of the correlated seismicity z_1 and z_3, and the local logarithmic seismicity x, or from the incomplete machine learning. Thus, it should be checked that whether the latent space maps of above global prediction case are plausible for the detection of large earthquake in the period from 2023/3/30 to 2024/3/30.

The results of the latent space mappings of NEJ, Chi, and Him of the final dense layer having 16 nodes are shown in Figs. 6.50, 6.51 and 6.52. As noted previously, the preferable set of the latent space parameters in the dense layer of 2dCNN modeling is available for identifying the clusters in the map. It shows that, in the latent space map, the two clusters of labels 1 (earthquake on) and 0 (off) are distinctively separated, thereby showing the availability of the full prediction introduced previously.

In Fig. 6.50, the result of NEJ is shown, and it is obvious that the large earthquake over M6 near the initial period (shown by arrow) is sharply predicted in the timeseries, judging from the plotting in the cluster of labels 1 in the latent space mapping, though there is no signal in the predicted timeseries of output of 2dCNN mentioned above. In addition, in the case of Chile, it is sure that the prediction by means of the latent spacing clustering method proposed above is clearly plausible for the real occurrence of the large earthquakes over M6, as is seen in Fig. 6.51. It is noteworthy that the above figure displays a gradual increase of the latent space parameter showing gradual increase of probability of large earthquake in the timeseries before several ten days.

On the other hand, the investigation of the prediction by the same method was carried out for the Himalayan plate boundary region (Fig. 6.52), and it shows that the timeseries of the predicted earthquake labeling shows the appearance of the large one over M6 at the time of the middle of the studied period, thereby suggesting the plausible prediction by the present method.

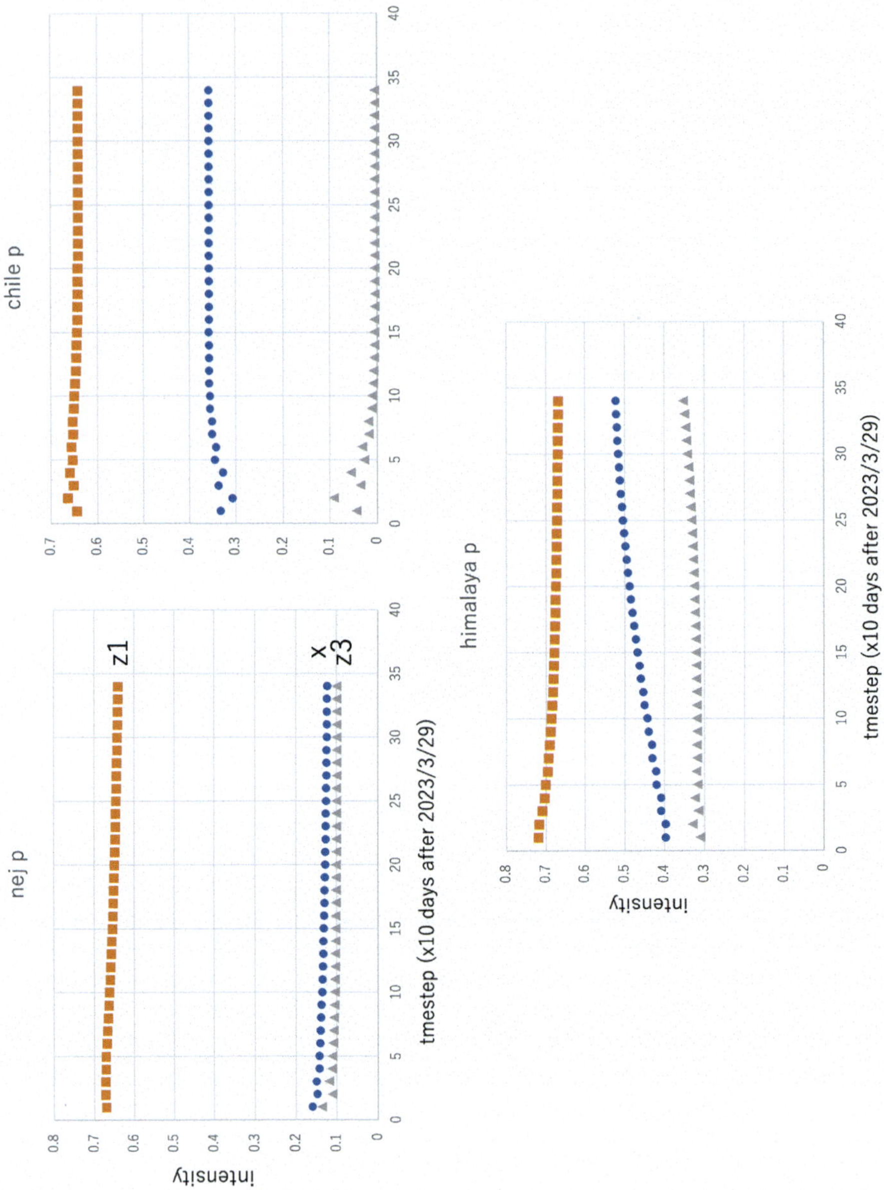

Fig. 6.48 Prediction of multivariate timeseries of global correlated seismicity z_1 and z_3, and logarithmic seismicity x of northeast Japan, Chile, and Himalaya regions from 2023 to 2024

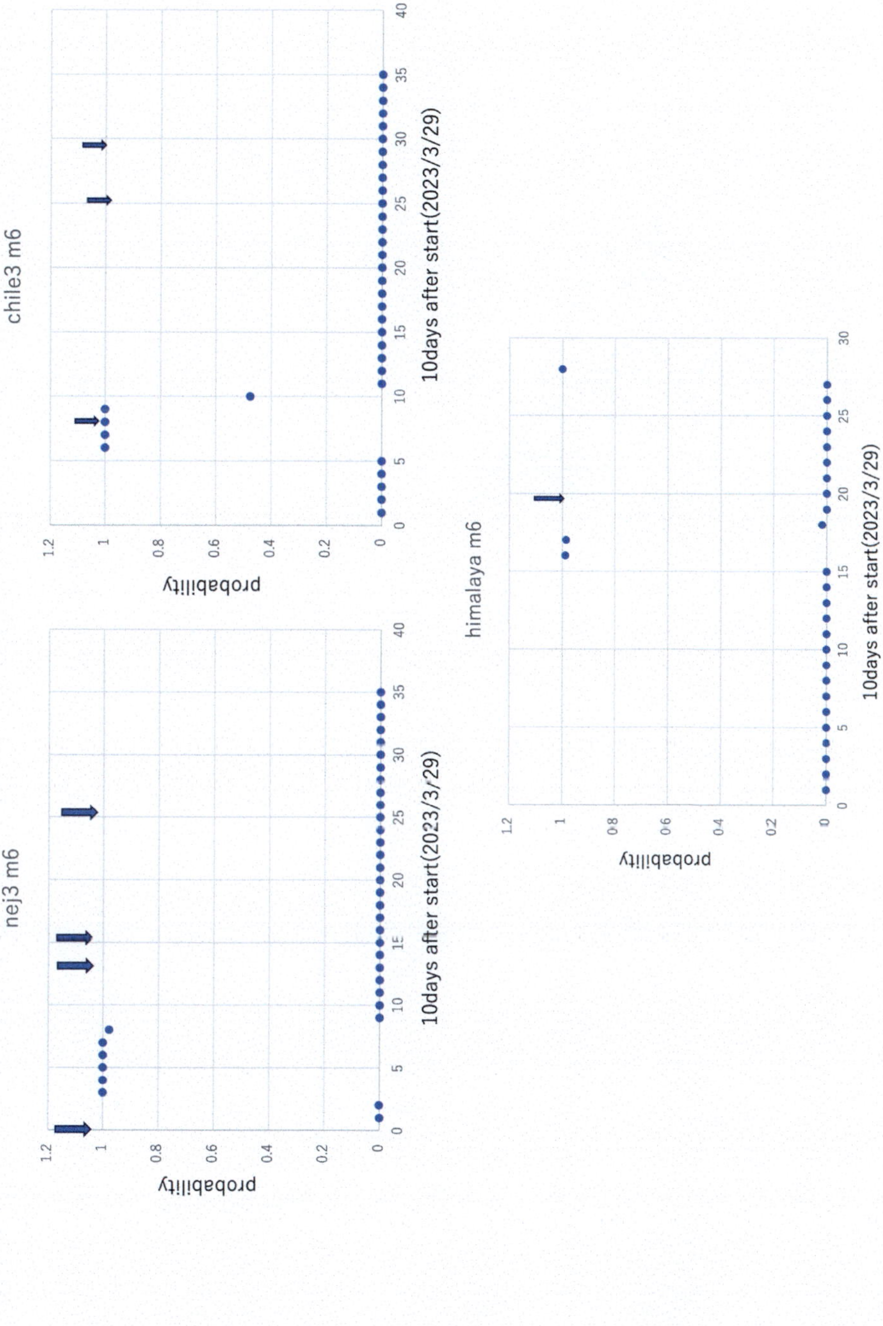

Fig. 6.49 Full-prediction results of large earthquakes from 2023 to 2024 in the regions of northeast Japan, Chile, and Himalaya

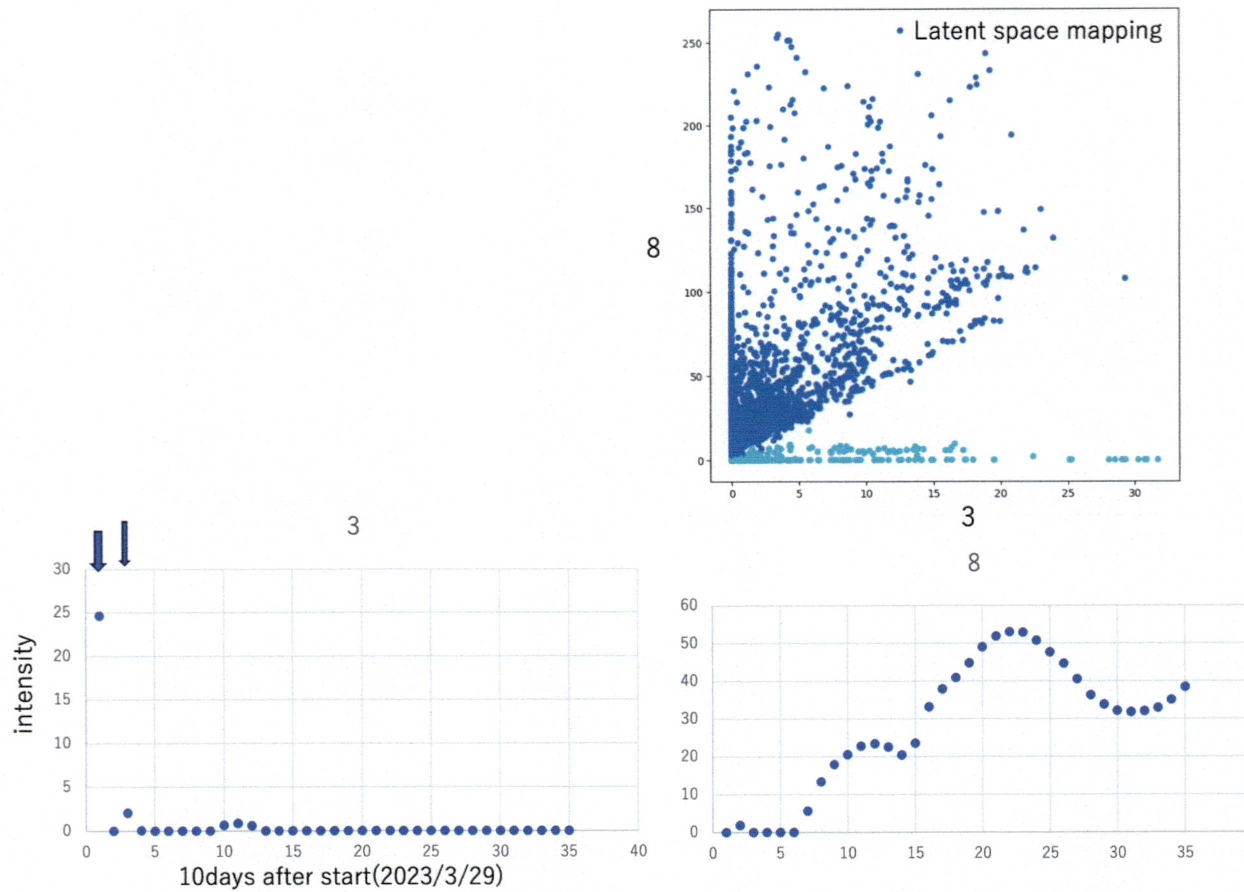

Fig. 6.50 Prediction of large earthquake over M6 of northeast Japan region by latent space mapping method from 2023 to 2024. Right, latent space map showing large earthquake cluster (light blue) and no earthquake cluster (blue). Arrows show occurrence of M6 over earthquakes predicted in the figure

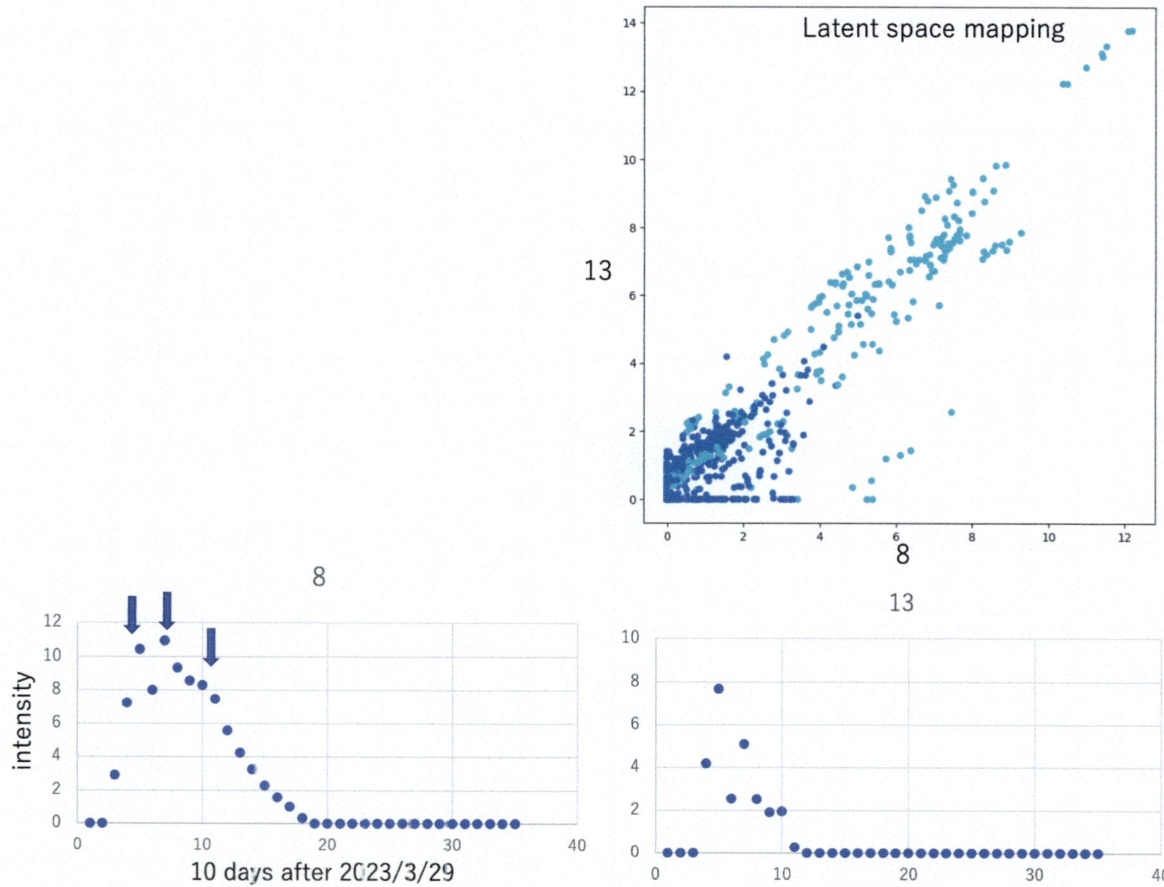

Fig. 6.51 Prediction of large earthquake over M6 of Chile region by latent space mapping method from 2023 to 2024. Right, latent space map showing large earthquake cluster (light blue) and no earthquake cluster (blue). Arrows show occurrence of M6 over earthquakes predicted in the figure

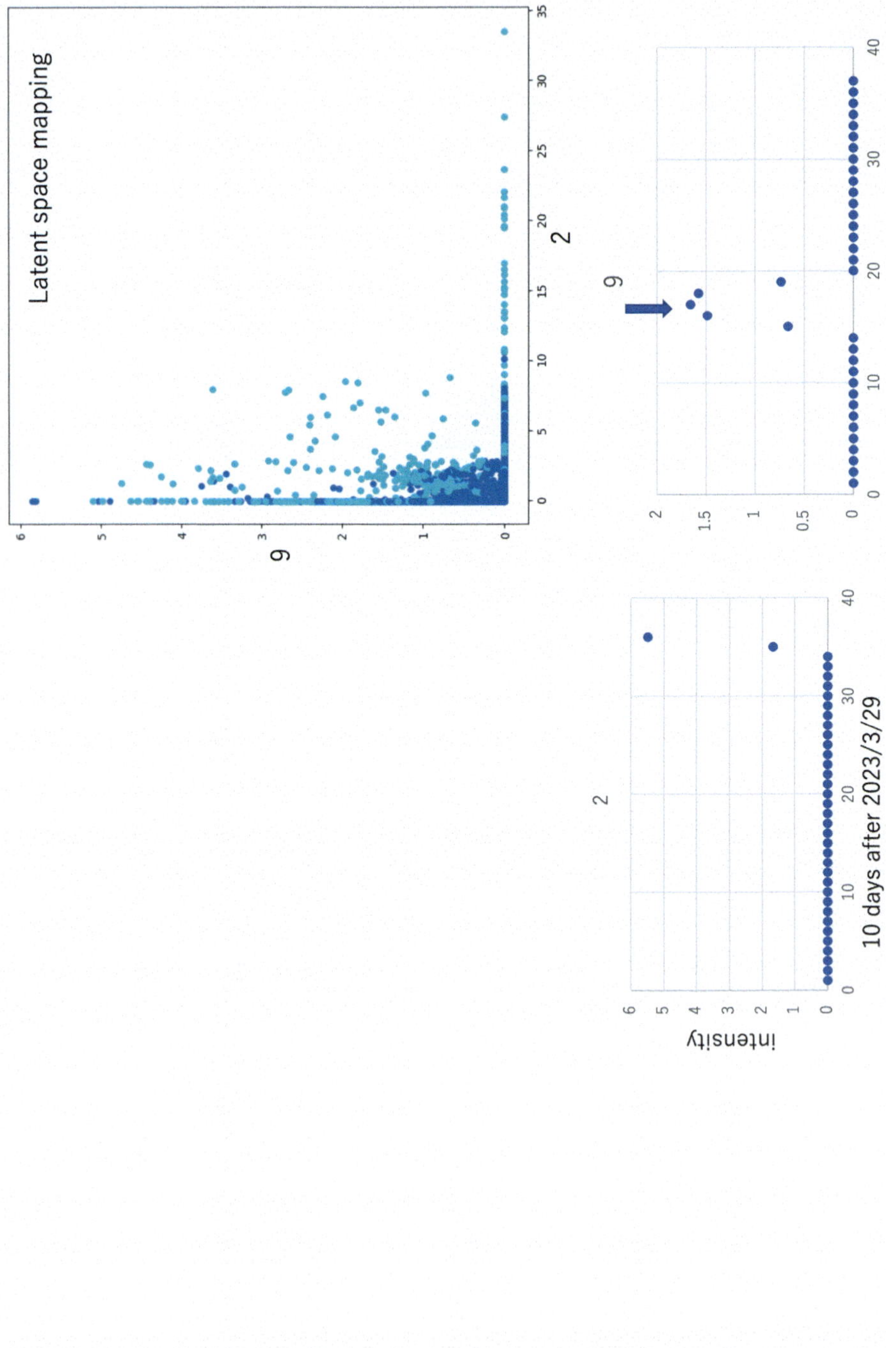

Fig. 6.52 Prediction of large earthquake over M6 of Himalaya region by latent space mapping method from 2023 to 2024. Right, latent space map showing large earthquake cluster (light blue) and no earthquake cluster (blue). Arrows show occurrence of M6 over earthquakes predicted in the figure

References

Chollet F, Allaire JJ (2018) Deep learning with R. Manning Publishing, USA
Haken H (2006) Information and self-organization, a macroscopic approach to complex systems. Springer
Kitagawa G (2009) Introduction to time series modelling. CRC Press
Ohnaka M, Matsu'ura M (2002) The physics of earthquake generation. University Tokyo Press 2002, 378pp. in Japanese
Schreiber T, Schmitz A (2000) Surrogate time series. Physica D: Nonlinear Phenom 142(3–4):346–382
Takanohara D (2023) Diffusion models. Iwanami Shoten Publication
Toriumi M (2021) Global seismicity dynamics and data-driven sciences. Springer
Toriumi M (2022) Geochemical mechanics and neural network model. Springer

Chapter 7
Concluding Remarks

Abstract Plate boundary metamorphic schists contain characteristic features formed by slow to rapid shear slip deformation in the very wide range of space and time scales. Shear crack jog on the slip surface of which size and spacing range from 10^{-3} to 10^{3} mm and growth velocity do from 10^{-3} to 10^{3} mm/s. Then, the grain growth of metamorphic minerals in the growing jog volume filled with aqueous fluid takes place in the very fast to very slow velocity because of various oversaturation degrees. On the other hand, the jog on the shear crack surface behaves as the vamp of the slip interface at the seismic and aseismic slippage of the plate boundary zones. The periodically arrayed jog seems to be source of the non-volcanic tremor associated with slow slip event and low frequency seismic slip. It seems that the background oscillation of the earth in the range of 10^{-3}–10^{-4} Hz may be derived from the resonance seismicity of vamp oscillators with very slow slip and plate subduction slip.

Keywords Shear crack jog as vamp on slip interface · Non-volcanic tremor and low frequency earthquake · Background oscillation of the earth

7.1 Geochemical Mechanics and Seismic Activity of the Earth

Until those decades, it has been thought the seismicity of the plate boundary zone is strongly dependent on the fluid pressure and fluid flux of aqueous solution in the deep-seated rocks which should be experienced by mineral reactions in the multi-component system including aqueous fluid. Classically, in the field of metamorphic petrology, it has been long recognized that the aqueous fluid has a role of enough mobile phase and component in the situation of mineral reactions, and thus the fluid pressure is approximated to be equal to solid pressure, for the convenience of phase-petrological modeling of the multi-component and polyphase system, using the pseudo-section method of metamorphic rocks (Holland and Powell 1998).

M. Toriumi, *Physics of Geochemical Mechanics and Deep Neural Network Modeling with Diffusion Augmentation*, Advances in Geological Science,
https://doi.org/10.1007/978-981-97-9376-1_7

On the other hand, the pressure–temperature path of metamorphic rocks inferred from pseudo-section method and differential thermodynamic method is very useful to investigate the materials composed in the deep plate boundary of which seismic activity is extremely high. Especially, in the studies of the metamorphic petrology applying to the estimation of content of aqueous fluid in the plate boundary zone, it has been clarified that the upper layers of the subducting slab change their H_2O content to decrease with increasing their metamorphic pressure and temperature (Hacker et al. 2003), and that in the prograde and retrograde metamorphism (i.e., subduction and exhumation of the metamorphic rocks), pressure and temperature together with H_2O content of subducting oceanic crust change along the specific path depending on the thermal state of the oceanic plate (Hacker et al. 2003; Okamoto and Toriumi 2004; Inui and Toriumi 2004). Furthermore, the H_2O release rate in the anhydrous garnet formation from hydrous chlorite and quartz during the increasing temperature was estimated from the pressure, temperature, and content of H_2O at every portion of the garnet grain that can calculate the molecular weight per unit advance of pressure of product garnet from the core to rim region (Toriumi and Inui 2001). From these studies, it is clarified that there are two different stages concerning the magnitude of H_2O release rate in the subducting process of the metamorphic rocks along with oceanic plate subduction: one is very large releasing rate of H_2O, and another is low rate of H_2O release from the pelitic schists during the process of low to high pressure conditions. On the other hand, during the process of the exhumation of the boundary metamorphic rocks, it was concluded that the H_2O absorption by amphiboles reaction which can be estimated from the retrograde chemical zoning of growing amphibole grains from core to the rim, thereby suggesting the hydration of the plate boundary rocks during the exhumation stage (Okamoto and Toriumi 2004; Uno et al. 2015).

By the way, fluid flux in the deep-seated metamorphic rocks should be passed through microcracks and grain boundary as advection. In the previous chapters, it is manifested that the open microcracks are very abundant, and these are filled with metamorphic minerals such as albite, chlorite, and actinolite, thereby showing that the aqueous fluid flow took place through the clusters of microcracks. These microcracks are aligned perpendicular to the shear slip plane (i.e., schistosity plane), and therefore, it should be obvious that the fluid flux by dehydration in the subduction and by hydration in the exhumation is possibly controlled by the shear crack growth and propagation in the whole process of metamorphism along the plate subduction zone. Accordingly, it is suggested that abrupt increase in the rate of H_2O release should be derived from the abundant active shear cracks motion in the subduction process of the metamorphic rocks. Furthermore, the growth amount of garnet per unit pressure change in the dehydration reaction is estimated to be very larger in the high flux stage rather than in the low flux stage (Toriumi and Inui 2001). These results show obviously that the metamorphic reaction and mineral growth were surely coeval with the large shear deformation of the rocks, and they are controlled by the intensity of aqueous fluid flux. Therefore, it strongly suggests that the metamorphic rocks should be composed of many minerals grown in the wide range of time scales and synchronous with brittle shear deformation in the various slip velocities and shear crack sizes. It seems that the present metamorphic rocks should be suffered from annealing and fracturing in the various degrees after successive formation of minerals and their grain boundaries.

On the other hand, it should be noticed that the mode of shear deformation changes from brittle type to ductile one with increasing metamorphic temperature just over the temperature of 350–400 °C by the detailed study of metamorphic quartzose rocks (deformed radiolarian metachert) by Toriumi (1990), thereby indicating that the low-grade metamorphic schists maintain the microstructure of cooperative shear crack slip and related crystal growth during their exhumation. In the present book, the author intends to introduce the mechano-chemical characteristics of the plate boundary metamorphic rocks, which is called as the geochemical mechanics of the plate boundary by the author in the previous book (Toriumi 2022).

By the way, the aqueous fluid flowing in the open cracks should be highly oversaturated for metamorphic minerals in the matrix and precipitates them in the high growth velocity. During the long period of metamorphism, the short term and high growth velocity crystallization should take place along the shear slip zone at the various periods, and thus, it can be thought that the minerals grains crystallized in the earlier stage should be suffered from longer time annealing. Therefore, it seems that there are possibly various microstructures of minerals annealed during various duration in the high-temperature metamorphism. The change of mineral microstructure by various duration of annealing time has been discussed in the previous book (Toriumi 2022), by means of spherical shape transformation of inclusion minerals in the temperature conditions above 500 °C. However, in the lower grade metamorphic rocks suffered from 350 to 400 °C, the shape change of mineral grains by such low-temperature annealing seems not to be enough, thereby suggesting that the many microstructure of minerals formed by rapid crystallization in the shear crack jog may keep during low-temperature metamorphism. In addition, it showed in the previous chapter that the secondary fluid inclusions formed by microcrack healing manifest various duration of annealing in the albite- and quartz-sealed cracks, thereby suggesting the short and long time annealing after crystal growth of the metamorphic minerals.

Moreover, it seems that various degrees of annealing manifest plural phases of the shear crack growth/propagation and associated rapid growth of minerals during the prograde and retrograde metamorphism, thereby showing that the deep seismicity along the subduction boundary of the slab should be accompanied with rapid precipitation recovery of shear crack and fast flow of aqueous fluid through the plate boundary rocks. Furthermore, judging from the spatial and temporal concentration of the shear and open crack slippage in the metamorphic schists as shown in the previous book (Toriumi 2022), extremely rapid extrusion and exhumation proposed previously is probably responsible for the local and episodic activity of the concentrated shear slip zone with large-scale hydration and precipitation of minerals. On the other hand, the present geochemical shear fault activity should bring the metamorphic rocks to be highly brecciated by rapid decompression and precipitation of metamorphic minerals in their matrix, as described in the previous chapter. It seems that the above process should control the rapid and extremely rapid extrusion of the metamorphic mass driven by the very low frictional resistive force against buoyant force.

On the contrary, it is probably said that, in the prograde stage of metamorphism, metamorphic rocks should subduct into the upper mantle, being associated with the subducting slab, and the boundary between them is mechanically glued with each other. The above strong coupling between the metamorphic rocks and subducting slab is probably due to the strong frictional force of the boundary zone, which is the fault zone. Accordingly, it may be concluded that, if the concentrated shear crack zone will be suffered repeatedly from decompression brecciation and hydration resulting in decoupling, and precipitation-recovery resulting in strong coupling of them, the metamorphic rocks should experience repeated subduction and extrusion as discussed in the previous chapter. In that chapter, it was concluded that the plate boundary metamorphic belt consists of high-grade metamorphic rocks containing coesite- and diamond-bearing eclogite blocks and surrounding amphibolite and greenschist, and their isotopic ages of these rocks range from 40 to 260 Ma, thereby suggesting their episodic extrusion and mechanical mixing at the stagnant region in the rather deep plate boundary zone. The boundaries between exotic high-grade blocks must be the brecciated and recrystallized shear fault zones. The final extrusion of the Alpine metamorphic rocks was presumably 38–40 Ma in the rapid mode of extrusion under the condition of regional extension along the plate boundary zone as noted previously.

These conclusions suggest that the plate boundary metamorphic rocks are the products of the geochemical mechanics processes involving geochemical reaction, fluid flow, geomechanical processes, global tectonics, and celestial mechanics discussed later.

7.2 Free Oscillation of the Earth and Non-Volcanic Tremors from Arrayed Jogs

Now, in the present book, it has been clarified that the high-speed co-seismic slip of shear cracks generates the sudden pressure drop by formation of the shear crack jog, and that the aqueous fluid in the matrix emits into the very low-pressure jog. In addition, it has been clarified that the jog along on the shear crack interface appears periodic and has systematic spacing as illustrated in Fig. 7.1, thereby suggesting that at the time of shear crack active, the many arrayed jogs on the shear crack interface should behave as the periodic sources of the elastic oscillation of the surrounding rocks, that is the seismic oscillators on the shear plane as illustrated in Fig. 7.2. Otherwise, as discussed in the previous sections, the expanding jogs should lead the influx of the intergranular aqueous fluid in the matrix by rapid decrease of fluid pressure, and then the surrounded rocks may become dry-up in some degrees. These two processes make the resonance and modulation of elastic waves generated from the periodic jogs having a similar wavelength in the surrounding rocks and the far-field ground motion as proposed in the previous sections.

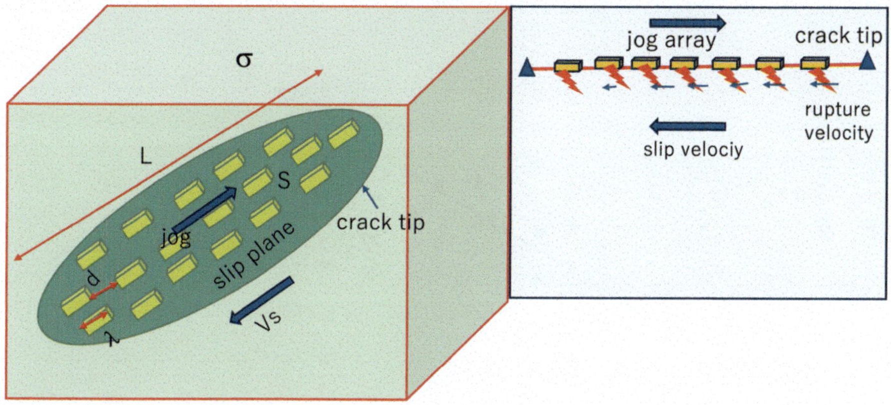

Fig. 7.1 Illustration of arrayed jog on large shear crack surface. L; length of shear crack, d; jog spacing, λ; width of jog (wavelength), v_s; shear slip velocity, σ; external stress, v_r; rupture velocity of jog

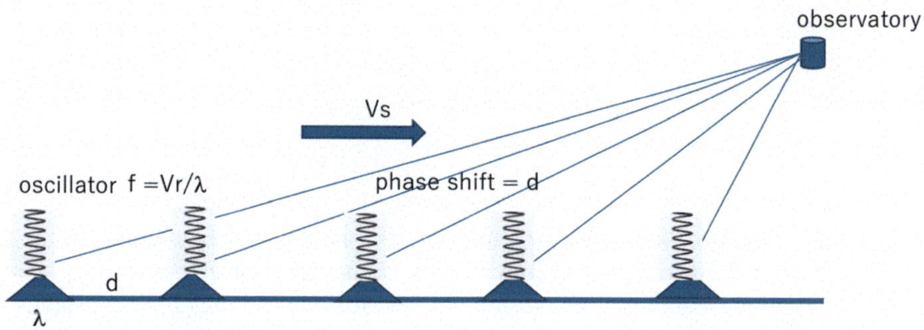

Fig. 7.2 Illustration of arrayed oscillators of jog on shear slip surface showing the superposition of seismic wave at observatory

Now, let us consider the seismic wave emitted from arrayed jogs on the various shear slip planes discussed in the previous sections. First, it has been clarified in the previous sections that the average width of the arrayed jog is approximately proportional to the spacing of them as shown in Fig. 7.3, but the height (amplitude) of them is not always related directly to the wavelength. Then, it is available that the stepwise jog gives a simple oscillation against the surrounding rocks by slip motion of the single shear crack.

If the shear crack jog velocity is v_j, which is approximated to be rupture velocity of the earthquake crack v_r, and the wavelength and amplitude of the stepwise jog are λ_n and a at the position r_n, respectively, the displacement of the shear crack interface holds,

$$Xo(r_n, t) = a \sin(v_r t / l_n + r_n). \tag{7.1}$$

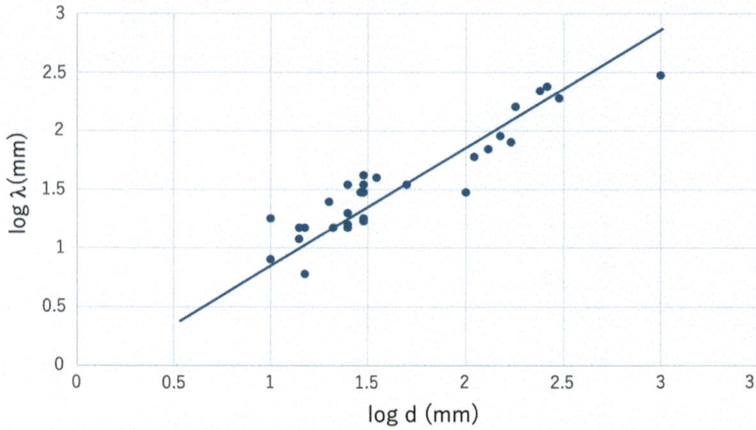

Fig. 7.3 Logarithmic relation of spacing of jog d and jog width λ of low-grade Sanbagawa basic schists and Muroto gabbroic mass in Shimanto belt

In the far-field approximation of the ground motion by Saito (2009), the displacement vector at the distance R from the arrayed multiple jogs in the spherical coordinates becomes,

$$u_x(x, t) = S_n 1/4pr\left(1/V_p^2 R\right)KXo(t) \tag{7.2}$$

in which K is the function of the orientation from the jog to the ground station, ρ and V_p are density of rocks and seismic velocity of primary wave, respectively. The modulation and resonance of these seismic waves are shown in the previous chapter, but in this chapter, the basic frequency of the seismic wave from jogs observed in the ground state seems to be approximated to,

$$f_n = v_r/l_n. \tag{7.3}$$

In this equation, the wavelength varies even in the single shear plane as like as Gaussian distribution function. Thus, it is probable that the mean width of the stepwise jog in the single shear plane probably depends on the shear slip velocity, for the wavy shear slip interface in the rather ductile environmental case, has the wavelength applicable to Kelvin–Helmholtz (KH) instability model, and it seems to depend on the slip velocity as shown in the previous section. In the case of the KH instability of the ductile shear slip zone, the wavelength of various shear crack interface appears from several mm to several meters as described in the previous chapters.

Furthermore, it is important that the logarithmic slip velocity of the seismic shear crack is proportional to the logarithmic rupture velocity in the cases from slow slip events, rapid slow slip events, low frequency earthquakes, and regular earthquakes by Fukao et al. (2021), and in the cases of simulations by Bizarre (2012), as shown in

Fig. 7.4. The above diagram between them is inferred from the relationship between the stress drop of the earthquake and the rupture velocity in Fukao et al. (2021) as shown in Fig. 7.5, and the experimental relation proposed by Ohnaka (1996).

Now, let us apply to the slow slip $v_s = 10^{-7}$ m/s, that corresponds to $v_r = 10^{-1}$ m/s and the wavelength is about 100 m, and then we have

$$fc = 10^{-3}(\text{Hz}). \tag{7.4}$$

The above frequency is near the observed free oscillation of the earth in the seismic quiet periods (Nishida et al. 2000). If this is the case of the free oscillation exited from the plate boundary arrayed jogs source of 100 m length, it seems,

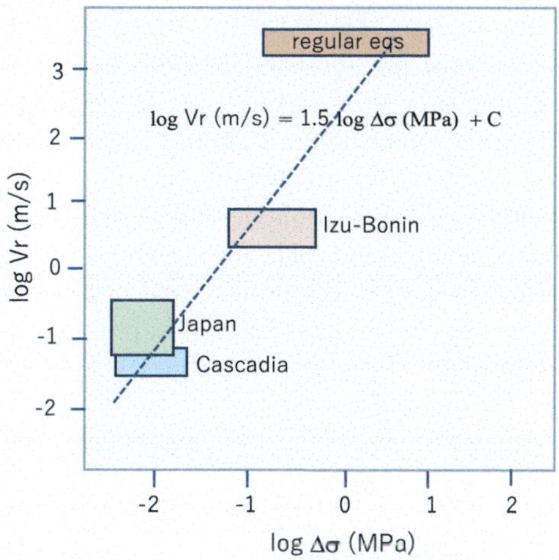

Fig. 7.4 Logarithmic relation of stress drop $\Delta\sigma$ and rupture velocity v_r of earthquakes, modified from Fukao et al. (2021)

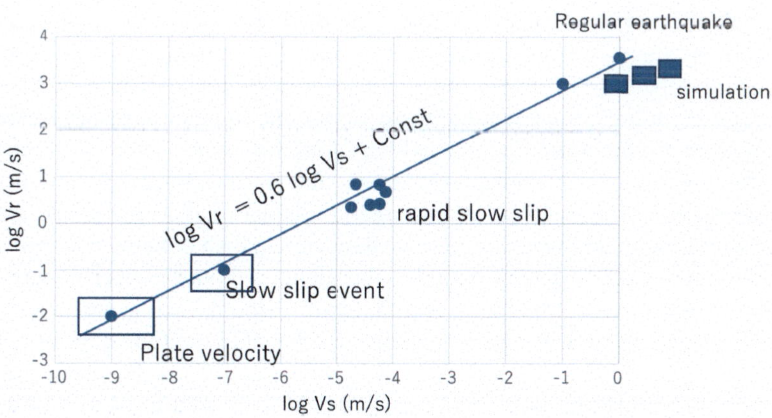

Fig. 7.5 Logarithmic relation of slip velocity v_s and rupture velocity v_r of various seismic and aseismic slip and simulated seismic slip (closed square; Bizarre 2012)

$$
\begin{aligned}
v_s &= av_r \\
&= 10^{-7} \times 10^{-2} \\
&= 10^{-9}\,\text{m/s}.
\end{aligned}
\tag{7.5}
$$

Therefore, it may be suggested that the abundant shear crack arrayed jogs are derived from the shear slip velocity about 10^{-9} mm/s which is near the plate subduction velocity, and the free oscillation of the earth (HUM) is responsible for the arrayed jog source on abundant shear slip planes which appear in the plate boundary metamorphic schists in this model for the HUM.

Nishida et al. (2000) clarified that the spectrogram of the 1–10 mHz of the free oscillation of the earth displays the intensity peaks with near annual but not regular periodicity, being compared with the atmospheric exiting cases. Furthermore, the source regions of HUM are localized in the high latitude ones of northern and southern hemisphere, but above HUMs may be one of atmospheric excited one, but not of the multiple very slow earthquakes of the plate boundary region.

On the other hand, it should be noticed that the very low frequency oscillation may be excited from the arrayed jogs on shear slip interface of rather rapid slow slip events ($v_r \sim 1$ m/s) found by Fukao et al. (2021) which yields probably the oscillation of frequency about 1 Hz for jog about 1 m size, and it is just in the frequency region of low frequency earthquake and non-volcanic tremor, thereby suggesting the low frequency earthquakes derived from the harmonic arrayed jog with finite height along on the shear slip plane of the slow slip events.

7.3 Future Problems on DNN Predictions of Large Earthquakes

In this book, the author intends to propose a prediction model of the large and giant earthquakes by means of deep machine learning using neural network modeling with multiple labeling. To obtain the accurate probability prediction of the large earthquakes, it should be needed that there would like to be enough large amounts of seismic data in the target regions and very long timeseries of seismic activities surrounding the regions. As being well-known, the large earthquakes, which occur mainly surrounding the Pacific Ocean along the subduction plate boundary, are rare and short-term event, and bring large-scale damages against human community and surrounding system during long periods. The giant earthquakes over magnitude 8 occur in the about several hundred years interval without known precursors until present. However, as far as the large and giant earthquakes emit the huge amount of seismic energy in the very short duration, attaining the 10^{19-20} J/s, some physical processes, that govern the large earthquake should have continued as a precursor before the events, and thus the process should be formulated by several basic but complicated mechanisms.

In the previous chapter, it seems to be clear that the repeating large earthquakes are associated with the non-volcanic tremor with low frequency during and before the events. The author has proposed the model that the repeating low frequency earthquakes and non-volcanic tremors are associated together, and thus the non-volcanic tremors before large repeaters are a potential precursor for the use of the practical prediction of the large earthquakes. If this is the case, it may be obvious that the characteristics of the non-volcanic tremors as a precursor should be intensely and comprehensively studied, especially the distribution and the change of spectrogram with approaching the initiation of the large earthquakes.

On the other hand, it is available that the recent-developed augmentation method can synthesize huge amounts of timeseries of rare phenomena such as large earthquakes, together with associated characteristic timeseries. If this is available for practical prediction of the earthquake, it seems that the augmentation method has a great advantage for increasing the resilience from the large earthquakes. In the present book, the author proposed a new method of augmentation of multivariate timeseries of the correlated seismicity and local seismicity rates of the target region by using the successive Gaussian density distribution function of the signals in the short time span. The author called this method as 'diffusion augmentation method (DAM)' of the multivariate timeseries in this book. This method is based on the facts that the global correlated seismicity timeseries contain various Gaussian noises in the high dimension space of the logarithmic seismicity along the plate boundary zones, as clarified in the previous book (Toriumi 2021). Furthermore, it shows that these timeseries, such as z_1 to z_8 in the previous chapter, are obviously divided into several stages from 1990 to 2023, and all stages are characterized by the patterns of the seismicity network connecting some localities along the subduction plate boundaries, thereby suggesting that the global correlated seismicity dynamics is represented by the transition of the characteristic network structures of them.

The Gaussian noise in the correlated seismicity probably comes from the fluctuation of the shear crack propagation in the inhomogeneity of rocks and aqueous fluid concentration, and the various orientations of the shear crack planes with applied shear stress in the solid earth interior. It follows that the large earthquake corresponding one by one to the segment of assemblages composed of the correlated seismicity and local seismicity timeseries should be related to the values shifted randomly in some distances toward the mean value of the local Gaussian distribution function. In other words, the labeling method of the large earthquake for the correlated seismicity and local seismicity assemblage should be applied for the random-shifted augmented seismicity assemblages with earthquake labeling.

On the other hand, in the previous book, it has been clarified that the bifurcation of the global correlated seismicity rate against the earth rotation rate is derived from the two-state nonlinear dynamics system. Further, it suggests that this type of nonlinear dynamical behavior should be derived from the transition kinetics in the mechanical network of the global plate boundary zones, which are constituted with the three-armed chain-network of nodes in the global subduction plate boundary. The simulation study of the simple slider-spring connection of above network clarified that the periodic external force makes arising the multiple bifurcation of correlated oscillation mode of the nodes as like as the bifurcation in the global correlated seismicity by means of change in earth rotation rate.

It is noteworthy that the giant earthquakes over M9 occurred in the transitional state between the single stable mode to the high stable one of the bifurcated modes in the diagram of global correlated seismicity and earth rotation rate as shown in Fig. 7.2. It suggests that the giant earthquake, which emits very large moment release in a short period, occurs in the unstable situation of the network mode derived from the external periodic force such as change of earth rotation, which is probably derived from the change in zonal flow of the fluid earth.

Well, it will be said that what is the reason of the relation between the earth rotation change and the global correlated seismicity. As discussed in the previous chapters, the change in the global correlated seismicity z_1 and z_2 displays weak periodicities of about 1 and 3 years, thereby suggesting the synchronicity with the change of earth rotation due to the change of the atmospheric zonal wind correlated with the sun–earth distance. However, it is still unknown that the secular trend of the global correlated seismicity z_1 shows general increase from 1990 to 2023. It looks like the present secular trend of z_1 may be coherent with the general trend of the length of day (LOD) which displays a gradual decrease from 1995 to 2023, and the LOD seems to a decrease of ellipticity of the solid earth by centrifugal force. In the previous chapter, the author tries to estimate the elastic strain energy release rate is surely larger than the moment release rate by large earthquakes, thereby showing the possible model for the recent high activity of the global seismicity. Furthermore, considering the annual change of the distance between sun and earth, which is responsible for the annual change of the LOD, it seems that the climate change in these decades possibly gives an effect on the high activity of the global seismicity as like as the earth rotation.

7.4 Conclusion

In this book, the author discusses physical processes of geochemical mechanics that involves fracturing and deformation of micro- to global scale solid earth, and data-driven scientific analyses of the regional to global seismicity dynamics. It, furthermore, has been discussed that the full-prediction model of large earthquakes needed for the resilience in human society against the giant earthquake should be composed of the global correlated seismicity timeseries together with the local seismicity, corresponding to the large earthquake event soon, from observed data sets. However, for our human society, it shows that the giant earthquakes are very rare and surely takes the severe damages on the society, but the rareness of the giant earthquake leads us to difficulty in practical prediction of them. To avoid this difficulty, it is obvious that the augmentation of the timeseries of the characteristic features and large earthquakes in the target localities is very needed. In this book, the available augmentation model, that is the diffusion augmentation method, is proposed for the set of global and regional correlated seismicity and local logarithmic seismicity timeseries associated with large earthquake labels.

On the other hand, it shows that the possible augmentation method using Gaussian distribution function of the prior and the timeseries of the local mean and variance of the characteristic global correlated seismicity and local seismicity timeseries, showing the stochastic variation of these multivariate timeseries, is not enough accurate yet for full-prediction method by CNN modeling to large earthquake. Thus, it needs that how we can obtain the signals of precursor of the large earthquakes in the various layers of the CNN modeling by taking the data matrices stored in the intermediate layers. As a result, it seems that some nodes in the final dense layer in the 2dCNN modeling contain available data matrices for the identifying the clusters of labels of large earthquake event-on and -off, showing the somewhat signals of precursor of earthquake. In this book, it seems successful in part that the method using the 2dCNN output and final dense layer vector matrices as discussed in Chollet and Allaire (2018) is applicable in some degree for the full prediction of the large earthquake at the target locality of the plate boundary zone. Soon, the more available and practical method of the large earthquake prediction should be found, and the author wants to develop the more feasible model for the accurate prediction of the large earthquakes.

References

Bizarre A (2012) Rupture speed and slip velocity: what can we learn from simulated earthquakes? Earth and Planet. Sci Letts 317–318:196–203

Chollet F, Allaire JJ (2018) Deep learning with R. Manning Publishing, USA

Fukao Y, Kubota T, Sugioka H, Ito A, Tonegawa T, Shiobara H, Yamashita M, Saito T (2021) Detection of "rapid" aseismic slip at the Izu-Bonin trench. J Geophys Res Solid Earth 126:e2021JB022132. https://doi.org/10.1029/2021JB022132

Hacker BR, Peacock SM, Abers GA, Holloway SD (2003) Subduction factory 2. Are intermediate—depth earthquakes in subducting slabs linked to metamorphic dehydration reactions? J Geophys Res 108(B1). https://doi.org/10.1029/2001JB00129

Holland TJB, Powell R (1998) An internally consistent thermodynamic data set for phases of petrological interest. J Metamorphic Geol 16:309–343

Inui M, Toriumi M (2004) A theoretical study on the formation of growth zoning in garnet consuming chlorite. J Petrol 45:1369–1392

Nishida K, Kobayashi N, Fukao Y (2000) Resonant oscillations between the solid earth and the atmosphere. Science 87:2244–2246

Ohnaka M (1996) Nonuniformity of the constitutive law parameters for shear rupture and quasistatic nucleation to dynamic rupture: a physical model of earthquake generation process. Proc Natl Acad USA 93:3795–3802

Okamoto A, Toriumi M (2004) Optimal mixing properties of calcic and subcalcic amphiboles: application of Gibbs' method to the Sanbagawa schists, SW Japan. Contrib Min Petrol 146:529–545

Saito M (2009) The theory of seismic wave propagation. University Tokyo Press

Toriumi M (1990) The transition from brittle to ductile deformation in the Sambagawa metamorphic belt, Japan. J Metamorphic Geol 8:457–466

Toriumi M (2021) Global seismicity dynamics and data-driven sciences. Springer

Toriumi M (2022) Geochemical mechanics and neural network model. Springer

Toriumi M, Inui M (2001) Pressure-temperature-water production rate paths in the subduction metamorphism. Bull Earthquake Res Inst Univ Tokyo 76(3):367–376

Uno M, Iwamori H, Toriumi M (2015) Transition from dehydration to hydration during exhumation of the Sanbagawa metamorphic belt, Japan, revealed by the continuous P-T path recorded in garnet and amphibole zoning. Contrib Mineral Petrol 170:33. https://doi.org/10.1007/s00410-015-1185-9

Appendix

In this appendix, the R-code of the augmentation of multivariate timeseries is given as follows. The input file "filename.csv" is the multivariate timeseries containing correlated seismicity rates z_1, z_3, and local seismicity rate x, and their mean and variance of segmented Gaussian distribution of them. Thus, the input data matrix is composed of column (z_1, z_3, x, label, vacant, vacant, $z_1$mean, $z_1$variance, $z_3$mean, $z_3$variance, $x$mean, $x$variance).

```
df<- read.csv("filename.csv")
ds<-matrix(0,9201,4)
dr<-matrix(0,9201,3)
dr2<-matrix(0,30,1)
df[,3]<- df[,3]-4
n<-1
m<-0.5
dr[,1]<-rnorm(9201,mean=0, sd=0.7)
dr[,2]<-rnorm(9201,mean=0, sd=1)
dr[,3]<-rnorm(9201,mean=0, sd=1)
for (i in 1:920) {
for (j in 1:10){
for (k in 1:3){
dd<-df[i,k]
m1<-df[i,2*(k-1)+7]
11<-(dd-m1)/v1
1<-(4-k)
ds[j+10*(i-1),k] <- dd - 0.8*(n/v1)*(1+dr2[j])*11*exp(-0.5*11^2)
+ m*dr[j+10*(i-1)+1,k]
}
}
for (l in 1:920){
for (m in 1:10){
ds[m+10*(l-1), 4]<-df[l,4]
```


M. Toriumi, *Physics of Geochemical Mechanics and Deep Neural Network Modeling with Diffusion Augmentation*, Advances in Geological Science,
https://doi.org/10.1007/978-981-97-9376-1

```
}
}
for (i in 1:9201){
ds[i, 3]<-ds[i,3]+4
}
write.csv(ds, "augfilename.csv")
```

The manufacturer's authorised representative in the EU is Springer Nature Customer Service Centre GmbH, Europaplatz 3, 69115 Heidelberg, Germany. If you have any concerns regarding our products, please contact ProductSafety@springernature.com

Printed and bound by CPI Group (UK) Ltd, Croydon, CR0 4YY
30/09/2026
02225298-0007